Christmas 2001

To Nick —
May this book add
Knowledge to this dark
period of human history.
Love,
Dad + Mom

Grenadiers

Kurt Meyer

Translated by
Michael Mendé and Robert J. Edwards

Grenadiers

Kurt Meyer

Translated by
Michael Mendé & Robert J. Edwards

Published by
J.J. Fedorowicz Publishing, Inc.
104 Browning Boulevard
Winnipeg, Manitoba
Canada R3K 0L7
Web: www.jjfpub.mb.ca
E-Mail: jjfpub@escape.ca
Telephone: (204) 837-6080
Fax: (204) 889-1960

Printed in Canada
ISBN 0 - 921991 - 59 - 2

Printed by
Friesens Printers
Altona, Manitoba, Canada

Publishers' Acknowledgements

We wish to thank you, the reader, for purchasing this book and all of you who have written us with kind words of praise and encouragement. It gives us the impetus to continue translating the best available German-language books and producing original titles. Our catalog of books is listed on the following pages and can be viewed on our web site at www.jjfpub.mb.ca. We have also listed titles which are near production and can be expected in the near future. Many of these are due to your helpful proposals.

John Fedorowicz, Mike Olive and Bob Edwards

Editors' Notes

Readers who read the original English edition published in 1994 will note a number of changes to the text in this edition. In addition to reviewing the original translation alongside the the most recent German edition of the book, the entire style has been changed from British English to American English to benefit our North American customers, who make up the bulk of our readership.

Modern American Army terminology is generally used wherever an equivalent term is applicable. In cases where there may be nuances where we think the reader might enjoy learning the German term, we have included it with an explanation

In cases where the German term is commonly understood or there is no good, direct English equivalent, we have tended to retain the original German term, e.g., *Schwerpunkt* (point of main effort), *Auftragstaktik* (mission-type orders) etc.

In an attempt to highlight the specific German terminology, we have italicized German-language terms and expressions. Since most of the terms are repeated several times, we have not included a glossary. There is a rank -comparison table at the back of the book listing German Army, *Waffen-SS* and US Army equivalents.

Unit designations follow standard German practise, i.e., an Arabic numeral before the slash (e.g., *1./SS-Aufklärungs-Abteilung 1*) indicates a company or battery formation. A Roman numeral indicates the battalion within the regiment.

Other Titles by
J.J. Fedorowicz Publishing

In Preparation (Working Titles)

J.J. Fedorowicz Publishing

Table of Contents

SS-Obersturmbannführer Kurt Meyer in the spring of 1943. An official portrait after his award of the Oak Leaves to the Knight's Cross. (Roger James Bender)

Preface

I often fought alongside *Waffen-SS* formations as an armor commander; I found I could rely on them.

The *12. SS-Panzer Division "Hitlerjugend"* mentioned in the second part of the book was under my operational control during five hard weeks on the Normandy invasion front. Its commander was the author of this book, Kurt Meyer, *Generalmajor der Waffen-SS*. At the end of the war we spent several months together in a camp at Enfield in England.

In December 1945 I was flown to Kurt Meyer's Canadian court-martial at Aurich. I was the sole German soldier allowed to be a witness in his defense. Some of his comrades and I were also given the opportunity to be with him for a short period of time after he had been sentenced to death.

After his sentence had been commuted to life imprisonment, I got in touch with him and his wife as soon as possible. We remained friends until his far too premature death.

As a result, I knew *Generalmajor der Waffen-SS* Kurt Meyer and his *12. SS-Panzer-Division "Hitlerjugend"* rather well. I knew them in good times and, even more, in the bad ones.

The book *Grenadiers* chronicles the fighting of the *Waffen-SS* units during the Second World War under the command of *Panzermeyer* — as the author was known to his troops — in Poland, France, the Balkans, Russia, and on the Normandy invasion front. The courage, comradeship, chivalry and patriotism of the troops described are also representative of the military discipline, the selfless devotion, and performance of all other Waffen-SS Divisions and, indeed, the entire German Army.

Kurt Meyer wrote this book after his release from nine years of prison. It was important to him to memorialize through this book those soldiers of his who were still living — and who looked up to him as a father figure — and to the dead of all of the divisions of the *Waffen-SS* and the army.

The *12. SS-Panzer Division "Hitlerjugend"*, to which a portion of this book is dedicated, fought in Normandy for ten long weeks, mostly as the *Schwerpunkt* of the counterattack against Montgomery's army groups' continuous assaults and massive material advantage. The division was nearly destroyed. Its performance was always more than could be expected of it. Such exceptional accomplishments would have been impossible if the soldiers had been drilled to zombie-like obedience. The young soldiers were trained to act on their own initiative, thanks to the exemplary training that had grown out of the practical experience of war. Behind all this was a love for the Fatherland.

The success of the *12. SS-Panzer-Division "Hitlerjugend"* was frequently due to the personal intervention of its 34-year-old commander. His analytical skills, coupled with a sixth sense for danger and his ability to make the correct decision, enabled him to intervene personally at the right place and time. His determination and personal example gave the soldiers — and not only of his division — the strength to persevere and to counterattack. He suffered the deaths of his soldiers as if they were his sons.

Kurt Meyer's courageous bearing as he stood before the victors' court-martial at Aurich at the end of 1945 far exceeded mere warrior mentality, as did his composure when he listened — unjustly condemned — to his death sentence. I take my hat off to the courage then required and the chivalry of the Canadian General who did not sign this sentence but commuted it to life imprisonment. I also take my hat off to our Kurt Meyer who remained the same German officer in the death cell and in prison amongst criminals that he had been on the battlefield.

An additional heavy burden was his anxiety over his wife and five children, who had to survive just on social security benefits during his nine years in prison.

Neither during this time, nor after his release, did Kurt Meyer feel any hatred. With the help of his old comrades he soon built a new life for himself. In spite of his wounds, illness and imprisonment, he also felt obliged to support the reputation of his fallen comrades and their widows and children as well as those who survived. That is how this book came into being. Consequently, he shouldered the burden of being the first spokesman for the *HIAG* (Editor's Note: *Hilfsgemeinschaft auf Gegenseitigkeit der Angehörigen der ehemaligen Waffen-SS = Waffen-SS* Mutual Aid Society).

Nine hundred thousand soldiers served in the 36 *Waffen-SS* Divisions. About four hundred thousand were killed or are missing. Of the survivors, every second one had been wounded, often several times. These numbers speak for themselves. If one adds the families, the former members of the *Waffen-SS* total several million German citizens. In the long run, no democracy can do without the willing participation of so many people without severe problems. They had clearly proven their willingness to sacrifice themselves for Germany.

As the first spokesman for the *HIAG*, Kurt Meyer led his old comrades by his example and words in a deeply felt involvement in our democracy. His involvement stemmed from his former love of fatherland, coupled with the insights he had won during and after the war. He did this although the former members of the *Waffen-SS* and their families do not receive the same state benefits as other German combatants of World War II. Even today, Kurt Meyer's widow receives no pension.

After 1945 a flood of hatred was poured on the *Waffen-SS*. The things that were said about this component of the German Armed Forces do not, in the main, stand up to detailed inspection. Not only foreigners, but also many of our own population, lump the soldiers of the *Waffen-SS* with the members of the *SD* and those of the *Allgemeine SS*. This book has also been written to set truth against libel. In this way the contribution of the *Waffen-SS* may be seen objectively. Furthermore, the book will show the children of the *Waffen-SS* soldiers their fathers' deeds in an undistorted manner. They will be proud of their fathers' courage, constancy, decency, and their love of the fatherland. They also will read about the terror of war.

In *Panzermeyer* our German soldiers lost an armor commander of exemplary courage, chivalry and responsibility. His human stature, proved before his judges, in prison, and after his release, is an example to our people, whom he loved all his life.

Heinrich Eberbach
General der Panzertruppe a.D.

Poland

"Achtung! Panzer, Marsch!"

We had been standing, waiting for this moment, our eyes glued to the faces of our watches. Now it started to turn crazy.

The vehicle motors roared in the dawn. We increased speed, faster and faster, up to the border. I listened intently in the half-light. The first rounds would start their death-dealing flight any moment, opening up our way east. The hissing, wailing and shrieking was suddenly above us, enhancing the perception of our own speed that we were sensing with every nerve. During the approach we caught a brief glimpse of our assault groups as they dashed towards the frontier barriers and destroyed the obstacles with demolition charges. Machine-gun fire lashed down the street and short fiery flashes of exploding grenades illuminated our target. The armored cars entered the village of Gola at top speed. Infantry assault troops captured the bridge over the Prosna — it had already been prepared for demolition — and it fell into our hands undamaged. In a few minutes the village was occupied. The Polish soldiers crawled out of their positions, baffled and dazed, and approached us with their hands raised. They could not believe that, barely ten minutes after it had started, the war was already over for them.

I was suddenly standing in front of the corpse of a Polish officer. A round to the throat had killed him. The warm blood was spurting from the wound. Yes, this was war! This initial sight of death impressed grim reality onto my brain with great clarity.

But it was time to move on! Uprooted trees and smoldering houses made it difficult to advance. We could hardly see. Ground mist mingled with the smoke of destruction.

I could not stay with the regimental staff. I moved forward to the outskirts of Gola and followed the reconnaissance patrol. Of course, as the company commander of the *SS-Panzerjäger-Kompanie*, I had completely different duties. An enemy tank attack was not expected and my company had also been dispersed among the individual battalions. This kind of warfare didn't suit me and so I secretly followed the tanks. Since 1934 I had been following the development of the tank as a weapon at Döberitz-Elsgrund and later at Wunstorf-Zossen. Now I suddenly saw myself in a dead-end occupation as a *Panzerjäger*.

Whirled up dust was still hanging in the air as I came across two of our heavy armored cars and a motorcycle platoon just beyond Chroscin. The armored cars moved slowly into the fog. Visibility was less than 300 meters. Suddenly, the eerie silence was broken by the whiplash round of a Polish anti-tank gun. The first armored car rolled to a smoking halt. Its wheels had hardly stopped when the second one was also destroyed. Both armored cars were about 150 meters in front of the antitank gun. The position was well camouflaged and difficult to find.

Round after round penetrated the vehicles; machine-gun bursts swept down the street, forcing us to take cover. We heard cries from the *Panzeraufklärer* trapped in the armored cars and were forced to watch without being able to go to their aid. Each time a round penetrated the armored car's interior the shrieks of our mortally wounded comrades grew louder. We tried to reach the armored cars to help our comrades who had scrambled out to escape the antitank gun's field of fire, but it was impossible. Hostile machine guns hammered down the street. The machine-gun fire mowed down the *Panzeraufklärer* who managed to get out of the armored cars. The moans in the vehicle grew weaker. I was lying behind a pile of gravel. Spellbound, I watched blood dripping from the fissures in the first vehicle. I was paralyzed. I had not yet seen a live Polish soldier, but my comrades were already lying dead, right in front of me.

Polish cavalry came galloping out of the smokescreen. They were charging directly towards us, and wouldn't be stopped by the fire from my machine pistol. It was only when the motorcycle platoon opened fire and brought down some horses that the fierce cavalry troop galloped back into the fog. Artillery was engaging the hill in front of us, while a battalion of *Panzergrenadiere* assaulted the enemy positions. The young grenadiers were moving like they were in a training area. They could not be stopped either by machine gun or artillery fire. The battlefield looked deserted, however, innumerable soldiers were advancing towards the enemy.

I watched astonished as the attack was carried out almost noiselessly in front of me. *Panzergrenadiere* dashed forward with mechanical precision. As their attack gathered momentum, the Poles were swept from their positions. The attack rolled on irresistibly; it could not be stopped either by the enemy or the difficult terrain. Each of these fabulous soldiers was convinced about the justice of this war and had no scruples about giving his life for the rights of his people. Still, no cheers rang out over the battlefield. The faithful young soldiers carried out their duty and made unequalled sacrifices with earnest expressions. For these men the war against Poland was no war of aggression but the elimination of a scandalous injustice. They wanted to expunge the rape of the German people at Versailles. Their strength came from the purity of their aspirations. These were no ordinary soldiers, nor were they political mercenaries, risking their lives for their people's future.

These young people belonged to the elite of the nation. They had been selected from thousands of volunteers and had been intensively trained for four years. The *Leibstandarte SS "Adolf Hitler" (mot.)* consisted of men who had just reached their 19th year at the outbreak of war; the noncommissioned officers were about 25 years old. These young men obviously had had no influence on the political events of 1933. In 1933 they had been mere schoolboys who had sought ideals and wished to serve those ideals with devotion. How have they been repaid, with what infamy were they tortured, and how are they being treated even now? But on 1 September 1939 the *Panzergrenadiere* could not have known that they were to become scapegoats for spiteful politicians. They were soldiers, fulfilling their duty according to the traditions of the Prussian soldier.

At about 1000 hours the town of Boleslawez surrendered to the tempestuous assault after fierce street fighting. Enemy artillery fire rained down on the town, causing casualties among the population. By nightfall we were near Wieruszow and were planning the following morning's attack. The *Leibstandarte SS "Adolf Hitler" (mot.)* was attached to the *17. Infanterie-Division* and had to defend the right flank against a Polish cavalry brigade's attack.

The approaching darkness hid the day's destruction. The battlefield's misery was only visible in the illumination of nearby fires. The horizon was marked by burning hamlets and the thick smoke wallowed over the violated earth. We sat silently behind the remains of a wall trying to make sense of the first day's battle. We gazed earnestly into the glow of a former farmhouse and listened to parts of Hitler's historic speech. "I have decided to solve the question of Danzig and the Polish Corridor, and to find a way to make sure that a change in the relations between Germany and Poland will make peaceful coexistence between us possible." His words echoed in our ears for a long time.

The regiment was employed at the Warthe River as part of the *17. Infanterie-Division*, and advanced towards Pabianice. On 7 September, at around 1000 hours, we reached the outskirts of Pabianice and received orders to establish a blocking position to the south along the ridge running through Rzgow-Wola, Rakowa and Lodz. Stronger enemy forces with antitank weapons had occupied Pabianice. The attack of the *I./Panzer-Regiment 23* had just been repulsed by the Polish defenders. Damaged and destroyed tanks were on the battlefield. They had been rendered combat ineffective by Polish antitank rifles.

The regiment took over the mission of the tanks and immediately carried out an assault. The *1.* and *2./Leibstandarte SS "Adolf Hitler" (mot.)* broke into the town and the battalions followed up. This violent attack forced the Poles to withdrew to the town center. Strong Polish counterattacks were then made against the regiment's exposed flank.

The firing positions of the *II./Artillerie-Regiment 46* were desperately defended against persistent attack by Polish infantry. The frontline was everywhere. The Polish units came flooding back from the west attack without thought for casualties. The regimental command post suddenly became the focal point of the attack. All of the clerks and drivers fought for their lives. The Poles approached the command post through a potato field and, as the foliage offered excellent cover and camouflage, we could not see them until they came within grenade range. We could not stop the enemy infantry from winning more and more ground.

I leapt to my feet and fired, standing, into the field. This was the only way to hit the Poles. On my right a grenadier from the *13. Kompanie* was firing at them as if he were on a firing range, round after round. Our "target shooting" did not last long. Suddenly, I found myself back at the bottom of the trench, thrown there by a bullet grazing my shoulder. My neighbor had been killed with a bullet wound in the neck. Never again would I try to stop an attack standing up for all to see. The attacks continued with determination on both sides and only in the late afternoon was the momentum of the Poles broken. Hundreds surrendered and started the long march into captivity. Meantime, the *XVI. Armee-Korps* had advanced to the gates of Warsaw and was grappling with the Polish units defending the city, as well as those fleeing eastwards from the west. The Commanding General, *General* Hoepner, greeted the spearhead of the regiment at Nadarzyn. We would be attached to the *4. Panzer-Division*.

The *Leibstandarte SS "Adolf Hitler" (mot.)* received orders to secure the Kaputy — Oltarzew — Domaniew line and block the enemy retreat from the west towards Warsaw.

While on the march, the *I./Leibstandarte SS "Adolf Hitler" (mot.)* was ordered to change direction northwards towards Oltarzew. Motorized infantry followed the motorcycle troops and armored cars. They vanished into the night.

General Hoepner felt confident of the outcome of the war in Poland, but he predicted heavy fighting for the *XVI. Armee-Korps*. He thought the Polish forces still west of Warsaw would make every attempt to break through our blocking positions. After a few kilometers, it became obvious to us that the oncoming night would bring us some hard fighting. We had to work our way through the suburbs of Warsaw to reach the main street. Loud sounds of fighting could be heard from Oltarzew. The *I./Leibstandarte SS "Adolf Hitler" (mot.)* had reached the main line of retreat and was fighting strong enemy forces. On the road the columns had driven into each other and it was completely jammed. They were totally destroyed overnight. Hundreds of dead were lying in the rubble. Artillery, weapons and ammunition covered the road. The merciless fighting lasted until the morning, and both sides waited, exhausted, for daylight to reveal the true situation.

First light showed a grim situation. Not only had Polish soldiers been killed on this straight road but boxed-in refugee columns had also been shot to pieces. Dead and wounded horses hung in their tack awaiting the coup de grace. Women and children had been blown apart in the fury of war. Whimpering children clung to their dead mothers or mothers to their children. The wounded crawled out from under the rubble and cried for help. The field dressing station was soon overflowing. Poles and Germans worked together to relieve the suffering. Not a shot was heard. The war had been suspended. The refugees were bitter; they were from Posen and had been incorporated into the column to provide protection for the Polish troops.

This night had revealed the naked face of war to us for the first time. There was no longer any difference between soldier and civilian. Modern weapons destroyed them all. I did not see a single German soldier laughing on the "death road" at Oltarzew. The horror had marked them all. The September sun shone brightly on the blood-covered road and changed the destruction into a flytrap. More than 1,000 prisoners were ordered to remove the rubble. Six hundred were sent over to the enemy lines with the message "Warsaw has fallen."

A single antitank gun destroyed an enemy armored train; the exploding ammunition cars flew into the air with a loud crash and destroyed the train completely. In the next two days, strong enemy attacks ran into positions held by the *II./Infanterie-Regiment 33*, the *II./Panzer-Regiment 35* and the regiment. Their attacks were in vain.

In vain, I asked the commander for different duties so I could take a more active part in the fighting. I was fed up with commanding a company that was scattered around the regiment, platoon by platoon. I reminded the commander at every opportunity that I was a tank and motorcycle man, and felt totally superfluous in my present position. But it was no use; I remained a *Panzerjäger* for the present.

During the night of 12/13 September a strong enemy unit penetrated the positions of the *II./Leibstandarte SS "Adolf Hitler" (mot.)*; a total breakthrough seemed imminent. Early in the morning we received a message that the *6./Leibstandarte SS "Adolf Hitler" (mot.)* had been overrun and the company commander killed. I had always felt very close to him; we had belonged to the same regiment since 1929. We found the message of an impending breakthrough incredible. We simply did not believe that the enemy cold break through our defensive positions.

I received orders to go and find out if there was any truth to the report. I leapt into the driver's seat of a motorcycle/sidecar combination accompanied by *SS-Obersturmführer* Pfeifer, and we vanished in the direction of Blonie. Pfeifer died a soldier's death some years later commanding a company of *Panther* tanks. We moved out at speed along the "death road" to get past the insects as quickly as possible. The horse carcasses stank dreadfully.

A few hundred meters outside of Swiecice I saw two Polish soldiers and a member of the *6./Leibstandarte SS "Adolf Hitler" (mot.)* crouching behind a small bridge. The behavior of the three soldiers seemed so peculiar to me that I braked sharply, jumped off the machine, and walked towards the group kneeling in the ditch. It was only when I was standing on the edge of the ditch that I understood the reason for the German soldier's strange behavior. He was a prisoner of the Polish soldiers and was looking at me flabbergasted, as I had walked alone towards the group. Damnation, was I ever lucky again! Only Pfeifer's machine pistol had prevented the Poles from advancing me into the great beyond. It was true, the company had been overrun; the company commander was lying dead in a trench a few hundred meters away. Pfeifer and I worked our way further towards Swiecice and soon found our fallen comrade. He had been shot through the chest. Seppel Lange died an exemplary soldier; we would never forget him.

The enemy units that overran us were destroyed during the day; the front line was reestablished to its former position.

The *Leibstandarte SS "Adolf Hitler" (mot.)* and the *4. Panzer-Division* were employed in the Bzura sector to stop retreating portions of the Polish Army from crossing the river. The Poles attacked with great stubbornness and proved repeatedly that they knew how to die. It would be unjust to deny the courage of these Polish units. The fighting on the Bzura was desperate and intense. The best Polish blood was mixed with the river water. The Poles' losses were terrifying. All their attempts to break through were broken by our defensive gunfire.

Polish strength was broken on 18 September, and we were ordered to attack the fortress at Modlin. Heavy fighting developed in the forest area south of Modlin. The *I./Leibstandarte SS "Adolf Hitler" (mot.)* was attacked and surrounded by superior forces.

At 0700 hours on 19 September *Generalleutnant* Reinhardt ordered an attack to relieve *I./Leibstandarte SS "Adolf Hitler" (mot.)* and break through to the Weichsel River. The attack was supported by the *II./Panzer-Regiment 35.*

The deep sandy roads made movement very difficult and wheeled transport could only advance very slowly. Again the fighting was bitter and, although the Polish situation was hopeless, they did not consider surrender. They fought to the last round.

During the attack we discovered the remains of *SS-Obersturmführer* Bruchmann and an *SS-Unterführer* of the *I./Leibstandarte SS "Adolf Hitler" (mot.)*. Both had been captured after being wounded during the encirclement and had been badly mutilated. Bruchmann had been a platoon leader in my company and married only two weeks before the outbreak of war.

The battle for the old fortress at Modlin started with a heavy artillery barrage and *Stuka* attacks. We experienced the destructive impact of our dive-bombers for the first time and could not understand how the Polish garrison could withstand such a storm of fire. Contrary to our expectations, the Polish units in Modlin resisted stubbornly and defied every attack. In fact the fortress only fell during the final phase of the campaign.

On 25 September Adolf Hitler visited the front, to include the *15./Leibstandarte SS "Adolf Hitler" (mot.)* at Guzow.

Infantry divisions relieved the armor and motorized formations around Modlin. The mobile forces were readied for the attack on Warsaw that started with a bombing attack and concentrated fire on the fortifications and military strong points. The main bombardment of the city only started on the evening of 26 September. The Poles did not consider surrendering. It would be a fight to the finish; there were still 120,000 Polish soldiers fighting in the city.

The Poles only offered to surrender the city on the afternoon of 27 September. In the afternoon all fighting along the front ceased. The campaign for Poland was over. On 28 September the capitulation was signed by the Commander-in-Chief of the Eighth Polish Army and *Generaloberst* Blaskowitz. We listened to the generous conditions in astonishment. The officers were to keep their swords, and the noncommissioned officers and soldiers would only be held as POW's for a short time.

Very soon, on 1 October, the *Leibstandarte SS "Adolf Hitler" (mot.)* received orders to move west. We were all convinced that we would be marching to the banks of the Rhine. We were wrong. We reached Prague on 4 October and were allowed to stay in the golden city for a few weeks. The regiment received a tremendous reception from the German populace in Prague; thousands cheered us as we arrived in Wenceslas Square. *Freiherr* von Neurath, the venerable *Reichs* Protector, said words of praise.

I reported once more to the regimental commander in Prague and pleaded earnestly for another duty appointment. My experiences in Poland had left me dissatisfied, and I was afraid that I would remain commander of the *SS-Panzerjäger-Kompanie* of the regiment for the rest of the war. I must have gotten through to him because, at the end of October, I took over command of the *SS-Kradschützen-Kompanie*. That meant I would be at the spearhead of the regiment. Although I had long wished for that appointment, I was sorry to leave the *SS-Panzerjäger-Kompanie*. I had formed it in 1936 and felt attached to my *SS-Panzerjäger*. Still I was very pleased to know that I was allowed to take a platoon leader and several *SS-Unterführer* with me. In addition, my dependable driver was also allowed to accompany me to the *15./Leibstandarte SS "Adolf Hitler" (mot.)*.

At last I was in my element. We trained hard daily. The motorcyclists participated enthusiastically and gave me all their support. My slogan — "The engine is our best weapon" — was fully accepted and obeyed by the lads. In the space of a few weeks I had won the trust of my new company and I knew I could rely on every single *Kradschütze*. We awaited further developments on the Western Front with interest.

From Prague To The Western Front

The *Blitzkrieg* in Poland had made the soldiers hope the politicians could end this ill-fated war and a campaign against the Western Allies might be avoided. Our "pipedream" was rapidly shattered when we heard that the Allies had rudely rejected Adolf Hitler's peace offer at the beginning of October. From then on, it was obvious to the soldiers that a decision could only be achieved by the sword.

The question of how the military defeat of the Western Allies could be brought about occupied the thoughts of the youngest grenadier as well as the most experienced troop commanders. All agreed that to remain on the defensive would not result in a military decision. Seen from the soldier's point of view, only a major offensive would force a military decision, if a political understanding were impossible.

In November we moved into the Koblenz area and came under the command of *General* Guderian. We used the experiences gained from the Polish campaign and trained the soldiers in their new tasks. Planning and exercises and maneuvers followed in quick succession. The enthusiasm of the soldiers under my command encouraged me anew. Neither the hard training nor the icy weather of the winter could dampen their zeal. Training continued under the slogan — "Sweat saves blood. Rather dig a 10-meter trench than a 1-meter grave."

My company was billeted in empty buildings in Bad Ems. The rough terrain there was very suitable for our training, as we knew that our advance would be through the Ardennes forests with Guderian's corps, and we would find similar terrain problems terrain to those in the *Westerwald* region.

Guderian inspected each and every company. His planning exercises were of special interest to us. All his opinions served as great guidance for us. He said, "The tank engine is your weapon, just as much as its main gun". Under this very experienced armor commander we prepared ourselves for the inevitable attack in the West.

On 24 December 1939, Adolf Hitler visited us in Bad Ems. He talked to the regiment and told us of the trust he put in us. He hinted we would soon be marching across the battlefields that were drenched with our fathers' blood, fighting for a lasting peace and a strong Europe.

In February 1940 we were placed under *Heeresgruppe von Bock* and moved to the Rheine area. This sudden move came as a surprise; we would rather have stayed with Guderian.

The move to Rheine started a new phase in our training. We were attached to the *227. Infanterie-Division*. Our orders, as a motorized unit, were to cross the Dutch frontier, breaking through the frontier defenses, and advance to the Ijssel line. These orders would require the troops to move at top speed so they secured the many road bridges over the canals, and especially over the Ijssel, undamaged. We practiced river and canal crossings continuously. Soon we had explored every possible combat situation and felt confident in mastering our task.

My unit was quartered at Salzbergen and I was staying at the vicarage. It was there on 1 May that I got to know the well-known Bishop *Graf* von Galen, who would plead for my life a few years later and would draw my judges' attention to the principles of Christian-based justice. *Graf* von Galen insisted upon giving my company his blessing.

With the approach of the favorable season, the day we were to be employed drew irresistibly close. For days we had been waiting for the code-word, "Study Anton". On 9 May 1940 the code was given and operational readiness established. At 0205 hours, the next code — "Danzig" — was given. It was the final order to attack the Dutch frontier fortifications. We left Salzbergen in the dead of night and moved silently into the darkness. The people were standing on both sides of the road and waving. They wished us luck and a speedy and healthy return.

The final attack preparations were complete at 0400 hours. Once again I collected my young *Kradschütze* around me to remind them of the combat fundamentals. At the dawning of the day — that fateful 10 May — I promised my soldiers that an officer of our *Kampfgruppe* would always be in the front, thus affirming the leadership principles that we had been preaching. In the presence of my soldiers I shook hands with all my officers to lend emphasis to my promise.

The attack started at 0530 hours sharp. An assault party ambushed the outpost near De Poppe and took the surprised Dutchmen captive. The bridge fell into our hands undamaged, the assault party had cut the demolition cables.

Over our heads a countless stream of *Ju 52's* was flying westwards. Comrades of the *22. Luftlande-Division* and *Fallschirmjäger-Regiment 1* flew towards their objective. Fighters whirled through the air like hawks and plunged on their designated targets.

It was as if we were in the grip of a fever. Hardly had the frontier barrier been lifted and the bridge immediately behind it secured, then we were sweeping down the smooth asphalt road like race-car drivers. Max Wünsche, the platoon leader of the 1st platoon, raced ahead of his men and pulled them forward with his enthusiasm. I moved after Wünsche's platoon, and was surprised that we had met no resistance. Our advance towards Oldenzaal and Hengelo continued with utmost speed. The concrete tank traps and bridge barricades were undefended. Some bridges were only slightly damaged by explosives; we could bypass them.

Bornerbroek was reached without a shot being fired. The Dutch population stood by the road and watched the rapid advance of our troops. Enemy combat engineers had blown up the bridge leading over the canal just beyond Bornerbroek. We had hit the first enemy resistance. The canal was crossed in minutes. Barn doors and other materials were used for bridging. Speed was of the essence. All motorcycles were sent in pursuit of the enemy's demolition party to prevent it from destroying the next bridge. *SS-Obersturmführer* Kraas, platoon leader of the second platoon, took up the pursuit of the enemy combat engineers. Meanwhile the temporary bridge was solid enough to allow the crossing of the motorcycle/sidecar combinations as well. Antitank guns were towed by the motorcycles. The wild chase continued. Unfortunately, the armored cars had to stay behind. They were providing security for the pioneers who were working in exemplary fashion and were busy throwing an assault bridge over the canal.

Unfortunately we could not prevent the enemy demolition party from causing damage. The bridges already prepared for demolition were blown up. But these demolitions could not seriously hamper our progress. We advanced on Zwolle without great delay.

At about 1130 hours the advance guard was waiting on the outskirts of Zwolle, which meant it was 80 kilometers deep into enemy territory. The lead platoon (led by Reuss) moved towards the railway embankment directly south of the town and then decided to dismount behind a road embankment and proceed on foot. What a surprise the next minute brought! The wonderful chestnut trees along both sides of the road had been cut down to block entry into the town. But what use was the best obstacle if combat-ready soldiers did not guard it?

North of the barricade, only a few hundred meters away, we saw machine gun and antitank pillboxes and — wonder of wonders — the defenders were calmly sitting in their shirt-sleeves on top of the pillboxes and having lunch. They were enjoying the May sunshine that had clearly seduced them into leaving their somber bunkers.

The tree barricade stopped us from breaking straight through to the line of bunkers and taking the Dutch by surprise. Surprise fire on the crews sitting on top of the bunkers took care of the fortifications and allowed the grenadiers to cross an area free of cover.

Before the Dutch even knew what was happening our *Kradschützen* have reached the bunkers and disarmed the defenders. The trees could only be removed with difficulty. Armored cars were dragging the gigantic trunks off. The removal of the barricade was taking too long for my liking. The enemy could not be allowed to come to his senses. We had to take advantage of the element of surprise. Without a moment's hesitation I jumped into a Dutch vehicle and moved out quickly into Zwolle with *SS-Obersturmführer* Wünsche and *SS-Grenadier* Seelenwinter. *SS-Oberscharführer* Erich accompanied us on a Dutch motorcycle. I intended to take the town commander unawares and make him agree to a ceasefire.

Dutch soldiers were standing transfixed on the street as we shouted at them and pointed towards the tree barricade. They threw down their weapons and went towards the barrier. The further we went into the city, the more uncomfortable I felt about this "excursion". I would have liked to turn back, but it was too late, we had to play this game out to its end.

The sound of firing at the bunkers had not reached the center of town. Hubby and wife, out enjoying the beautiful May day, were scattering like frightened hens threatened by a hawk's shadow. Despite the extremely uncomfortable situation, we had to laugh at the Dutch reaction. An imposing civic building in the city center and the sight of a number of uniformed people coming and going made us try fate there. We moved right into the middle of the crowd. Amidst the squealing of brakes the car seemed about to overturn. In a split second we pointed our weapons at bewildered men in uniform. The Dutch stood transfixed. A respectable elderly gentleman in civilian clothes introduced himself as "the Queen's representative" and told us that he would order the Dutch troops in Zwolle to cease resistance. He kept his promise. Not another shot was fired in Zwolle.

With several captive officers we hurried back to the tree barricade. Zwolle was ours but, unfortunately, we could not prevent the destruction of the large bridges over the Ijssel. Both bridges had already been blown up in the early hours of the morning.

I nearly had a stroke when I reached the dismantled barrier — my men and some Dutch youths were amusing themselves on a carousel with hardly anyone on guard.

Meanwhile, the *III./Leibstandarte SS "Adolf Hitler" (mot.)* had forced a crossing over the Ijssel 800 meters to the south of the destroyed railway bridge at Zutphen. It was attacking Hooen. Under the command of *SS-Sturmbannführer* Trabandt, the village was taken at about 1400 hours. Four officers and 200 men of the "Gendarmes" Regiment had been taken captive. Our own losses were negligible. The missions of the regimental *Kampfgruppen* had been fulfilled. The Ijssel had been reached and partially crossed. My *Kampfgruppe* had only a single wounded man to show for it. *Kradschütze* Fleischer had been shot through the leg at the tree barricade.

During the night our regiment was detached from the *227. Infanterie-Division* and placed under the operational control of the *18. Armee.* The commander of the *227. Infanterie-Division, Generalmajor* Zickwolf, recognized the regiment's fast and successful advance. As the first officer so honored in this campaign, *SS-Obersturmführer* Kraas received the Iron Cross, 1st Class from the hands of the general. Kraas advanced about 60 kilometers beyond the Ijssel with his reinforced platoon and took seven officers and 120 men prisoner.

Operations Against Rotterdam

After the fighting on the Ijssel line, the regiment received orders to advance to Geertrnidenborg via 's Hertogenbosch and establish contact with the *9. Panzer-Division*. After skirmishes with Dutch infantry we reached Geertruidenborg late in the afternoon of 13 May. Contact with the *9. Panzer-Division* was established.

Next morning at 0400 hours we started our advance on the bridge across the Meuse near Moerdijk. As a result of the paratroops' efforts, the bridge had fallen into German hands intact.

Parachutes were dispersed in the broad meadows on both sides of the bridge embankment. Many a brave paratrooper had been killed in front of the numerous pillboxes, but the element of surprise had also won here. The enemy did not have a chance to destroy this all-important bridge. The way into Fortress Holland was open.

The *9. Panzer-Division* had advanced to the port of Rotterdam and made contact with the *11./Luftlande-Regiment 16*. A company had been set down near those bridges by gliders and had defended them against continuous Dutch attack.

The regiment's mission was as follows: "The reinforced *Leibstandarte SS "Adolf Hitler" (mot.)* in conjunction with the *9. Panzer-Division* and following behind it will push through or bypass Rotterdam to relieve the surrounded airborne troops in the area Delft/Rotterdam and then continue towards Gravenhage (den Haag)."

The regiment prepared to attack south of Katendrecht. Preparations were complete by 1300 hours.

Rotterdam was to be attacked at 1440 hours after softening up by artillery and *Stukas*.

My advance guard had been deployed ahead to the port of Rotterdam and was positioned near a big Dutch passenger liner. The ship had been burning since 10 May. Its cargo consisted of American cars.

At 1400 hours word got around that the Dutch were negotiating a capitulation. The negotiators were *General* Student, *Oberstleutnant* von Choltitz of the *22. Luftlandedivision* and the Dutch Colonel Scharro. During the parley *General* Student was shot in the head and taken away seriously wounded.

It had yet to be confirmed whether the surrender demand would be accepted by the Dutch high command.

At 1525 hours the corps issued orders not to attack Rotterdam. General Winkelman was expected as representative of the Dutch high command. I watched from the bridge with a group of officers as several waves of *He 111* bombers approached Rotterdam. Dutch antiaircraft guns fired at the planes. The ceasefire was broken. We tried vainly to attract the pilots' attention by firing red flares in order to stop the attack. We were standing in the middle of their target area. We believed we could prevent the attack up to the last moment but, as we heard later, the pilots could not see our flares through the haze. The dense clouds from the burning ship threw a pall over the city. As we heard the whistle of falling bombs we vacated the bridge and hurried into nearby cellars. That was it. The attack could no longer be stopped. Rotterdam was a sea of flame. The last bomb fell at 1545 hours.

We looked at the raging fire in horror and experienced the enormous violence of a bombing attack for the first time. The fire in front of us was building up into an impenetrable wall. The streets were almost impassable. Our doubts about negotiating the streets of the burning city were dispelled by an order to move out immediately. My advance guard was supposed to make contact with troops of the *22. Luftlandedivision* at Overschie.

We approached the maze of blocked streets and searched for a route to Overschie, pushing deeper and deeper into burning Rotterdam with our faces covered. People were fleeing towards the port area to escape the inferno.

My motorcyclists were moving through the narrow streets as if possessed by the devil. Shop windows exploded about our ears. Burning decorations and clothed mannequins presented an unearthly picture. The further we moved into the city the emptier the streets. There were soon no Dutch to be seen; the blazing incandescence had driven them all away.

Two heavy armored cars were moving through dense smoke clouds and their taillights showed the way. There was no room for mistakes; it was impossible to stop. The heat was unbearable. After we had passed through the shopping quarter and reached a tree-lined avenue, I ordered a short break to let the motorcyclists catch up with us. Soot-caked, with singed hair, but laughing faces, the last section came out of the burning city. Behind us things were closed off tighter than a drum. The fire had formed a convincing barrier for us. We could not turn back, so — onward!

We moved cautiously towards Overschie in the protection of a canal embankment and were met by infantry fire. The drawbridge over the canal had been raised and proved an effective barrier. We quickly blew up the bridge-operating mechanism. A heavy vehicle pushed on the bridge and it descended slowly. The road to the north lay before us. But how did this straight stretch of road look? Plane after plane sat there on the broad concrete road, destroyed, shot up, or burnt out. They were the transport aircraft of the *22. Luftlandedivision* which had used the road as a runway when they could not use their designated landing zones. They had fallen victim to the Dutch artillery. The air-landed troops had held out against all enemy attacks for three days. The fighting was especially fierce in Overschie. We worked our way forward down both sides of the road. Dutch machine gun and rifle fire failed to stop us. We searched Overschie in vain for survivors of the *22. Luftlandedivision*. Apart from traces of battle and dead comrades we could not find any German soldiers.

Only after advancing further towards Delft did about ten soldiers and a *Leutnant* come running towards us. The young officer threw his arms around my neck in exhaustion. At about 2100 hours we reached Delft and made contact with elements of the surrounded *22. Luftlandedivision*. The regiment took 3,536 Dutch prisoners on 14 May.

The disarmament of Dutch troops in Gravenhage and Scheveningen was completed on 15 May without enemy resistance. The regiment took 163 officers and 7,080 soldiers captive. With the occupation of the Ministry of War, the war in Holland was over for us.

Into France

The regiment moved into northern France by way of Arnhem and Namur and, near Valenciennes, it tackled French troops for the first time.

The regiment's mission was to prevent a French breakthrough to the south. All enemy attempts at a breakthrough were frustrated by our soldiers' defensive fire. The width of our regiment's allocated front was about 30 kilometers.

Near the old fortress of Les Quesnoy a freshly harvested field gave me a supernatural feeling. A few hours before, thousands of French must have been camped here. Now not a single French soldier was to be seen. But innumerable French helmets were lying in the great field as if arrayed for a parade. The neatly arranged helmets, so it seemed to me, expressed the helplessness and weariness of the French Army. It was an army without spirit and drive. It no longer consisted of "Verdun Soldiers". It was fighting without faith in its cause and goals.

The battles of World War I were still firmly embedded in the French soldiers' minds. They believed in their Maginot Line and therefore in the invincible force of the greatest defensive line on earth. France not only had the Maginot Line but also a superior tank force. The Allied armed forces had more than 4,800 tanks at their disposal. This quantity of armor was faced at the beginning of the attack by 2,200 German tanks and armored cars. The reason for the rapid breakdown of the French was surely due to their old-fashioned leadership principles.

On 24 May the *Leibstandarte SS "Adolf Hitler" (mot.)* came under the operational control of *Panzergruppe von Kleist* and was apportioned to the *1. Panzer-Division*. A few days before the rapidly advancing *Panzergruppe von Kleist* had reached the scarred battlefields of the Somme from World War I. The division, having advanced to the Channel coast by way of Cambrai, Peronne, Amiens, and Abbeville, was ready to take Boulogne, together with the *2. Panzer-Division*. On 24 May the *1. Panzer-Division* was positioned on the Aa canal at Holque and had orders to attack Dunkirk. Within the framework of this operation our regiment was attacking Watten to lend more weight to the attack of the *1. Panzer-Division*.

I brought the advance guard up to the canal and on to Watten Hill in a night march. Watten Hill is 72-meters high, which was enough in this flat marshland to command the surrounding area. The hill was east of the canal; the bridges over the canal were destroyed and its bank was defended by English and French troops. Under these circumstances a surprise attack on the hill was impossible. It could only be taken by a deliberate attack. That night, the *III./Leibstandarte SS "Adolf Hitler" (mot.)* made ready for the attack on the hill.

Shortly before the start of the attack, the crossing of the canal was forbidden by an order from Hitler. Dunkirk was to be left to the *Luftwaffe*. All offensive operations by *Panzergruppe von Kleist* were immediately brought to a halt. We were left speechless by this order, because we were now out in the open on the west bank of the canal. We sighed with relief when we heard of Sepp Dietrich's decision to go through with the attack despite Hitler's order. After effective preparatory fire, the *10./Leibstandarte SS "Adolf Hitler" (mot.)* succeeded in crossing the canal and entering the outskirts of Watten east of the canal. Stubborn resistance by the English and French hindered the progress of the units that had made the crossing. Only the attack of the *III./Leibstandarte SS "Adolf Hitler" (mot.)* finally brought the high ground into our hands.

The hill was crowned by castle ruins that afforded us excellent observation to the east. We were standing on the ruins when suddenly the Commanding General of the *XIV. Armee-Korps* appeared and demanded an explanation from Sepp Dietrich for advancing on his own. Sepp Dietrich answered: "The area west of the canal is in full view from Watten Hill. Those bastards were able to look right down our throats. That is why I decided to take the hill." *General* Guderian approved Sepp Dietrich's decision. A few seconds after this conversation we were all lying in the dirt and having to crawl for our lives. Enemy machine gun fire forced us to take cover. The dexterity with which the tank veterans Dietrich and Guderian disappeared behind the ruins was amazing.

In the face of this "ambush", Guderian ordered a continuation of the attack in the direction of Wormhoudt-Berques. For the duration of this attack our regiment would be attached to the *20. Infanterie-Division (mot.)*. On our right *Infanterie-Regiment 76* was attacking; on our left it was the reinforced *Infanterie-Regiment "Großdeutschland"*.

The start of the attack on 27 May was delayed because the bridging of the canal was not completed in time. At 0745 hours there was an enemy attack from a small patch of woods two kilometers east of Watten Hill that was repulsed by our artillery. At 0828 hours the regiment went on the attack and rapidly gained ground. At 1000 hours the regimental command post came under heavy enemy artillery fire that continued into the early afternoon.

In Bollezelle the *I./Leibstandarte SS "Adolf Hitler" (mot.)* met with strong resistance and, in addition, came under heavy fire from the sector belonging to *Infanterie-Regiment "Großdeutschland"*. *Infanterie-Regiment "Großdeutsch-*

land" was lagging behind and it could only deal with this threat to our battalion's flank after some time had passed.

The *Kradschützen* waited in readiness for the result of the attack. Once Bollezelle had been taken, it was planned to shoot my advance guard like an arrow from a tightly drown bow and seize Wormhoudt from the English by surprise.

I could not bide my time and tried to get a hand's-on view of the situation in the area of the *I./Leibstandarte SS "Adolf Hitler" (mot.)*. A solo motorcycle seemed to me the "right horse" for this purpose. Shelling on the crossroads forceed me into racing along the road at top speed. Knocked down telephone wires laying on the road turned my racecourse into an obstacle course. Suddenly I felt a jerk and could only just begin to perceive that I was flying past a tree like a rocket. From that point on I couldn't remember a thing.

Someone must have picked me up and brought me to the regimental command post. The not-so-friendly voice of Sepp Dietrich called me back to reality. In accordance with his orders I was promptly packed onto a litter and received doctor's orders not to get up on any account. The crash had given me a concussion. Some time later I heard in my dazed condition that my unit was beginning to move and saw the *Kradschützen* moving off in the direction of Bollezelle. The deep growl of the *BMW* machines was music to my ears. My crash was a thing of the past — I had to lead my troops.

Without anyone noticing me, I jumped up from the stretcher onto the road and scrambled onto a dispatch rider's machine. I quickly reached the lead elements of the company. Questioning glances from Wünsche, the leader of the spearhead, greeted me but he did not have the opportunity to ask questions. I thundered up to the lead platoon and raced towards Bollezelle. My troops followed me; they had no idea that I had just gotten up from a stretcher.

We were met with rifle and machine-gun fire from the outskirts of Bollezelle. Mortar fire landed on either side of the road. A halt was not recommended under these circumstances! So — we moved out at top speed towards the town's entrance. The machine seemed to be flying over the cobblestones; I knew that only a few seconds were necessary to cross the danger zone and that my men were following me without hesitation. To the left of the road I saw a machine-gun emplacement. It could no longer fire at our lead elements. A small hut denoted the limit of its field of fire. At full throttle we flew past the first houses. A small barrier consisting of farm machinery was being constructed beyond a slight bend in the road. Without firing a single round, Erich's group succeeded in disarming the French at the barrier.

Behind us the *Kradschützen* fired into the gardens to their left and forced the surprised defenders to give up the fight and gather in the street. Fifteen officers and 250 enlisted personnel set off down the road to captivity. We had two casualties to report. *SS-Unterscharführer* Peters was killed and *SS-*

Oberscharführer Erich was shot through the thigh during the approach. Our bold stroke was a success, but I had to delegate my advance guard to the second-in-command for a couple of days and obey doctor's orders.

On 28 May the regiment, the *2. Panzer-Brigade* and the *11. Schützen-Brigade* went into the attack on Wormhoudt. At 0745 hours the tanks started to move taking the grenadiers with them. Heavy enemy artillery fire attempted to stop our tanks. The enemy was superior to us in artillery. He was also strong in infantry. Just in the sector of the *II./Leibstandarte SS "Adolf Hitler" (mot.)* two enemy regiments were identified.

I was at the regimental command post and was not allowed to leave without permission. My *Kradschützen* were waiting for things to develop in Wormhoudt. They would be employed after the town was taken. The ring around Dunkirk was becoming tighter and tighter.

Sepp Dietrich and Max Wünsche moved to the *I./Leibstandarte SS "Adolf Hitler" (mot.)* to get a clear picture of the situation. At 1150 hours a dispatch rider returned with the disastrous message that Sepp Dietrich and Max Wünsche had been cut off on the eastern outskirts of Esquelberg on their way from the *I.* to the *II./Leibstandarte SS "Adolf Hitler" (mot.)*.

The *2./Leibstandarte SS "Adolf Hitler" (mot.)* tried to free its commander from this critical situation but was prevented by heavy machine gun and artillery fire. The attack of the *15./Leibstandarte SS "Adolf Hitler" (mot.)* was also repulsed by English defensive fire. A reinforced platoon of the *6. Kompanie* of the *2. Panzer-Brigade* under the command of *Leutnant* Corder lost four tanks and was unable to cross the open ground. *Leutnant* Corder and *Feldwebel* Cramel were killed a couple of hundred meters outside of Esquelberg.

Sepp Dietrich's encircled position was clearly visible. It was 50 meters in front of the enemy's position. His car was positioned at a road barrier. The staff car was burning and, from the ditch, thick smoke clouds were rising. The fuel had leaked into the ditch and the dry grass had started to burn. While all this was going on, Dietrich and Wünsche were lying in a narrow culvert, covered with mud from head to toe to protect themselves from the fire.

Five *Panzer IV's* and a platoon of *Panzer II's* advanced to the outskirts of Esquelberg. Those tanks to the left of the road advanced through a park that was stubbornly defended by the English. When they pulled back, the English ignited the fuel they had poured on the park paths so that further forward movement of the tanks became impossible. The regiment's entire sector was under heavy enemy artillery fire. The *III./Leibstandarte SS "Adolf Hitler" (mot.)* succeeded in breaking through to the southeastern part of Wormhoudt at about 1500 hours.

The commander was finally freed from his predicament at 1600 by an assault team from the *I./Leibstandarte SS "Adolf Hitler" (mot.)* commanded by

SS-Hauptsturmführer Ernst Meyer. Unfortunately, the leader of the assault troops, the courageous *SS-Oberscharführer* Oberschelp, was killed. Oberschelp was the first noncommissioned officer of the regiment to receive the Iron Cross, First Class during the Polish campaign.

The *II./Leibstandarte SS "Adolf Hitler" (mot.)* pushed irresistibly on against the toughest of enemy resistance. They were defending tenaciously. Our grenadiers dashed from house to house and, at about 1700 hours, they succeeded in reaching Wormhoudt's market place. The enemy's counterattacks were repulsed. During a surprise advance by enemy tanks, the commander of the *II./Leibstandarte SS "Adolf Hitler" (mot.)*, *SS-Sturmbannführer* Schützek, was wounded. Two tanks were shot up in flames; the regiment captured 11 officers and 320 enlisted personnel. A great many guns and vehicles and ammunition were seized in Wormhoudt. At 2310 hours the regiment recommenced its attack with tank support and forced the English to retreat. During the night six English officers and 430 enlisted personnel were taken captive.

By dawn the regiment had advanced to the Ost Cappel — Rexpoide road without meeting serious resistance. The enemy forces facing the regiment's sector were completely scattered and tried to escape to the north, leaving all equipment behind.

The roads to the north were completely blocked. Endless columns of English trucks, tanks and guns left them useless for any other traffic. The amount of equipment left behind was enormous. The English retreat had taken on the aspect of uncurbed flight. At 1545 hours, the *XIV. Armee-Korps* ordered the attack broken off and immediate preparedness to move. The *Leibstandarte SS "Adolf Hitler" (mot.)* came under the operational control of the *9. Panzer-Division* and was to pursue the enemy fleeing to Dunkirk. At 1800 hours this order was cancelled for reasons unknown. Once again we watched the English and remained inactive. We were not allowed to continue the attack. We had to watch the English evacuating Dunkirk and see them vanish across the Channel. How the war would have turned out, had *Panzergruppe von Kleist* been allowed to continue the planned operation against Dunkirk and had taken the British Expeditionary Force captive, can scarcely be imagined.

The fighting against the English had ended; we did not participate in the final phase of Dunkirk. The regiment came under the command of *Armeeoberkommando* 6 and found itself in the Cambrai sector on 4 June.

The battle of the Somme had started and the Somme position had undergone heavy attacks that led to various breakthroughs. The regiment stood by to be directed towards Amiens or Peronne via Bapaume. The enemy had brought up fresh forces. Their orders were possibly to stop the deep breakthrough designed to hinder the retreat of those French forces fighting in the northern part of the front line and to buy time to construct a new line behind the Oise. But it was also possible the enemy would try to escape to the south during the night.

It was therefore intended to attack with four divisions and break through in a southwesterly direction on 8 June. The regiment was placed under the operational control of the *3. Panzer-Division*. The attack started out on schedule. The breakthrough was forced. On 9 June we were suddenly transferred to the *XXXXIV. Armee-Korps* sector and received orders to advance towards Soissons, Villers Cotterets, and then in a southeasterly direction. By the time my *Kradschützen* crossed the Aisne to the west of Soissons they were dead tired. But there was no time for sleep. We were supposed to move through the forest of Villers Cotterets during the night and then advance to La Ferte Milon.

Deepest night surrounded us as we entered the dark forest and moved slowly along the road that had been severely damaged by mines and bombs. *Infanterie-Regiment 124* was bivouacked along both sides of the main road in the forest of Doxauiale. Soon we passed the last outposts and moved into no-man's-land. French stragglers willingly gave themselves up. They mostly belonged to the French 11th Division. The rustling of the tall beeches disturbed us. We expected to encounter the enemy at any moment and each sound seemed to proclaim his presence.

This movement through the woods must have been a special experience for Sepp Dietrich. It was there, during the First World War, he took part in his first tank action and destroyed his first enemy tank.

About 0400 hours we reached Villers Cotterets and took a number of surprised Frenchmen captive. The whole situation already bore the stamp of impending collapse. Scattered units of the French 11th Division offered only token resistance.

At 0500 hours we advanced in the direction of La Ferte Milon and, in the forest 4 kilometers south of Villers Cotterets, we took prisoner some more members of the 11th Division. A short distance from La Ferte Milon the leading elements came under hostile infantry fire. The village was then quickly taken.

The *I./Leibstandarte SS "Adolf Hitler" (mot.)* succeeded in entering Chateau-Thierry with its first attack and advanced as far as the destroyed railway bridge. Chateau-Thierry, a fateful town for the Germans, had been evacuated by the French, but heavy artillery fire was falling on the deserted streets, turning the sleepy town into a rather uncomfortable place.

On 11 June my advance guard moved through Brumets and Coulombe towards Montrenil and broke through several of the enemy's lines of resistance. I could not restrain my grenadiers. The dash for the Marne started. At 0530 hours on the next day we raced through Montrenil and surprised the French during their wake-up call. They eagerly threw away their weapons and congregated on the main street. At 0904 hours we reached the Marne at St.

Aulge. Our heavy weapon fire destroyed enemy columns on the south bank before we handed our position over to the *II./Leibstandarte SS "Adolf Hitler" (mot.)* and set out to pursue still further the already beaten enemy.

Although we had fulfilled our objectives by reaching the northern bank of the Marne, the *II./Leibstandarte SS "Adolf Hitler" (mot.)* forced a crossing at St. Aulge to create a bridgehead and so enabled an attack against the railway embankment in the bend of the Marne. At 1850 hours Moey was taken, the embankment was reached and the attack pressed. By creating this bridgehead, further operations over the next few days were made easier and the enemy was prevented from deploying defensively in the bend of the Marne.

During the night the regiment was relieved and, once again, came under the *9. Panzer-Division*. At 1245 hours of the memorable 14 June we listened to a special broadcast: "German troops have been marching into Paris since early morning…" Soldiers of the *11./Leibstandarte SS "Adolf Hitler" (mot.)* rushed into the village church at Etrepilly and started the bells ringing. Silently we stood on the advance route and listened to the solemn peal. Nobody cheered, there was no toast of joy, no bonfire. Deeply moved, we acknowledged this fact and our eyes followed the *Stuka* squadrons flying over the Marne, carrying death to the south. On the evening of the same day one of my best noncommissioned officers died a soldier's death. *SS-Hauptscharführer* Schildknecht was killed, an example to his whole platoon.

We continued south through Montmirail and Nevers, with orders to form a bridgehead across the Allier near Moulins. The road bridge was blown up by the French in front of our eyes. The detonation happened at the same time as a *Leutnant* of *Schützen-Regiment 10* was trying to reach the other bank. He was flung into the torrent of the Loire together with the bridge. In contrast to the road bridge, we succeeded in taking the railway bridge. It had been set on fire, but we were able to form a bridgehead, the fire being unable to destroy the iron structure of the bridge. The enemy offered only token resistance, although the French high command still had some 70 divisions available for battle. But the French army no longer wanted to fight. We only met isolated pockets of determined resistance.

On 19 June I received orders to reconnoiter the route from Moulins to Gannat via St. Pourcain. At dawn, my advance guard was moving through the tree dotted, undulating landscape, clearing the road with gunfire. The fleeing French units repeatedly tried to form lines of resistance to gain time and space for their retreat. These attempts didn't bother us any more. We had only one aim: To gain ground to the south. The flanks had become unimportant. We moved down the roads like a fire-spitting dragon. Halting was taboo. Firing was only conducted from moving vehicles. The advance was beginning to look like a wild hunt.

About 1030 hours we crested a small rise and looked down on St. Pourcain. I was moving with the spearhead and saw French soldiers eagerly endeavoring to erect a roadblock at the entrance to the town. The area on either side of the road was open, without cover, and it sloped down for about 800 meters in the direction of St. Pourcain. This unfavorable terrain prohibited an infantry attack on the town. I therefore decided to surprise the French and overrun them by a lightning attack on the obstacle. For this surprise attack I chose a lead element under the command of *SS-Obersturmführer* Knittel. I ordered the rest of the advance platoon to follow 100 meters behind and give covering fire to the leading element.

The French still had no idea that we were already so close to St. Pourcain. They were calmly collecting all kinds of things to block the entrance to the town.

The first motorcycle/sidecar combination then swept over the hill and down the road like lightning, firing as it went. The remaining motorcycles followed at top speed. The armored cars followed on either side of the road and fired their 20 mm rounds in front of the advancing motorcycles. Mortars were firing on the town. Within a few seconds all hell broke loose. I rushed after the lead element and saw startled soldiers running, panic stricken, from the houses into the street. Wildly gesticulating officers tried in vain to force their soldiers to fight. The surprise was too complete and an effective defense was no longer possible. Only single rifle rounds whistled over our heads. Within a short time the obstacle was breached and a gap made for our passage. A 75 mm field gun was positioned at the barrier. The gun crew did not have time to fire, our lead elements had been too quick.

The initial surprise was gone and it was no longer advisable to continue with a mounted attack. We pushed along both sides of the road dismounted. As we turned into the main street we encountered lively machine gun fire that made us more cautious. But we could not lose any more time; the attack had to be executed as quickly as possible to stop the demolition of the bridge that was at the far end of the town.

The advance-guard platoon leapfrogged towards the bridge. The French prisoners ran back trying to leave the danger zone. Fifty meters from the bridge the assault-group leader, *SS-Obersturmführer* Knittel, was shot in the thigh and had just enough time to roll behind a big elm tree before pieces of the bridge flew around our ears. As soon as the smoke cloud began to clear we were greeted with lively infantry fire. The other bank was somewhat higher and afforded an excellent defense. Under these circumstances we remained in the line we had secured and I asked the following battalion to bypass St. Pourcain and try to capture the bridge over the Sioule about 12 kilometers to the south.

Meanwhile, beyond the demolished bridge, the enemy was feeling secure and had no idea that complete destruction was already on the way. At 1420 hours the commander of the advance-guard company of the *III./Leibstandarte SS "Adolf Hitler" (mot.)*, Jochen Peiper, sent a message that the crossing of the

Sioule had been accomplished. In addition, an enemy company with all its equipment had been captured as it was retreating towards Gannat. The *III./Leibstandarte SS "Adolf Hitler" (mot.)* was quickly thrown across the river and sent to attack the French in St. Pourcain. The battalion attacked the French rear and was able to finish the fight without great loss.

My advance guard left St. Pourcain and chased the enemy towards Gannat. By 1600 hours Gannat had already been taken without a fight and we reconnoitered towards Vichy.

Strong barriers made from trees on the Gannat — Vichy road prevented us from carrying out this task before nightfall. Shortly before reaching Vichy we surprised a heavy artillery unit that had halted. Its ancient trucks were unable to climb the steep hill. The guns must have been relics from the First World War and were surely not fit for use. They had probably led a peaceful existence in some depot up to then.

The motorcycle company succeeded in disarming the French without loss and sent them marching back to Gannat. A demoralized French officer was standing on the road looking mournfully at the guns. I saw tears running down his cheeks. He stuttered: "For shame! The soldiers of Verdun would not have let that happen."

We found the bridge over the Allier intact and established contact with German troops at Vichy. On 19 June, 17 officers and 933 enlisted personnel were captured. All the prisoners looked worn out and demoralized.

On 29 June the *II./Leibstandarte SS "Adolf Hitler" (mot.)* advanced on Clermont-Ferrand and seized 242 planes of different types on the airfield. Eight tanks, innumerable vehicles and other equipment fell into the hands of our troops. In addition, the battalion captured a Major General, 286 officers and 4,075 enlisted personnel.

A French captain captured at Pont du Chateau after the start of the attack volunteered to be a parlementaire. Despite his white flag, he was shot and killed by French troops when he reached the French lines to ask the so-called "open city" to surrender.

On 23 June the advance guard set out in the direction of St. Etienne and, 2 kilometers north of La Fouillouse, came under brisk fire from a road obstacle. The obstacle was located behind a small outcropping and was therefore difficult to engage. A 37 mm antitank gun was prepared for firing. It was planned to push it round the outcrop to fire at the obstacle.

The lead element had moved into the undergrowth on both sides of the road and was trying to see beyond the far side of the obstacle. I moved up along the ditch with the second section and had just passed the antitank gun's position when lively shooting started and a tank rolled out from behind the obstacle. It moved around the outcrop, firing. We pressed ourselves into the bottom of the ditch like rabbits and gazed at the advancing mass of steel.

Spellbound, we watched the tracks as they came closer and closer to the edge and looked as if they were just about to slide down into the ditch over the masonry border. Finally, the tank stopped even with us at the apex of the bend.

Tank and antitank gun confronted each other, 20 meters apart. The antitank gun fired first and, after the ringing impact, we heard the shrill whistle of the ricocheting round. Nor did the second round penetrate the armor. The steel plates were too thick for the 37 mm rounds. We watched as the tank rolled directly towards the antitank gun and scored a direct hit on the crew. Only a few meters before reaching the gun's position, it turned around and retired behind the barrier. We were relieved to see that the second round had jammed the turret and the gunner could not aim his weapon properly. Unfortunately, three *Panzerjäger* from the antitank gun crew were killed. They were the last of our men to die in the French campaign of 1940.

From the lead element's position, I could see six enemy tanks in all positioned behind the obstacle. They were First World War veterans, which had been built for the planned 1919 offensive, but never saw action. The tanks were forced away half an hour later by our 150 mm rounds. The road to St. Etienne was open. Next morning the *I./Leibstandarte SS "Adolf Hitler" (mot.)* entered the town and took several hundred French captive.

At 2145 hours we heard a truce between Italy and France had been signed. The fighting in France had ended. Was this the end of the war?

We were concerned to hear of the arranged demarcation line and we had to withdraw from occupied territory on 4 July. Our regiment was then attached to *Armeeoberkommando 12* and, in the early hours, set out for Paris. We were to take part in the planned parade.

In spite of the defeat the French army had suffered, the French population was quite friendly. Shortly before we reached Paris we heard of the sinking of the French fleet by British battleships in the port of Dakar. This action touched the French very deeply. Never before and never again have I seen so many people crying as that time in France. Churchill's deed was not regarded as an act of war but as a crime.

Paris was surrounded by a strong cordon from *Division von Briesen*. The city center could only be entered with permission and a pass from headquarters. I took the opportunity to see the sights of Paris with my *Kradschützen*. Because the planned parade for the *Führer* was first postponed and then finally cancelled, the regiment left Paris and marched towards Metz.

I asked Sepp Dietrich's permission to leave 24 hours earlier so that I could show the blood-soaked battlefields of Verdun to my soldiers. Permission was given and consequently there were some one hundred soldiers at Fort Dounaumont on 28 July 1940.

Together we climbed through the casemates that, 25 years previously, had been taken by *Hauptmann* von Brandis and *Oberleutnant* Haupt and their courageous Brandenburg grenadiers. Deeply moved, we stood before the great casemate, the gate of which was walled up, and behind which innumerable German soldiers were taking their final rest.

The scarred area around the destroyed Fort Donaumont told an unambiguous tale. Line after line of craters, a scene from a lunar landscape. The thin layer of grass was unable to hide the suffering of this tortured earth. Communication trenches cut through the landscape like deeply furrowed wrinkles.

Between Fort Donaumont and the charnel house we discovered the grave of a fallen comrade, who had lost his young life only a few weeks ago. With bared heads we stood by this forlorn grave and gazed at the innumerable graves to our left. Thousands of wooden crosses could be seen in front of the charnel house. Words lost their power. The invisible regiments, whose former existence was signified by the crosses, needed no interpreter, they spoke for themselves.

From the charnel house we walked slowly up the mount of Vaux and tried to picture those enormous efforts of the German and French soldiers who lost their lives on this hill in June 1916. We climbed up to the shattered hilltop of the fort and tried to follow the route of *Leutnant* Kiel who, on 2 June 1916, advanced towards the fort's center by the way of the eastern trench with about 40 grenadiers. We soon gave up. In this churned up earth nothing could be found any longer.

At this place, man's destructiveness had changed the face of the earth. In our imagination we saw the dark shadows of the advancing grenadiers, dashing through the roaring barrage and breaking through a battered gap in the outer trench wall. We imagined how the combat engineers put incendiary charges into the firing embrasures and incapacitated the crew in the armored turret. Today the armored cupola lay destroyed at our feet; its power had been broken.

As I described the plight of the French fort's garrison to my soldiers I seemed to hear the roaring thunder of the French artillery that tried to chase the German grenadiers off the hilltop. In the dark corridors of the fort we found scorch marks on the walls and ceilings of the vaults and recognized the effects of the German flamethrowers. Shaken, we stood by the cistern which was partly responsible for the downfall of the French garrison and sensed the agony of the French soldiers who suffered from unbearable thirst. But the Germans on the fort were in the same position. The owner's sweat and blood clung to each water bottle that passed through the storm of steel into the fort.

The visit to this historic place had turned my comrades into a silent audience. Without uttering a single word they surrounded me as I told them of the heroic battles of 8 June 1916. On that day the French attacked in seven waves, one after another, to recapture the fort. But the exhausted Germans fought like madmen. They were not prepared to surrender the fort.

At twilight we followed the route of the 21 soldiers and 2 German officers who passed through the French barrage and reinforced the forlorn group of defenders. These men were the remnants of two German companies. All the others remained on the battlefield.

In the shadow of the all-concealing night our vehicles moved eastwards. The visit to the battlefield had left us circumspect. Verdun had taught us that, in spite of two campaigns behind us, we had not yet experienced our fathers' appalling deprivations.

The first fighting in Poland.

Officers of *SS-Aufklärungs-Abteilung 1* in 1940. Left to right: Hugo Kraas, Max Wünsche, Hermann Weiser, Kurt Meyer.

"Der schnelle Meyer" in July 1940 in France. Note the Army officer cap which has been impressed into *SS* service by Meyer.

August 1940: Fort Alversleben (Metz). Taken after the *"Leibstandarte"* received the *Führerstandarte* from Adolf Hitler. Left to right: Sepp Dietrich, *SS-Sturmbannführer* Keilhaus, *SS-Untersturmführer* Ritz, *Panzermeyer*.

The Formation of the Reconnaissance Battalion at Metz

On 29 July we occupied Fort Alversleben at Metz. The fort lies west of the Moselle and commands a view far down the beautiful Moselle valley.

In the fort we discovered old batteries with Krupp cannon from the turn of the century. Even the appropriate ammunition was stored next to the guns, neatly arrayed. The inventory report originated from the Prussian gunners who had to surrender their guns to the French in 1918.

With a lot of effort and work we succeeded in making the fort fit for billeting troops. The fort was to become a training ground for the newly established *SS-Aufklärungs-Abteilung* of the *verstärkte* [reinforced] *Infanterie-Brigade Leibstandarte SS "Adolf Hitler"*.

I was entrusted with its formation in August. As a nucleus of units I had at my disposal the former *15. (Kradschützen-)Kompanie*, the *Panzer-Späh-Zug* [armored car platoon] and the *Kradmeldezug* [motorcycle messenger platoon] of the former regiment. I was allowed to select additional personnel from the Motorcycle Replacement Battalion at Ellwangen.

I did not have to search around Ellwangen for long. The young *Kradschützen* wanted to join combat units and were glad to leave their home barracks. Splendid, healthy young men surrounded me as soon as I asked for volunteers. These were young men who had just turned 18 and had been soldiers for six weeks. Within a few days the new battalion in Metz was complete and started an intensive training program. Nothing was too difficult for our young comrades. They eagerly followed their instructors' orders and welded themselves into a steel-hard team. The old battlefields of St. Privat, Gravelotte, and Mars la Tour became the training area for the *Kradschützen* and the *Panzeraufklärer*.

In the years since the defeat a lot of nonsense has been written about the composition of the *Waffen-SS*. I think it is necessary to give the reader a survey of the origins and sociological composition of the unit at this point. As an example, I quote from the records of the *2. (Kradschützen-)/Aufklärungs-Abteilung Leibstandarte SS "Adolf Hitler"*. The men practiced the following professions prior to their enlistment:

1. Technical professions: 42.73% (Their fathers: 10.9%)
2. Skilled trades: 21.69% (Their fathers 39.03%)
3. Self-employed: 14.16% (Their fathers: 26.08%)
4. Farming: 6.41% (their fathers 8.76%)
5. Unskilled: 15.01% (their fathers: 18.23%)

The average age of the soldiers was 19.35 years. The noncommissioned officers were 25.76 years. The average age of the entire company was 22.5 years. In all, the members of the unit had 452 siblings. All regions of the *Reich* were represented. It may be said with complete justification that these troops represented a cross-section of the German people and were neither a unit for bigwigs or mercenaries.

Forty-eight officers, noncommissioned officers and men were killed from within this splendid company during the six months from 10 July to 31 December 1941. A further 122 were wounded during the same period. In the tough defensive fighting at Rostow in December 1941 the company was reduced to platoon strength.

Where do today's public figures pluck the courage to call these faithful and self-sacrificing young men party soldiers? These young men fought for Germany and certainly did not die for a party.

*

In the autumn of 1940, I was detailed to a staff officer course at Mühlhausen in the Alsace. The Chief Instructor was excellent, the commander of the *73. Infanterie-Division*, *Generalleutnant* Bieler. During this course I got to know a couple of colleagues, *Oberst* Hitzfeld and *Major* Stiefvater, with whom l would later share some serious moments in Greece and Russia.

During this time the units were training for Operation Sea Lion and carried out amphibious landing exercises. The Moselle River was the favored place to train. Under secret conditions, the training was switched to mountain warfare. We rode our motorcycles at breakneck speed across the steep slopes of the Moselle Mountains. The area surrounding the fort, with its high walls and trenches, looked like a circus. We even practiced scaling up and down cliff faces with motorcycles and antitank guns. By the spring we considered ourselves to be a well-trained unit, ready for employment. The teamwork with the heavy weapons functioned as smoothly as a chronometer. *Generaloberst* Blaskowitz found words of high praise for our efforts. *General* von Kortzfleisch expressed himself similarly as we were inspected in Metz for the last time. The unit was ready for action and awaited orders.

The Balkans

During the First World War the German people fared badly on their southeastern flank. In the autumn of 1916, at the same time as we were making the bloodiest of sacrifices in the west on the Somme, in the east against Bruisow, and in the south on the Isonzo, the Allied Powers completed the encirclement of Germany as a result of Rumania's mobilization. For two long years the Central Powers fought the Allied Army of Salonika in the rugged mountains of Macedonia. Not until the autumn of 1918 did the Allied forces under General Franchet d'Esperey succeed in breaking through the German defense. With 29 Allied divisions he advanced to the Danube, thus sealing the fate of our allies.

Which role the Balkans were to play could only be guessed at in the spring of 1941. One fact was that Winston Churchill again held a decisive influence on British strategy. In his time he had organized the Gallipoli landings and then the Salonika operation.

In the spring of 1941 London held a Balkan expeditionary force in readiness in the Mediterranean and, subsequently, landed it in Greece. In the middle of February, Foreign Secretary Eden and General Sir John Dill, the Chief of the Imperial General Staff, spent time in Athens to discuss the deployment of British troops in Greece

In January the initial German units of *Armee List* marched into Rumania. The soldiers were regarded as a training cadre and the population gave them a warm welcome.

We also received marching orders at the beginning of February. Nobody knew where the move would lead. We crossed the Rhine near Straßburg and then moved through the magnificent southern German landscape into Bohemia. Passing Prague, we moved directly south and next morning saw the silhouette of Budapest. Continuing through the *Puszta*, our train approached the Rumanian frontier. We got to know the Transylvanian Saxons and their beautiful country in the Hungarian-Rumanian border area and were received with overwhelming hospitality and cordiality in Kronstadt, Hermannstadt and many other imposing settlements of the Teutonic Knights in the Carpathians.

Our troops were billeted in the Campulung district. This formerly tranquil small train station took on an entirely different aspect with lively activity reigning everywhere. In the very first hour of our road march I had an experience that cast its shadow far into the future. A Rumanian Lieutenant-Colonel was cursing the bad road conditions and asked me to drag his small car out of the mud. A lady, obviously in pain, was sitting in the car as the Lieutenant Colonel drove away. I quickly forgot the incident.

In 1943, however, while at in Döberitz/Krampnitz a Rumanian Colonel came up to me and greeted me warmly. He repeatedly told his comrades that I was the savior of his wife and son. It took a long time for me to realize what he meant. When the car was stuck, his wife was on the way to a maternity home and was in labor. He reached the maternity home just in time for her to give birth to a son. Naturally, we celebrated our reunion in Germany.

After a few weeks stay in Campulung we set out on our march south to Bulgaria along deeply rutted and muddy roads. The tank tracks cut deeper and deeper into the road and the repair columns worked ceaselessly. Bare, broad expanses with hardly any high ground or wooded areas could be seen on either side of the road. From time to time we passed through impoverished villages, with a well, a few mud huts pressed deep into the earth, a few windswept fences and nothing else. Then one morning we saw the wide, earth-brown ribbon of the slowly flowing Danube. South of the Danube, rising through the haze and mist, towered the mountains of Bulgaria.

The sun shone down unmercifully as we rolled into Bulgaria across the bridge built by our combat engineers. The Bulgarians give us a festive welcome. Many First World War memories were reawakened and Bulgarian peasants showed us their German decorations with pride. The march over the notorious Schipka pass was unforgettable. The dangerous hairpin bends were negotiated with élan as recovery teams stood by. When all else failed, the Bulgarians lent us their draught-oxen.

The long columns rolled inexorably southwards, past Sofia and into the Struma Valley. The jagged mountains threatened to crush us. The drivers just managed to move their heavy vehicles along the narrow mountain roads. A sea of dust lay on the roads. These roads, with their potholes, steep descents and sharp bends that demand the utmost of the vehicles, had been witnessing enormous quantities of traffic for many days. A traffic jam about 20 kilometers long developed in the Struma Valley and was an especially serious bottleneck.

The combat and construction engineers built a new road, continuously grading, dynamiting, and bridging. The danger of a bottleneck was soon over; the columns crossed the valley quickly and disappeared into the feeder valleys. The towering mountain ranges, hidden canyons and wide valleys effectively hid the enormous troop contingents. Large amounts of fuel, ammunition and other supplies were stockpiled by the roads. The approach march was over. The assault companies were ready.

Meanwhile, anti-German circles stirred up by the English had taken over in Belgrade. A revolt during the night of 26/27 March overthrew the government and Prince-Regent Paul was forced to leave the country. As a result, the Balkan situation changed dramatically. On the eve of the Belgrade coup Hitler had already decided to remove the Yugoslavian threat to our flank.

Besides *Generaloberst* von Kleist's *Panzergruppe 1* and *Generaloberst Graf von Weichs' 2. Armee* which were moving against Belgrade and northern Yugoslavia, *Feldmarschall* List was leading the *12. Armee* against southern Yugoslavia (Skopje) and Greece. The *12. Armee* had 16 divisions at its disposal, in addition to the *Infanterie-Regiment (mot.) "Großdeutschland"* and our regiment.

Hitler ordered the attack against Yugoslavia on 6 April, following the conclusion of a non-aggression and friendship pact between the Soviet Union and Yugoslavia on 5 April.

A hot spring day neared its end. The heat in the Struma Valley was almost unbearable. Because of the events in Yugoslavia we were moving north to Kustendil, which lies right on, the Bulgarian-Yugoslavian border. The *9. Panzer-Division* had already reached the border town and had orders to advance on Skopje and, if possible, take this important junction in a coup de main. We were to follow the *9. Panzer-Division* almost as far as Skopje, then turn south and head for the Greek border via Prilep.

My reinforced battalion was formed in a square before me. Darkness surrounded us as I told my comrades the essential details of our impending operation. They listened silently as I explained the mission of our advance guard and indicated expected problems. I also thought it appropriate to remind them of the fierce battles that our fathers fought during the First World War in the black mountains of Macedonia and in the occupation of Monastir, the city which cost no end of blood and which was our first objective. We wanted to take it swiftly and by surprise. At these words I sensed, for the first time, the boundless trust that united me with my men. I could lead them into hell, and they would follow.

The night was muggy, little was said and most of the men were smoking. Each man liked to be alone with his thoughts just before an operation. The silver sickle of the moon cast a ghostly light on the men crouched beside their motorcycles. The steep, bare slopes of the mountain rose ominously before us out of the first light of dawn. The white road snaked up it in steep curves, and we knew that pillboxes and dragons teeth awaited us at the summit.

The advance guard of the *9. Panzer-Division* set off at dawn on its way west across this natural frontier. At an elevation of 1,200 meters they came across the Yugoslavian fortifications. The heavy weapons spoke their first words. The 88 mm antiaircraft guns and the heavy antitank guns shattered the enemy pillboxes. The fortifications were turned into a smoking shambles within a few minutes. It was an unearthly picture. Far to the east a blood-red sun arose and, in the valleys, the morning mist boiled with dense dust clouds.

Red tracer rounds arced in short, flat streaks from the frontier mountains into the receding dawn. Machine guns lashed at the pockets of resistance, rattling down the valley. Suddenly, enemy planes appeared. Coming low over the mountains they struck at the valley road and attacked Kustendil. The roads were filled with columns of troops as bombs fell on the town. Thank God, the losses were few but, unfortunately, *SS-Obersturmbannführer* Mohnke, commander of the *II./Infanterie-Brigade Leibstandarte SS "Adolf Hitler"* was seriously wounded. *SS-Hauptsturmführer* Baum took over the battalion.

We were getting closer and closer to the border and finally reached it in the late afternoon. The *9. Panzer-Division* had broken through the frontier fortifications and was advancing deep into Yugoslavia. The armored spearhead rolled towards Skopje, fighting along the way. We moved down from the heights past destroyed frontier barriers, roadblocks and well–positioned pillboxes. We encountered innumerable prisoners, among them many Batschka and Banat Germans who welcomed us with loud shouts and shook our hands. Carcasses of dead horses, already bloated by the southern sun, lay in the ditch. Living horses trotted over the fields or stood apathetically by the side of the road. The rough landscape took on a different character. The mountains receded. Their snowy silhouettes remained behind us. We found friendly knocked-out tanks and fresh graves outside of Kuma Nowo indicating a fierce battle for the town. Darkness sank quickly over the line of march. We would soon have to reach the large road junction south of Skopje. From that point onwards we were to take the lead and strike southwards via Prilep.

We reached the last outposts of the *9. Panzer-Division* shortly after midnight and prepared to enter no-man's-land. Before the advance-guard platoon, under the leadership of *SS-Untersturmführer* Wawrzinek, moved out, I briefed the platoon once more on our situation and wished my comrades all the best. I sent the platoon into the darkness with the words: "Boys, the night belongs to the good soldier!"

The motorcycles forged on, slowly at first, then ever more quickly. It was reminiscent of the Dutch campaign. I soon found out that Wawrzinek had personally taken the lead and, without much ado, scrambled to the south. But these were not the smooth asphalt roads of Holland and France. Our advance had to cross narrow mountain paths and ravines. The road rose steeply. After a short while the first rounds whistled over our heads. The enemy was crouching somewhere in the mountains and trying to halt our advance. I was moving behind the lead section. A short call sufficed to get it moving again. Onward, ever onward. Our goal: Win ground to the south and take advantage of the enemy's confusion.

We came under fire below a small hill outside a village. Armored cars supported the already attacking *Kradschützen* and fired their tracers into the enemy. *Kradschützen* combed the village. Over a hundred bewildered Yugoslavs were the consequence of our battalion's first engagement. The enemy officers were cursing their tactical outposts in the mountains. They listened incredulously to our interpreter's explanation that the fire from their outposts did not bother us and that we simply continued our advance to the south. Half an hour later everything was over.

The *Kradschützen* pressed on. The men could not be restrained. On we went! In a breakneck move past the slopes and through ravines we surprised an enemy battery on the move. In a couple of minutes the excitement was over. Creaking and groaning, the guns tumbled into the ravine.

At dawn we arrived at Prilep and made contact with the advance guard of the *73. Infanterie-Division*. The commanding officer of the battalion was *Major* Stiefvater with whom I attended a training course at Mühlhausen. Stiefvater had advanced directly from east to west and reached Prilep without great loss.

We granted ourselves a sorely needed rest. It looked like it could have gotten hot that day. We had a long-range objective and advanced on the important town of Monastir. A light rain drizzled in the early dawn. It mixed with the dust and changed it to gray, glutinous mud. Anxiously, we gazed into the fleeing shadows of night. The road now led out onto a plain, with only the outline of a high mountain visible to the right. From behind an outcrop we looked out at the Zrna River and caught sight of a substantial bridge with steel arches — it had not yet been blown.

Some enemy trucks and horse-drawn carriages were approaching the bridge. I only saw the bridge — nothing else interested me. It had to fall into our hands intact. Automatically two armored cars dropped out of the line of march and fired their 20 mm rounds at the far approaches to the bridge. The lead elements raced to the bridge as if possessed by the devil.

Horse-drawn carriages and motor vehicles collided in a tangle, each one wanting to get across first. The lead elements were only 100 meters away. Stray shots winged past us. I already saw myself in possession of the intact bridge, but then, just before we reached our goal, a muffled bang echoed across the river valley. The bridge lifted into the air before my eyes and collapsed in on itself. Enemy horses, soldiers and vehicles were thrown into the air and then disappeared into the swirling waters of the Zrna. Clattering machine gun fire chopped into the wreckage.

First horrified, then angry, and finally, coldly assessing the situation, I approached the wreckage. The commander of the *2./SS-Aufklärungs-Abteilung 1, SS-Hauptsturmführer* Kraas was beside me. The situation was quickly assessed and a decision made. The enemy was not to be allowed any time to rest! He had to be hunted down.

We were lucky! The iron structure projected above the water and could be used as a foundation for a temporary bridge. Grenadiers clambered across the wreckage and secured a small bridgehead. Combat engineers and anyone else who was free gave a hand to lug over the beams and other building material.

Solo motorcycles were taken over the river and reconnoitered in the direction of Monastir. The combat engineers constructed a new crossing, as if they were in the training area. The bridge grew before our eyes and, soon afterwards, the first heavy armored cars crossed the river. The advance continued.

The *2. Kompanie* was put back in the lead. A railway line to Monastir ran to the left of the road on which we were advancing. Behind its embankment were crouching enemy riflemen who were trying — in vain — to delay our rapid progress. The armored cars merely let a few machine-gun rounds fly against the enemy lying behind and on the embankment. All the others had their eyes directed to the front. We wanted to take Monastir in a coup de main; everything else was irrelevant. The railway embankment got closer and closer to the road, crossing it a few hundred meters ahead of us. The lead section halted and the grenadiers leapt into the ditches on either side of the road. They opened fire on the occupants of a linesman's shack. An enemy machine gun spat its ammo down the road. The cottage was seen as a pocket of resistance. The 50 mm antitank gun took up position under enemy fire and fired a couple of rounds at the walls. The building burst asunder with a crash.

It was only then that I noticed the enemy on the embankment had become more active, emboldened by the halt of our spearhead. Machine-gun fire lashed down from the embankment and there was nothing else to be done but take out the opposition on the embankment. The matter was speedily resolved under covering fire from the armored cars. The survivors fled into the marshy ground beyond the embankment.

I was just about to jump onto the motorcycle of *SS-Unterscharführer* Weil when renewed fire from some Yugoslavs forced us to the ground again. My map board was shot up; its remains lay at the edge of the ditch. Enemy rounds pumped into the grass and tore up the soggy earth around us. A gurgling noise compelled me to look at Weil. He was squirming at the bottom of the ditch, his lower jaw hanging — shattered.

We couldn't allow ourselves to get hung up there! The enemy could not be allowed to nail us down outside the gates of Monastir and rake us from those towering heights. I shouted to the lead elements. The *Kradschützen* leapt like acrobats onto their motorcycles and dashed on. The leading elements pulled the remainder of the battalion like a magnet as it tore along the rain-soaked road. It had been raining all day; at that point, the sun began shining through patchy cloud cover.

Resistance stiffened; tracer bullets whistled venomously towards the sheltering haystacks and turned them into gigantic torches. Monastir was ahead of us. We could see the city sprawling between the mountains. On the right-hand slope I saw an enemy battery that was just moving into position. Onward! No time for firefights! We had to get into the city. We wanted to jump down the defenders' throats. We were intoxicated by the speed! Our machine gun bursts raked the enemy on either side of the road. In front of us was a half-finished roadblock. Fire! Bursting rounds from the armored cars and hand grenades flew through the air. The surprised and totally dazed defenders dove for cover.

We didn't advance on a broad front as the enemy expected. Instead we were like a flexible, lightning-fast dagger thrust — the battalion moved into town in column. It was only the artillery that wasn't with us. It was set up and fired its heavy rounds over our heads.

I saw neither the minarets nor any other buildings — only machine-gun nests, defended houses and determined enemies. The battalion bored deeper and deeper into the town. I had lost my maps but I knew where the barracks were. We wanted to reach them because it was there that we would find the strings that directed the enemy movements.

The troops formed up on the square scattered as soon as the motorcycles came roaring around the corner. Enemy fire was hitting us from all the windows, roofs and vegetation. The armored cars proved their worth. Their weapons dusted down every suspicious-looking corner and forced the enemy riflemen into cover. Two heavy infantry guns took up position under the covering fire of the armored cars. They were less than 200 meters from the barracks into which they fired their 150 mm rounds. The result was convincing. Within 20 minutes Monastir was no longer being defended with the exception of a pocket of resistance at the train station. It was engaged by a combat-engineer element in a fight that was over within the hour.

Our advance had proved what we had learned during years of training: The motor is a weapon.

In the course of the next few hours apathetic prisoners were brought in and disarmed. An entire artillery battalion fell into our hands without a shot being fired. But the battle had to be carried forward; we had no time to rest.

Serbian forces were positioned at Lake Ochrida and had occupied the Javat Pass, 20 kilometers west of Monastir. We knew a strong British force had come up from the south and pushed into the Florina area southeast of us on the Greek border. I was hard pressed to make a decision on our next move. We were alone in Monastir and could not reckon on support for the next 24 hours. I had to advance in two directions and also hold Monastir with the staff, artillery and trains drivers.

Kompanie Kraas got the job of forcing the Javat Pass with a reinforced company and making contact — via Ochrida — with the Italians who were in the mountains west of Florina. *Kompanie Schröder* was ordered to reconnoiter the British lines and to stay in contact with the enemy. If possible, the British were to be prevented from leaving the Klidi Pass. Both company commanders looked at me in astonishment as they received these orders. Hugo Kraas shook his head in disbelief.

The situation was not too bad for Schröder. He had a large area in which to maneuver and, with the roads being in good condition, he could exploit every opportunity for reconnaissance. The companies moved out. My *Kradschützen*, *Panzerjäger*, *Pioniere* and *Grenadiere* moved past me laughing. They moved into the darkness, into uncertainty. The headquarters staff set up for defense and monitored the radios.

We remained in constant contact with the companies and monitored the reports from the armored cars. We knew every move that was made. Within a few minutes Kraas attacked a battery positioned in an orchard west of Monastir. It was still waiting for orders to fire. The complete battery marched off to captivity.

By midnight Kraas had advanced through several villages and was at the foot of the Javat Pass. Reconnaissance reported the pass was occupied and a well-developed defensive position had been constructed along the ridge. Enemy reconnaissance patrols were taken prisoner. The attack on the pass was to take place at dawn.

Schröder made good progress and soon reported from Florina and Vevi. The company had a bizarre experience between those two villages. As the old saying goes, all cats are gray when the candles go out. Schröder gave the following account the next morning.

I dispatched several reconnaissance patrols from the road junction and slowly followed the first one that was reconnoitering towards Florina. Before long, two reconnaissance cars loomed up out of the darkness coming towards us. Unsuspecting, we continued to move on. I thought they were our own cars. We realized our error only a few meters ahead of them. The two English reconnaissance cars stopped in front of us then moved slowly on. They didn't know who was next to them either. They must have taken us for Serbs. Relieved, I moved the company forward a few hundred meters and awaited the return of the English armored cars. A half hour later they fell victim to our obstacle on the road. We could tell the enemy's intent from the captured maps. Australian troops had occupied the high ground and closed the valley with extensive minefields.

Schröder remained in contact with the British and continued his aggressive patrolling. Our infantry guns and mortars must have deceived the enemy concerning our true strength. He did not advance beyond his own obstacles.

Kraas' attack on the Javat heights began early in the morning. The road led steeply into the mountains so a surprise attack was out of the question. Hairpin bends alternated with steep curves. Steeply dropping precipices, barren ravines, overhanging cliffs and bare, treeless expanses completed the picture. The pass was more than 1,000 meters high; it bordered on insanity to risk an attack there with only a reinforced company. But surprise was on our side. No one had suspected such a rapid advance and absolutely nobody could have conceived that a single company might risk an attack on the pass.

I went forward to *Kompanie Kraas* before dawn. I was anxious and wanted to experience the attack myself. We passed a war memorial from the First

World War just north of Monastir. There, on the heights, innumerable German soldiers slept in foreign soil.

The sun broke through as we heard the first sounds of fighting.

We could clearly see the impact of the heavy infantry guns. The 150 mm rounds must have had a terrifying effect in the mountains. The 20 mm tracers rose into the mountains like pearl necklaces. I came across the heavy guns and vehicles in the valley. Only the armored cars could accompany the company in the attack.

The *Kradschützen* become mountain troops. During the night they had climbed up the pass along both sides of the roads and were in front of the positions on the ridge. Several of the positions had been bypassed and were attacked from the rear. With considerable élan, the company commander led his soldiers onto the ridge and rolled up the Serbian positions. The psychological effect of the heavy guns had been of considerable unexpected benefit. The heavy-caliber rounds created a hellish noise.

I had attached myself behind an armored car and participated in the last fighting for the high ground. *Gruppe Tkocz* eliminated the last pocket of resistance. I found Hugo Kraas behind a small chapel and congratulated him on his success. A few hundred prisoners lay, stood and crouched in front of us. An entire battery had surrendered. The result was beyond our comprehension. Our battalion in this formidable position would have been able to resist an attack by regiments. The enemy battalion commander gave us an explanation. He said, "When my men heard yesterday evening that German troops had already reached Monastir and would appear in front of our positions tonight, their will to resist was considerably weakened. The fact of having to fight German soldiers in itself unnerved my battalion. Your "bomb throwers" [meaning the heavy infantry guns], did the rest."

From the highest peak we looked down on shimmering, blue Lake Ochrida and brightly illuminated Florina. The city had to be ours before the enemy noticed his battalion in the blocking position on the high ground had surrendered. We had no motorcycles available, but a number of armored cars was ready to roll into the valley and surprise the enemy. While we slowly groped down around the serpentine bends, *Kompanie Kraas* assembled and awaited its vehicles. We reached the valley floor without encountering the enemy and dashed towards the city. I was in the armored car of Bügelsack, this *SS-Oberscharführer* was the leader of my best reconnaissance team and had an excellent nose for sniffing out the situation.

The fleeing Serbs rushed from the road and sought cover in the undergrowth. Others threw away their weapons and moved towards the pass. At that point, it was impossible to stop; we had to enter the city and take advantage of the confusion. Bursts of machine-gun fire cleared the road. The surprise was complete. Within a few minutes we were standing on the Kirchberg and firing red flares into the sky. *Kompanie Kraas* dashed into the city soon after and sent out reconnaissance teams towards the mountains west of Lake Ochrida to make contact with the Italians. Contact was made in a few hours. The first mission of the reconnaissance battalion was carried out quickly, successfully and without excessive casualties. I was proud of my men, and knew that I could stake everything on them.

I reported the latest success of my battalion to the commander of the *Leibstandarte* in Monastir and accompanied him to *Kompanie Schröder* where we met the commander of the *I./Infanterie-Brigade Leibstandarte SS "Adolf Hitler"*, *SS-Sturmbannführer* Witt. His battalion received orders to take the key British positions defending the Klidi Pass thus enabling the breakthrough of the brigade and the *9. Panzer-Division*. New Zealanders and Australians were deeply entrenched on the mountain slopes of the Klidi Pass. The enemy had had time to develop an impressively extensive defensive system.

Their artillery observers could observe far out across the plain over which our troops would approach. Monastir was the floodgate of the Klidi Pass, and the pass was the Yugoslavian gate into Greece. The enemy had every advantage on his side. A deep minefield in the pass ruled out an armored attack. It was our infantry that had to take the heights in hard fighting.

Into Greece

A sunny summer day turned into a rainy, changing mountain evening, and later into a freezing night. Snow covered the slopes. The men of the *I./Infanterie-Brigade Leibstandarte SS "Adolf Hitler"* faced a vastly superior enemy. They were dug into makeshift trenches, awaiting the attack.

It happened at dawn on 12 August. The screaming of heavy shells broke the silence. The heavy antiaircraft guns started to destroy identified pockets of resistance and the *Sturmgeschütze* rolled forward. I was standing at a scissors telescope, watching the attack of the *1./Infanterie-Brigade Leibstandarte SS "Adolf Hitler"* under the command of *SS-Obersturmführer* Gerd Pleiß.

There was still a hail of fire coming down the mountain; the entire summit was veiled in smoke and the air smelled of earth and sulfur. All of a sudden, the artillery fire ceased. The infantry leapt up and worked their way up the mountain. The heavy *Sturmgeschütze* climbed the slopes from the bottom of the valley. We watched the guns advance in amazement. They climbed higher and higher, and then joined the fight. Nobody thought it possible to use them, but now they were up there giving valuable assistance to the infantry.

Completely shaken by the impression German shelling had made on them, British prisoners came down the mountain. They were tall, strong fellows and formidable opponents. Our infantry advanced deeper and deeper into the defensive system. Combat engineers pushed into the minefield to clear a path for the armor. But, even there, the infantry had a hard job throwing the British out of their positions. Only then could the engineers clear the mines. Shaken, *SS-Sturmbannführer* Witt stood in front of the mortal remains of his brother Franz. His younger brother was trying to negotiate the minefield and was torn apart by the mines.

Pleiß was now leading from the front and was fighting just below the summit. The assault guns could no longer provide assistance, only men counted there. We could not hear the crash of the hand grenades but could see the clouds of smoke from their detonations. Nests of riflemen were taken in hand-to-hand combat and the summit stormed.

The brave men of *Kompanie Pleiß* had defeated the opposition. More than 100 prisoners, 20 machine guns and other equipment was taken. Gerd Pleiß himself had been wounded, but he stayed with his grenadiers. The gateway to

Greece had been kicked open. The fighting continued. The *I./Infanterie-Brigade Leibstandarte SS "Adolf Hitler"* attacked the withdrawing enemy at a furious pace. Enemy tanks were destroyed by antitank and assault guns. Enemy planes tried to halt our advance, but their bombs did not have the desired effect.

SS-Hauptsturmführer Fend, commander of an 88 mm battery, had been taken captive and spent the night in a British column. At dawn our infantry freed him. Additional New Zealanders started the way into captivity. Early in the morning the southern exit of the pass was taken. Strong British and Greek forces tried to turn the tables and push the Germans back into the pass. The British had a large number of tanks at their disposal and seriously threatened our spearhead. The *I./Infanterie-Brigade Leibstandarte SS "Adolf Hitler"* had already reached open country and our assault guns were still in the mountains, a dangerous situation. The first enemy tanks were already in the midst of the lead company when, suddenly, SS-*Obersturmführer Dr.* Naumann appeared with two 88 mm guns, opened fire and put an end to the nightmare. Tank after tank flew into the air or came to a smoking standstill. The attack was quashed in fire, death, and ruin.

While the *9. Panzer-Division* pushed south, my reconnaissance battalion dashed on in the direction of Lake Kastoria. The shadows of night were already encircling us when we recognized the threatening, dark mountains of the Klisura Pass. Our objective was Koritza, headquarters of the Greek III Corps, but before this was the Klisura Pass, a mountain that presented a great obstacle, if only from the standpoint of being able to negotiate it, even without opposition. Rising to almost 1400 meters, the summits seemed to be pressing down on us. The advance was rapid; two of the ridges extending in front of us were taken in the next 30 minutes.

Broad and massive, the mountain lay before us, the road wound upwards in a series of tight curves. There was no longer any way back; to turn around would have been physically impossible. To the left the terrain steeply sloped away into sheer, inaccessible ravines while, to the right, the vertical cliffs towered up. Small mountain villages appeared dead and deserted. In the last village the inhabitants gazed at us fearfully. Their faces were both questioning and expectant. A terrific tension filled us. The smell of sulfur filled the air. Cliffs stood out from the mountain like mock pillboxes.

The next ridge appeared in front of us in tiers. The road bore slightly to the right and then had to bridge a narrow but deep ravine. We maneuvered carefully towards the bend. At any moment we expected a hail of fire or the rocks to be blown down on us. We felt as though we were walking on hot coals. The spearhead halted. The men took cover and moved into firing position. What was happening? There was still no firing. Full of tension, I ran up front. Before us yawned a void in the road. The bridge across the ravine had been blown; the massive stone span was in a heap of rubble on the ravine floor and formed a low saddle.

We were surprised to see that the obstacle was undefended, with no sign of any enemy positions. We worked our way forward cautiously towards the broken span; the ravine was perhaps 15 meters wide and could easily be crossed by foot soldiers. It was impassable for motorcycles, however. The leading platoon was ordered to secure the far side of the bridge and cover the building of the intended provisional crossing. The grenadiers had hardly reached the rubble of the bridge when machine-gun bursts flew around our ears.

We could make out the enemy emplacement to our right on the mountaintop. The muzzle flashes of the machine guns indicated where their positions lay. Grenades came whistling through the air and exploded in the ravine behind us. Mortars tried to drive us away from the obstacle. My battalion had gotten itself into a very unpleasant position — it could neither advance nor retreat. There were no alternatives. We were on the only road leading across the mountains and into the rear of the Greek III Corps.

The conquest of such a massif was really a job for mountain troops, not for an armored reconnaissance battalion — but those reflections were overcome by events. There were no mountain troops available, so we had do the job, and we would do it, even if we had to sell our souls in the process! Both motorcycle companies would assault the enemy's position at dawn, while I continued along the road with the drivers, the staff, and the armored car company faking the main attack. The heavy weapons and artillery would only be available later.

Meanwhile, darkness had fallen. Weak harassing fire continued to play over the obstacle occasionally. The engineers bored holes for explosive charges to flatten out the rough approaches to the old bridge. Minutes later, masses of earth and rocks tumbled down into the ravine on top of the remains. My swift reconnaissance battalion then became a labor battalion. Strong grenadiers hauled boulders over and hurled them down on to the remains of the bridge, a living conveyor belt passed stone after stone into the sheer abyss. Shortly thereafter, the first antitank gun crossed into the bridgehead. Our bridge was holding. Just after the new bridge had been finished, the two motorcycle companies began to scale the massif. *Kradschützen* had become *Gebirgsjäger*! The grenadiers had to climb some 800 meters before engaging the enemy pockets of resistance. Both companies advanced like storm troopers. Separated from each other by the ravine they had to fend for themselves. Although on separate routes, they had a common objective: The peak!

We were then facing the enemy; the men's tiredness blew away. Nerves were tense, all adventurous instincts aroused. My *Kradschützen* were confident. They knew they would be successful. They used traditional tactics for rough terrain, passing one another, groping stooped from boulder to boulder. *Kompanie Kraas* also disappeared out of the ravine to the right and climbed up the mountain; it has the longest distance to cover. I took charge of the section that was to advance along the road. We were about 30 strong and had a few armored cars, antitank guns and a section of 88 mm *Flak* with us.

The road snaked higher and higher and we had no contact at all with the other companies. All was quiet. Nothing broke the stillness of the night, not a single round. The moon had vanished behind the mountains and the night became darker and blacker. According to the map, we had reached the big bend that had to lead around the last rock face and into the enemy's rear. The enemy's position had to be high above us. Our concept was to move around his flank and cut off his retreat.

The road curved around the mountain and continued some 400 meters in a northerly direction before it bent due west again into a group of farm houses. Near these houses the road crossed the crest of the mountains and then sloped down to Lake Kastoria. I dared not advance any further. I had the feeling there was something wrong. We needed to wait for first light.

It was windy and cold on the crest; we pressed close to the rock wall. Naumann's platoon manhandled an 88 mm *Flak* into position so it could take the farm buildings and the ridgeline under fire.

It had gradually become bitterly cold. Since we had neither coats nor blankets and were soaked through with sweat, we suffered a lot. We were shivering with cold. Sleep was out of the question. If we could only have had a smoke! A radio car moved up slowly. Under its cover I had a smoke and studied the map again. The longer I looked at it, the worse my shivering became. At first I thought my teeth were chattering because of the awful cold but then I realized I was very frightened. The more time passed, the tenser I became. I could no longer stand being in the car. The radios with their endless "beep, beep, beep," got on my nerves.

Outside I shied away from talking to a man, I was worried he might hear my teeth chattering and realize I was afraid. We all crouched silently behind the rock wall staring into the dark. Were my young comrades also afraid? I couldn't tell. *SS-Kradschütze* John, from the *1./SS-Ausklärungs-Abteilung*, arrived with a report from his unit. His company, unknown to the enemy, was just below its position, waiting for dawn. John had a bullet graze on his head. Yet he seemed unafraid. He gave me a short clear report, and was then given a drink from the medic's canteen.

It got lighter. We could soon make out the silhouette of the village. The attack by all three groups was to start with the firing of the 88. I crouched behind the gun and tried to penetrate the darkness with my binoculars. The closer it got to the time to open fire, the more I believed in the success of the attack. It simply had to succeed. I counted on my opponent having learned his lessons in the military academy and anticipated which measures he might take in this case. From all that the Greek commander had learned, he would expect me to advance along the road with my motorized unit. That was why I would attack him across the two ridges, and only conduct a feint down the road.

The outlines of the houses could be seen as the shadows faded. Pressed to the ground, I gave Naumann the order to open fire. Within a few seconds we found ourselves in a witch's cauldron. The 88 fired round after round into the ridge on our right; mortars and infantry guns let loose their rounds which then hailed down on the defenders. High above us the *Kradschützen* stormed the enemy defenses. I could not see the two motorcycle companies attack but I could hear their furious machine-gun fire and the crashes of their grenades.

The commander of a heavy field howitzer battery informed me he could no longer support the companies without endangering our personnel. The guns were positioned along the mountain road, one behind another. But, because the road was so narrow, they could not dig in their trail spades. The commander refused to accept responsibility. This sort of nonsense was the last thing I needed. Angrily I ordered him to open fire. We had to do it. The heavy rounds roared over the first ridge and smashed into the enemy's positions on either side of the little mountain village.

Enemy machine-gun fire hacked and sprayed down the road and into the rocks above us, bringing stones rolling down the slope and crashing among us. There was nothing to do but to go forward. We rushed the first bend in leapfrog fashion and took cover a few meters further on behind the rock wall. At the next bend we would be directly below the enemy position 100 meters above the road. I collapsed, exhausted, behind a block of stone and gasped for breath. Our forward movement was hampered by having to leap from one scrap of cover to the next in order not to give the enemy snipers an aiming point.

Above us we heard screams and the raging sounds of battle. Elements of the *2./SS-Auklärungs-Abteilung 1* had broken into the enemy's positions on the first ridge. We dashed on. At the final big bend we encountered some men who had become separated from their company above by a crevasse. Among them was *SS-Untersturmführer* Wawrzinek who gave me a brief report on the operation on the ridge. From the statements of prisoners, we were up against a reinforced infantry regiment on the left wing of the Greek defenses. It had the mission of holding the Klisura Pass for the retreat of the Greek III Corps from the Albanian Front. They were pulling back to avoid capture by German armored forces and continue the fight for southern Greece in conjunction with the British forces. The Greek plan could not be allowed to succeed. Not only had the retreat to be prevented, but it also had to be turned into a catastrophe. We had to cross the mountains and block the valley beyond Kastoria.

We moved along the road. Suddenly the ground in front of us heaved upwards. I couldn't believe my eyes. Where the road had been, there was a vast crater at that point. The road had plunged into the ravine. Sweat left bright tracks on our faces. We were terrified. Were we also to fly into the air in the next few seconds? A hundred meters further on the mountain shook anew and, after the dust settled, there was yet another hole in the road.

We hid behind the rocks not daring to move. Nausea almost choked me. I yelled at Emil Wawrzinek to press the attack. But good old Emil looked at me as if he doubted my sanity. Machine-gun fire splashed against the rocks in front of us; our lead element was only about ten men strong. Damn it! We certainly couldn't remain there, while craters were being blown in the road and machine-gun fire was pinning us down in the rubble. But I, too, was crouching in full cover and fearing for my life. How was I to order Wawrzinek to move first? In my distress, I felt the smooth roundness of an egg grenade in my hand. I yelled at the group. They all looked at me thunderstruck when I showed the grenade, pulled the pin and let it roll it behind the last grenadier. I had never seen such a unified leap forward. As if bitten by a tarantula we dashed around the rocks and into the fresh crater. The paralysis was broken; the grenade had done it. We grinned at each other and dashed forward into fresh cover.

On top of the ridge the companies had penetrated deeper and deeper into the Greek positions. The 88's were surrounded by clouds of dirt and shell bursts from Greek mountain artillery, but Neumann's section continued firing. The antiaircraft rounds prepared the way for us, burying the pockets of resistance under heaps of rubble.

We were just below the summit. The sweat was burning my eyes. I could observe the fighting through a film of dust and dirt. We rushed the ridge like madmen. The Greeks scrambled out of their positions holding up their hands, no longer defending themselves. Their line of retreat was already under fire from the *2./SS-Aufklärungs-Abteilung 1*, whose machine guns fired from the highest point directly into their positions. We broke the mountain battery's resistance with hand grenades. We had forced the crossing of the mountains. My grenadiers had achieved what others thought impossible and what, even today, was considered insanity. Klisura Pass belonged to us! There was no time for a break. Only pursuit would bring us the fruits of victory.

Our combat engineers blew masses of rock into the craters on the road. The artillery changed position and fired into the fleeing enemy. Whole columns pulled out to the west, onto the plain. The resistance of the Greeks who, here and there, bravely held their positions to the last breath, was broken. Over a thousand prisoners were taken, including the regimental and three battalion commanders. It was only then that the vital importance of the position became clear to us. From the pass we could look directly down onto the Greek Army's line of retreat at which all our weapons fire was then directed.

I wanted to push on after the fleeing enemy but, once more, the steeply descending road exploded around our ears. Precious time was lost in filling up the craters. The *2./SS-Aufklärungs-Abteilung 1* tentatively moved down the road and into a small village. It had been evacuated. I wanted to regroup my battalion there and then push on along the main Greek line of retreat. I was waiting for the *1./SS-Aufklärungs-Abteilung 1*. The young soldiers appeared shortly. Their faces told me everything. On a blood-soaked shelter-quarter they carried the remains of their company commander. Rudolf Schröder lay before me, his chest torn to pieces. He had achieved a unique military success. As leader of the first assault group, he was killed in the initial breakthrough of the enemy defensive system.

We reached the plain in the late afternoon and reconnoitered towards Kastoria. I wanted to see the lay of the land and followed a reconnaissance section. We slowed the pace before reaching a small bridge. Beyond it was Hill 800, commanding the approach to Kastoria as well as the route of the retreating Greek III Corps. No movement was to be seen on the bridge; it had not yet been demolished. Suddenly, machine-gun fired opened up on us. Franz Roth, the war correspondent, began screaming. A bullet had split open his skull. He was returned to his colleagues in the rear with a bloody head.

The *2./SS-Aufklärungs-Abteilung 1* reached the bridge as night fell, forming a small bridgehead. The company reconnoitered north of Lake Kastoria and met stiff resistance. The attack on Hill 800 southwest of Kastoria started at daybreak.

Again the shells howled over us and bored into the masses of stone, but the Greek artillery was stronger. The bridge collapsed under a direct hit. We lay there dumbstruck and pressed hard into the filth at the bottom of the ditch. The intensity of the artillery fire told me that a surprise assault would not succeed. What was required was a deliberate attack. At about midday the attack was repeated with the support of heavier artillery and the *III./Infanterie-Brigade Leibstandarte SS "Adolf Hitler"*. The battalion moved out to envelop from the left and was intended to push on to the Greek's main line of retreat in the course of the afternoon. To eliminate the strong Greek artillery and help reduce the enemy positions on Hill 800, a *Stuka* unit was called up to support my battalion.

The operation was executed with a precision second to none. The *Stukas* struck the enemy positions like birds of prey, flying in wide curves around the mountain, and then diving, screaming, into the depths. They began their descent into Hell with full bomb loads. There were crashes and flashes on the heights and within the massif. Giant mushrooms of dust and rubble shot into the sky, merging into each other and drifting as a dark haze across the lake. In the scorching light of the sun, a thick veil covered the mountain, showing the devastating effect of our bombs and artillery shells. All hell had broken loose up there.

When the first bombs fall, the *Kradschützen* stormed out of their trenches and ran across the open field with lungs wheezing. The excellent firing of the 88's completed the work of the *Stukas* and artillery. It would be a long time before the Greeks recovered from the *Stuka* attack. By then it was too late. The *2./SS-Aufklärungs-Abteilung 1* climbed up the mountain and had a firm foothold in the tumble of rocks.

The rest of the battalion dashed into Kastoria over the temporarily repaired bridge. Unsuspecting Greek companies and batteries withdrawing from the mountains were so surprised that they did not fire a shot and marched willingly into captivity. One of their batteries continued to fire and was shot to pieces. The armored cars roared past the Greek columns into the center of Kastoria. Chaos was complete. In the market square the local priest greeted me. I will never forget his brotherly embrace. For hours afterwards I still stank of garlic.

At twilight, my brave comrades took over the security of the northern approaches. Greek units were still coming from there, having fought against Italian troops. It was raining steadily. A heavy thunderstorm combined with the thunder of shells and bombs. We were at the end of our strength. We fell asleep right where we stood. The extent of our success was only clear the following morning. During the previous 24 hours the reconnaissance battalion had taken 12,000 prisoners and captured 36 guns. I was awarded the Knight's Cross for this performance by my brave grenadiers.

The fight against the trapped Greek Army continued. The *Leibstandarte*, after overcoming considerable difficulties, forced the Metsovon Pass and the capitulation of 16 divisions. The surrender was signed on 21 April in Larissa. Late in the afternoon of 24 April, while in Joanina, I was ordered to pursue the defeated British forces. My comrades had just spent their first quiet night since the start of the Balkan Campaign. They were shaken awake from their dreamless sleep. Their fuel tanks were filled to the brim from fuel cans taken from a Greek depot. Nobody had bothered to gather up the Greek machine guns positioned in front of the Ali Pasha mosque in the old Turkish fortress or, for that matter, the untold mountains of weapons in the town.

The Greek soldiers, who were still coming out of the Albanian mountains, left their weapons leaning against the walls of the houses, shaved off their dark beards, strode into the nearest bakery and, coming out with loaves of fresh bread, a bundle of leeks and, if they were lucky, a few fish threaded on a stick, wandered off to the south.

We overtook them. This act of overtaking once more vividly drew our attention to the difference between the paths of the victors and the vanquished. These men, be they uniformed herdsmen, fishermen, farmers, shopkeepers or officers, had really shown us in great numbers a conduct that demanded our respect, but their returning home was chaos. They poured through the valleys and down the slopes in a thousand rivulets. The war ended in hopeless confusion, even if you saw the occasional colonel sitting erect in his saddle with a trumpeter by his side. This was disintegration.

On, on we went. We were going to have to bump into the British sooner or later. Without stopping, the battalion asked brief questions in every village and town. Whoever was able to cut himself a slice of bread while on the move, smeared some fat drippings on it and, until it had all been eaten, held his hand over it so he didn't swallow a coating of dust and dirt along with it.

Only once, on the Gulf of Arta, did I allow a short halt. I found the orange groves too seductive. The men filled their helmets with the aromatic fruit; we wanted to taste them to prove we were in the South! In a narrow mountain pass stood a pitiful Greek Army nag, a white horse, with blue shadows showing between its ribs. Lost and unharnessed, at the end of its patient strength. It didn't stir; it stood as a monument to the collapse. Every vehicle and motorcycle made a detour around it. It was a war veteran, an overworked, pitiful creature.

Further south, we moved past a gurgling mountain river and saw the disarmed soldiers in it enjoying the cool water. But we, covered in dirt and sweat, were not allowed to enjoy a single drop. We saw thousands lying in the shade of the olive trees. Instead we had to concentrate on fuel levels and the bends in the road. We had to avoid the potholes and hang on tight as we jolted over the ruts. We could certainly remember the miserable Polish roads, but this road was the devil's "cheese grater" which seemed intent on completely flogging us to death. Evening and night arrived and we still had not reached our objective. British stragglers and demolition squads were rushing away in front of us. Greek farmers told us the British were scattering nails by the bundle on the roads to delay our advance, and they did not seem to be entirely wrong. The young drivers swore; the old drivers showed them a trick or two when yet another flat had to be repaired. We took a short break in a one-horse town; the battalion had to close up.

The hunt continued at dawn. The trail led steadily south, up hill and down dale, through deep gorges. Ruins of classical Greece greeted us. Somebody mentioned Lord Byron, who was killed here while fighting the Turks in 1824. But we did not have any time to think about history. Mesolongion appeared in front of us. The Isthmus of Corinth would soon be reached. We would then be able to nab the British. Carefully, the spearhead moved its way toward the town and into the narrow streets. A jubilant Greek population welcomed us. The last British troops had only just left the town and were on the eastbound coast road, disappearing in the direction of the Isthmus of Corinth.

The Crossing to Peloponnesus

We had survived the 250-kilometer move through mountainous country and stood opposite the dark mountains that towered over Peloponnesus. We had no radio contact with the regiment. We were alone.

English reconnaissance planes flew above us and circled the harbor of Patras on the other side of the gulf. We could make out ships in the harbor and saw a British destroyer turning away to the south. We followed the tracks of the English demolition parties and advanced toward the Isthmus of Corinth. But this sort of thing was no longer to my taste. This kind of pursuit had become uninteresting to me. Yawning craters in the road reduced our speed. I thought I was most likely to gain lots of experience in road building but that I wouldn't nab any more British. The mountains on the other side attracted my attention more and more. On the coast road on the far side the British units were rolling from Corinth to Patras to reach the evacuation ships. I had to get there! But how was the gulf to be crossed?

I was standing on the mole at Navpaktos, a small, poor harbor with defensive towers from the Middle Ages, when a dive-bomber formation attacked the harbor of Patras. Clouds of smoke and explosions shot into the air from the convoy of ships. I caught sight of a telephone. It was still connected, it worked and Patras was answering! Startled, I replaced the handset back down on the ancient apparatus.

But I was fascinated by the idea of crossing the gulf and spoiling the British plans. I sent for an interpreter and asked him to call the Greek commander in Patras and request a situation report. The commander was still under the influence of the dive-bomber attack and answered all questions readily. Within a few minutes I had a precise report on English troop movements between Corinth and Patras. I told the commander to send a liaison detachment to Navpaktos.

Before long I observed a little motorboat on course to Navpaktos. The devil then proceeded to take over. Another *Stuka* squadron howled over us and repeated the attack on the British ships in the harbor. The discomforting aspect of this was that the town commander thought I had ordered the attack.

On top of everything else, the pilots attacked the boat carrying the liaison officer on their return flight. The boat immediately made a 180-degree turn and an insulted Greek officer reported by telephone that, under the circumstances, nobody wanted to make the trip across the gulf.

Sweat dripped onto my map. The entries had been out of date for a long time. Where were the English? On the left wing, following the conquest of Thermopylae, our troops must either have reached Athens or advanced further south to the Isthmus of Corinth. Therefore the British had either to defend Peloponnesia or make for the ports. I imagined German paratroops would land on the Isthmus to block the narrows near Corinth.

Had the British taken notice of our speedy advance? Had their reconnaissance worked well? Were destroyers standing by to prevent an attempted crossing? Nobody could give me an answer. My soldiers and officers watched me expectantly. They saw me standing on the mole, estimating the distance over and over again. More than 15 kilometers of water separated us from the British line of retreat. The next day, at the latest, the Isthmus of Corinth would be bitterly contested and I wanted to participate in that fight. I wanted to get across.

The moment had come: I would no longer act in accordance with the traditional conventions of war and when all responsibility rested in my hands. I would cross the Gulf of Corinth with the forces at my disposal. Whether that was a daring or a foolhardy move would only be revealed in the next few hours. My comrades were enthusiastic but practical objections were soon raised: The artillery could not support a landing; the distance was too great. Engineers drew my attention to the height of the waves and the miserable fishing cutters. The objections accumulated, but I had made my decision. The surprise attack had to succeed.

Two miserable fishing cutters were found in the harbor. Their crews were brought to the spot. The *2./SS-Aufklärungs-Abteilung 1* had to attempt the reconnaissance. Strong arms lifted the heavy *BMW* motorcycles and heaved them into the boats. The first boat took five bikes with sidecars and fifteen men. On the next cutter we put an antitank gun and some bikes. The mission: "Block the road and, in case of emergency, hide in the mountains."

Then the cutters left the harbor. I took my leave of Hugo Kraas and *SS-Sturmbannführer* Grezech. Those who remained christened the Kradschützen a "suicide patrol". A joker shouted: "Look out, mine on the port bow!" Everybody laughed. A young soldier shouted back: "What do you mean, a mine? This skiff isn't even worth a grenade!" The cutter started to pitch heavily. Breakers sprayed over me. Machine gunners were in position at the bow. The antitank gun was ready for firing.

All boats on this side were directed to Navpaktos. Soon the rest of the company was loaded onto boats. The first cutters could scarcely be made out any more. Two tiny points danced on the waves.

57

I stood on the marina again and watched the dark points on the water. A red flare was supposed to signal the failure of the mission and the existence of strong enemy forces. That's what I had agreed upon with my men. My eyes were burning. Soon, I was unable to make anything out anymore, but I didn't dare to put down my binoculars. By the time I was no longer able to see the boats, my clothes were soaked in sweat. We had been standing, waiting on the shore, for an hour. The tension had reached its peak. After an hour and a half two dots came back into view. Were they our boats? Had we been successful or were the boats bringing back shot-up bodies? They came closer and closer. Soon the outlines could be made out clearly and we could also see movement. A circle of half-smoked cigarettes lay around me as I stuck the next one between my lips. I calmed down and began to trust in the success of our operation.

Suddenly, a dust-covered staff car stopped on the shore and agitated officers jumped out of it. I recognized my revered commander, Sepp Dietrich, and reported my decision and the course of the operation so far. During my report, I started to notice the old daredevil gasping for breath and scrutinizing me from head to toe. Then a storm broke over me: "Are you crazy making such a goddamned decision? You should be brought before a court martial! How can you treat my soldiers that way?" I dared not reply in the face of this flood of undoubtedly justified reproof. I stood at the old harbor wall with my tail between my legs and wished for the world to end. Embarrassed silence all around. Only my men smiled at me surreptitiously, just as if they wanted to say: "Get a grip, don't be bothered about his ass chewing. He might be right, but take us across the gulf now so we'll have something to do again!"

Meanwhile, the cutters had come closer and, with binoculars, one could make out details on board. Both boats were full of men. There were more men coming back than I had sent over. I did not dare to say it aloud, but it was true. Both boats were bringing back captured English soldiers. Sepp Dietrich looked at me, turned and went. No more words were spoken.

I had no more reason for delay. Loaded boats sailed toward the returning ones. Tensely I awaited the report from the other side. What had happened over there? A *SS-Rottenführer* reported:

After an hour in those tiny shells the mighty mountainous coast of Peloponnesia stood out in front of the mast. Now came the ultimate test. All binoculars were searching the shore. 800 meters out, 700 meters, 600 meters, 500 meters, surely a machine gun would start chattering over there! Some forms could be seen between the houses and on the shore through our binoculars and with the naked eye. We no longer thought about anything. We lay flat in the boat, held our rifles and machine guns at the ready and we prepared to jump up as soon as we had touched land. We vaulted out and ran toward the houses. And, just at the very moment that we were running, a brown armored car appeared around a bend in the road about 50 meters away, revolved its turret and aimed the barrels of its weapons at the beach. We who had landed were paralyzed at first but then we waved at the armored car in a friendly way. Standing on the shore in shirtsleeves and without headgear we looked just like bandits. The Tommy vehicle growled, revolved its turret again and moved off.

What had happened? Didn't the guy recognize us as Germans? We stood among the first houses and clenched our eyelids — and some other parts. We stared back at the other part of Greece and saw nothing but water and, away over there, steep, bleak mountains. We had to act. We knew they were waiting. To the foot of the first range of the Peloponnesian mountains was only just over 100 meters, with only a railway track and a country road between the shore and the foothills. We ran up to the road and secured our eastern flank from where the English had come. We had scarcely reached the road when we again heard the sound of an engine. The platoon leader ordered us into full cover. Civilians, wine growers and fishermen, had already come out of their houses by then. When they saw the foreign soldiers suddenly disappearing between boulders and bushes they also threw themselves to the ground in fear. We heard our hearts drumming in our chests out of excitement.

Round the bend came an English dispatch rider with a truck behind him. They moved down the road without a concern since the armored car had already checked the area out. We let them approach until we could read their license plate, until the shield bearing the feather crested knight's helmet, the sign of the 4th Hussars, was above us. Then we jumped up and shouted: "Hands up!" The brakes squealed. The Englishmen's heads snapped up and they jumped down from the truck. The dispatch rider's feet were searching for the gravel. One Tommy cried out, his submachine gun flew into the ditch. "Hands up! Hands up!" They let their weapons fall and raised their arms. Then one of our men ran round the bend and shouted: "Another truck is coming!" In a fraction of a second an *SS* man got into the first truck and moved it across the road. The prisoners were quickly moved behind the houses. The second vehicle approached, again with a dispatch rider in front. Surprise and amazement were repeated. One Tommy was able to get out: "Germans?" Indeed the Germans were already there. Within a few minutes we had taken more than forty prisoners with three officers among them. They told us that they were on the way to Patras harbor. None of them realized that we had already crossed the gulf. Their unit was still fighting near Corinth."

Give me boats! All the small boats were assembled. The whole battalion had to cross during the coming night. My faithful driver, Erich Petersilie, had hung the last bottle of sparkling wine in the waters of the harbor. I clamped it under my arm and went to see Sepp Dietrich, who was having a conversation with the English officers. I invited the Englishmen to join us in a glass. We sat in the shade of a leafy tree. Before I could say a word, an English officer raised his glass and drank to the health of his sister whose birthday it apparently was. I'm sure we didn't look like a scholarly crowd during that round of drinking.

I took my leave and jumped into one of those damned skiffs. Half an hour later I was as sick as a dog. I didn't think this nutshell we were in would reach the other side, but it got us there. I greeted *Kompanie Kraas* suffering from complete exhaustion. In accordance with orders, the company had obtained vehicles and reconnoitered as far as Corinth. *SS-Sturmbannführer* Grezech had established contact with the town commander of Patras and ordered the larger boats to Navpaktos. The last Englishmen left the area of Patras and withdrew to the south.

Air reconnaissance in the afternoon reported an enemy regiment advancing between Corinth and Patras. The affair was getting interesting. The off-loading was completed with flying hands, so that the ships could return to the far shore as quickly as possible. As for heavy weapons, we had some antitank guns and a light armored car at our disposal. We prepared a "welcome" for the English. But the reported regiment did not arrive; perhaps it had already changed direction to the south, into the mountains.

Through the intelligence service, we learned about our paratroop operations near Corinth and that *Fallschirmjägerregiment 2*, under the command of *Oberst* Sturm, was deployed there. Contact with the paratroops had to be established immediately. The *2./SS-Aufklärungs-Abteilung 1* received the order to clear the south coast of the Gulf of Corinth of the enemy and advance to link up with the paratroops. The *1./SS-Aufklärungs-Abteilung 1* occupied Patras and reconnoitered south. The companies set off in captured and confiscated cars and motorcycles. An elegant limousine towed an antitank gun and mortars stuck out of a sports car. The combat engineer platoon was in a bus and gave the impression that the whole war was no longer any of its business.

Although the mass of our vehicles was still on the northern shore of the gulf, the battalion was nonetheless motorized and underway on the roads of Peloponnesia. I wasn't satisfied with its speed. I overtook the *2./SS-Aufklärungs-Abteilung 1* and sped toward Corinth. My instinct told me the British had already cleared off to the south. Our limousine bounced furiously along a washed-out coastal road. The little fishing villages lay sleeping and deserted in the brooding heat. Leaving a small village, just at a bend, I spotted a car disappearing off the road and racing at full speed toward a farmstead. We were even with them at that point. My comrades shouted: "Tommies!" I stepped on the throttle and the car accelerated like a rocket. I snatched another glance at the "Tommies" and recognized a German submachine gun. The last man was looking for cover behind his car when I spotted a German paratrooper helmet. The paratroops also recognized us and lowered their weapons. They took us for British and had also taken cover. A few minutes later my company arrived and established contact with *Fallschirmjägerregiment 2*. *Oberst* Sturm initiated the pursuit south.

We turned round immediately and sped back to Patras again. The shades of night had fallen. In the meantime, the *III./Infanterie-Brigade "Leibstandarte SS Adolf Hitler"* had also made the crossing and started its pursuit south. There was a train engineer in that battalion. He stoked up a railway engine in Patras and hauled the battalion southward. In the command post that day I encountered a now friendly commander. Everybody was silent while I reported to him about establishing contact. With a grin, he gave me his hand and said in his Bavarian dialect: "Hey, Kurt, yesterday I thought you had gone crazy. Now I take it back. It was brilliant. Come on, tell me how you got that crazy idea."

In the background, I saw my adjutant already making fresh entries on our map and looking at his watch again and again. I had hardly given Sepp Dietrich the information he wanted when he gave me a new mission. The battalion had to reassemble immediately and reconnoiter toward Kalumata via Pyrgos, Olympia and Tripolis and continue the pursuit.

My comrades were lying in ditches to both sides of the road and sleeping like dead men. More armor and vehicles had arrived in the meantime. The battalion was ready for operations once more. Just before dawn the move south began. Countless British vehicles lined the road. They had to be abandoned because of a lack of fuel. Our booty was very welcome. We even found the little Bren gun carriers intact. The Greek people in Pyrgos greeted us with wine and tropical fruit. I interrupted the advance in Olympia and took my soldiers into the stadium. The mayor of Pyrgos led us through the classical arena and, likewise, didn't forget to show us the memorial to Heinrich Schliemann [German architect]. For more than an hour we wandered around the rubble-strewn grounds admiring the wonderful mosaics and impressive construction of this historic site.

In Tripolis we linked up with units of the army that were bottling up the English in the southern harbors. *SS-Untersturmführer* Theede, the leader of an armored reconnaissance patrol, returned from Kalumata and reported: "The enemy was collapsing between fire and water." The battle in Greece was over.

The road led us through Patras and Corinth to Athens. In Athens we were supposed to participate in a parade held by *Feldmarschall* List in front of the palace. Moved by so many impressions, we crossed the deeply cut Corinth Canal and, that same evening, stood on the Acropolis. Many comrades, never having learned anything of classical antiquity, were astonished by the technical achievements that had been accomplished two and a half thousand years before. The classicists among us felt invigorated and, perhaps for the first time there on the Acropolis and Propylnea, were able to establish a true relationship with classical antiquity. For them the visit to ancient Hellas reunited them with their youth. They were seeing places which they had never visited before. Surrounded by a mysterious power, we gained strength from the Hellenic inheritance and were prepared to march toward further sacrifice for our homeland.

A Greek soldier stood guard over the memorial to the Unknown Soldier. Brave men were honored there who had sacrificed their all for their country. We moved via Thermopylne, Larissa and the Klidi Pass past burnt out wrecks of armored vehicles and fresh grave mounds, through Monastir, Belgrade and Vienna to the area east of Prague. In Gaya we regained our senses and started servicing our weapons and equipment. We had no idea of what awaited us. The troops were in a fever pitch. New equipment arrived and better weapons were issued. The experiences of the Balkan Campaign were analyzed and intensive training began once more.

Everything we did was governed by speed. We had learned that only the swiftest will gain victory and that only the most agile soldier will survive the fight.

The camaraderie of my battalion was like the comradeship of a big family. An iron discipline was the backbone of the community. We approached further training imbued with these values and forged an instrument upon which I could play all the symphonies of battle. Company commanders and platoon leaders mastered the tricks of the keyboard with finesse. And my young comrades had become soldiers whom I could lead on a loose rein and with few words. They were not uniformed dummies held together by slavish zombie-like obedience. No, before me stood young individuals who believed in themselves, their own values and their own abilities.

After the awarding of the Knight's Cross to Gerd Pleiß, Sepp Dietrich and Fritz Witt.

Struggling through the Klisura Pass with a *Sd.Kfz 222* armored car.

1941 in Greece. This photo appeared on the cover of *Illustrierter Beobacher* with the heading of "Heavy Artillery to the Front" and must be considered one of the most famous *Waffen-SS* images. (via Jost Schneider)

The end of the Greek campaign and a visit to the Acropolis. Kurt Meyer (middle) with Max Wünsche to his left. (Jost Schneider)

The modern soldier meets antiquity at the Acropolis. (Jost Schneider)

Right: The end of the campaign in Greece 1941. Left to right: Keilhaus (Operations Officer), Heinrich Himmler, Kurt Meyer (commander, *SS-Aufklärungs-Abteilung 1*), Surkau (*I./SS-Artillery-Regiment 1*), Mertsch (*II./SS-Artillerie-Regiment 1*). (Roger James Bender)

After the award of the Knight's Cross on 18 May 1941. **Right:** Posing with the
commander of the *Leibstandarte*, *SS-Obergruppenführer* Sepp Dietrich. (Jost
Schneider)

Military cemetary for *Leibstandarte* soldiers at the Klidi Pass.

70

Map of western Russia, the next theater of war for Kurt Meyer after the campaign in the Balkans.

Map of Southern Russia, the scene of the bold reconnaissance operations conducted by Meyer and his *SS-Aufklärungs-Abteilung 1*.

The Struggle Against the Soviet Union

The news of the attack against the Soviet Union struck us like a bolt of lightning. In Gaya, we heard Adolf Hitler's broadcast justifying his decision to do away with the worldwide threat of Bolshevism forever. With a gloomy foreboding that we might suffer the same fate as our fathers who fell victim to a war on several fronts in 1914/18, we prepared for the cruelest war that soldiers had ever had to fight.

On the morning of 27 June 1941, the battalion marched east through a jubilant populace via Olmütz, Ratibor and Beuthen. On 30 June we crossed the Weichsel near Annopol and reached the Russian border near Uscilug at 0800 hours.

Meanwhile, our forces had already advanced far to the east. The *Leibstandarte*, now officially redesignated as a division, was attached to *Heeresgruppe Süd* (*Feldmarschall* von Rundstedt), which had the task of advancing towards Kiev with motorized units and destroying the Russian forces west of the Dniepr. For this task the *Heeresgruppe* had at its disposal 26 *Infanterie-Divisionen*, 4 *Infanterie-Divisionen (mot.)*, 4 *Jäger-Divisionen* and 5 *Panzer-Divisionen*. *Luftflotte 4*, under *Generaloberst* Löhr, supported the advance of the *Heeresgruppe* against a strong force led by Marshal Budjonny. Budjonny was going into battle against *Feldmarschall* von Rundstedt with 40 Infanterie-Divisions, 14 motorized divisions, 6 armored divisions and 21 cavalry divisions.

The Russians fought tenaciously and obstinately over every inch of ground. There was bitter fighting over the entire front as we reached *Rollbahn Nord* at Luck on 1 July. On 2 July we received orders to advance on Rowno via Klevan and there to establish contact with the III *Armee-Korps* (*General* von Mackensen). It was reported the III *Armee-Korps* was surrounded at Rowno. Strong enemy formations were attacking our forces' extended flanks from the Pripet marshes.

This enemy situation was something completely new to us. Slightly confused, l looked at my maps to inform my men about the situation. The old criteria no longer applied. Where was the clear division between friend and foe?

Far ahead, the *Panzer-Divisionen* of the *III Armee-Korps* were fighting east of Rowno. Infantry units were fighting a few kilometers in front of us, and Russian forces were fighting to the north and south of us. After unsuccessful attempts to give my comrades a clear picture of the position, I summed up the situation with the following: "As of today, the enemy is everywhere!" I did not yet realize, as I was passing the word to my men, how closely it summed up the situation.

A couple of kilometers east of Luck, next to a railway line, we came across the last outposts of an infantry battalion. Dense smoke clouds showed us the location of a destroyed Russian tank. Dark forests were on either side of the road along which we were moving eastwards. We had to hurdle a distance of 60 kilometers to reach Rowno. Would we reach our objective in time and would we be able to give our comrades of the *25. Infanterie-Division (mot.)* tangible assistance? Calmly, I gave the lead elements the signal to start the move into the coming night. In the lead was Fritz Montag. Montag was a veteran of the First World War and was the platoon leader of a motorcycle platoon in the *1./SS-Aufklärungs-Abteilung 1.*

At first slowly, then more swiftly, the spearhead moved towards the dark forests. I wanted to shout out: "Stop! What I preached to you isn't true! The engine isn't your friend. Move slowly, or you will drive to your death!" But my lips wouldn't open. They were pressed together to protect themselves from the air stream. Squinting my eyes, I followed the lead elements of the column boring further and further into the forest.

Destroyed Russian tanks were on both sides of the road. Trucks and horse drawn vehicles were abandoned beneath the trees. At one place, we found twelve camouflaged Russian T26's. They were without fuel, deserted. Ready for combat, the battalion moved past a large area of heath that extended north. Suddenly, I saw the lead platoon disappear in front of me and the 20 mm cannon of an armored car fire at some bushes. Four or five of these "bushes" then moved towards us and started firing at a distance of about 150 meters. Well-camouflaged Russian tanks were attacking the line of march. In the twinkling of an eye, everyone took cover in the ditches and watched the duel between our armored cars and the enemy tanks. A few antitank guns entered the fray and put an end to the uproar. After a few minutes we resumed the march. The burning tanks illuminated the night sky to our rear for a long time. At 1930 hours we exchanged the first rounds with the Soviets. In order to make sure we would accomplish the mission, I sent combat reconnaissance south, with orders to advance on Klevan via Olyka and wait for the battalion at Klevan.

Extreme darkness surrounded us when we reached Stovek and moved further east. Shortly after midnight, I saw a truck crossing the road in front of me and more vehicles in a lane in the forest. Russians! Within a few seconds, the machine-gun fire of the *Kradschützen* was sent in their direction. Several enemy trucks burned like torches and high flames illuminated the junction.

The forest exerted an unearthly influence. The darkness had an oppressive effect and enormously increased the tension. I felt insecure, the battle was too mysterious for me and, above all, I had not yet had adequate experience with the Soviets. Hour after hour we moved through the darkness. More than 30 kilometers still separated us from Rowno. A couple of kilometers before we reached Klevan — right outside of the woods which we then, thank God, had behind us — we halted and the battalion closed up.

We could hear the noise of heavy fighting coming from Klevan. Flares rose up and pointed out the location of the fighting to us. Just as I was about to push onwards through Klevan, there was a fearful outcry behind me. Amidst grenade detonations, men swearing, grating tank tracks and the tumult of breaking iron and steel, I could make out the shape of a tank in the column. To my horror, I saw the tank steering its course over the motorcycles, tearing to the left across the road and disappearing into the darkness. Hardly had we recovered from this surprise, than the same game was repeated a little further back in the column. Two enemy tanks had slipped into our line of march in the darkness and only noticed their error during the halt. For our part, we had confused their shadows with our own prime movers and thus had moved through the forest in the company of the Russians. Additional tanks were observed to the left of the road.

The combat reconnaissance patrol I had deployed towards Olyka had made contact with part of the *III./SS-Infanterie-Regiment 1* and was under heavy pressure from strong Russian forces. The *12./SS-Infanterie-Regiment 1* called for help, as the Russians had surrounded it. We thus had to relieve the threatened units and continue the advance towards Rowno after daybreak. We turned quickly and established contact with the heavily pressured unit. The Russians were fighting for the road. In the meantime, it had become light and we could identify a long column of abandoned trucks. Enemy infantry were positioned in front of us in the fields of tall corn and had tried to overrun the *12./SS-Infanterie-Regiment 1*. So far the enemy had been unsuccessful. The company was reinforced in the nick of time by an armored car.

Without losing any time, we returned through the forest to Klevan. The village was taken and the advance on Rowno continued. The road ran dead straight in a southeasterly direction. A couple of kilometers outside of Klevan, the road descended and then rose again slowly outside of Broniki. On the horizon, clouds of smoke rose straight into the sky. I moved behind the advance-guard platoon and scanned the terrain with my binoculars. I thought I could make out an abandoned gun on the slope. Amidst the fresh green of the young grain, I spotted a couple of bright patches. The gun was a *leichte Feldhaubitze 18*, which was abandoned and in firing position. It made a depressing impression on us. For the first time we had found an abandoned German weapon on the battlefield. A few steps away from the gun there was a looted ambulance. Its doors were wrenched open and blood-smeared. Silently we observed the devastated area. Neither living nor dead soldiers were to be found. We moved slowly up the rise.

The prominent bright patches grew more and more distinct; we could make out a large and a small patch very clearly. I let my binoculars fall, rubbed my eyes and took up the binoculars again. My God! Was that really possible? Could that which I had just seen really be true? The last couple of hundred meters were traversed quickly. The advance-guard platoon dismounted and ran with me to the bright patches. Our steps slowed down. We came to a standstill. We dared not go further. Steel helmets were held as if in prayer. Not one word desecrated this place. Even the birds were silent. The naked bodies of a brutally butchered company of German soldiers were before us. Their hands were fastened with wire. Widely staring eyes gazed at us. The officers of this company had met an end that was perhaps even more cruel. They lay a couple of meters away from their comrades. We found their bodies torn to pieces and trampled underfoot.

Still, not a single word was spoken. Death's majesty spoke here. Silently, we filed past our murdered comrades.

My soldiers stood facing me. They were expecting me to give them an explanation or guidelines for their future conduct in Russia. We looked at one another. I searched the eyes of each and every soldier. Without a word, I turned and we moved out towards an unknown fate.

Up until 7 July, we fended off attacking Russians again and again north of Rowno. The enemy was suffering severe losses, our own were minimal. At 1400 hours we received orders to secure the flank of the *11. Panzer-Division* and reconnoiter to the northeast from Miropol. At noon the next day, we were engaged with strong enemy forces in the forest areas north of Romanov. Enemy artillery was laying down harassing fire. Through the activities of reconnaissance units and enemy deserters we identified a motorized battalion with several tanks and some batteries. Towards the evening, we lost a 20 mm antiaircraft gun to a direct hit. The wounded crew was evacuated.

The situation had become critical on *Rollbahn Nord* in the meantime. The *Panzer-Divisionen* of the *III Armee-Korps* had advanced further in the direction of Zhitomir and Kiev and the *25. Infanterie-Division (mot.)*, which was charged with securing the extended flank of the corps, was being attacked by strong enemy forces from the north. *Rollbahn Nord*, the lifeline of the corps, was threatened. So we received orders on 9 July to attack the enemy north of Romanov, advance north through the forest areas and establish contact with *Kradschützen-Bataillon 25* near Sokolov.

After a thorough artillery barrage, I wanted to carry out a motorized attack in order to immediately advance into the depths of the Russian defenses and take advantage of surprise. The *1. (Kradschützen)/SS-Aufklärungs-Abteilung 1* positioned itself for the attack behind a small rise. Fritz Montag took point again. The company commander, my old and trusted comrade Gerd Bremer, repeated his operations order once more and went to his vehicle. I had forbidden the company to engage the enemy or reduce speed before reaching the forest's edge. It was supposed to thunder at full steam through the enemy and leave everything else to the following battalion. Two 88 mm guns had been emplaced on either side of the road. They had the mission of opening fire as soon as the company set off and laying down covering fire in front of it. They would "shoot" the company forward.

At exactly 1730 the guns started roaring and smashed the forest on both sides of the road. The engines of the bikes howled; the bikes and sidecars with the men perched on them looked like beasts of prey. Pressed flat on the bikes, my comrades rushed down from the rise and raced towards the detonations of the rounds and the hammering of the enemy machine guns. Within a few seconds the company had reached the edge of the forest and disappeared. Peter stomped on the accelerator and rushed after the company. The artillery fire was still ranged on the edge of the woods. Not a round was fired at us. Small scruffy horses chewed on their bits. Escaping Russians ran north on both sides of the road. But then what happened? The company came to a halt. It started to fight with the fleeing Russians and with isolated pockets of resistance.

The company began to advance like infantry and wasted precious time. This could not be allowed to happen! We had to reach the crossroads a few kilometers further north and deny the Russians an orderly retreat out of the forest to the left of our line of advance. Angrily, I moved to the front to get the whole thing moving again. The armored cars and assault guns cleared the way for the *Kradschützen*. Within minutes, guns, prime movers and trucks were captured. The one thing not to do was stop — just continue to move and take advantage of the enemy's confusion! Exhausted Russians approached us, weaponless and crying. At first I couldn't understand their cries, but then I heard: "Ukrainski, Ukrainski! " They were as joyful as children and hugged each other around the neck again and again. The war was over for them.

The crossroads were reached at 1815 hours. Columns fleeing east were overtaken and disarmed. The Russians only defended themselves occasionally. They were completely unnerved by the battalion's lightning-like advance. Hundreds of prisoners were assembled along the road and countless small arms and other weapons were captured. The enemy had been caught in a situation favorable to us. Their units had just started a withdrawal. Unfortunately, we lost a light armored car to antitank fire during this engagement but, thank God, we only had one wounded comrade during the whole operation to complain about. Although we had adapted ourselves to the Russian style of combat and felt absolutely superior, I thought it wise to disengage and spend the night in a cleared area in the woods. Our division was still engaged elsewhere and could only reach us by dawn.

The forest was alive. We heard the Russian columns pulling away eastwards. They were the last units of the Stalin Line. They were looking for a way east between our lines of advance and along forest firebreaks and paths down which one could hardly move. We laid by our weapons ready for the fight and awaited the new morning. The rustling of the tall spruces mixed with the distant thunder of a few friendly guns somewhere to the east.

The sun was beneficent and the Lord presented us with a wonderful summer day. Reluctantly, I gave the order to move north. The *1./SS-Aufklärungs-Abteilung 1* took point again and was waiting on the road, ready to move off, as I gave the order to Bremer for the breakthrough to the *Rollbahn*. Overtaking my car, the men waved smiling at me as they stowed away the last remnants of breakfast. What would the new day bring?

I gave the commander of the *2./SS-Aufklärungs-Abteilung 1* (Kraas) a few quick instructions and followed the advance-guard company. It had a lead of about five minutes. Dark spruce forests extended along both sides of the road, interrupted every couple of hundred meters by firebreaks. We covered ground quickly. Apart from my command car (a converted medium-sized armored signals vehicle) a couple of dispatch riders made up the party. Heinz Drescher, my excellent interpreter, and I sat on top of the armored car as we crested a small rise and were able to survey the route for a few kilometers. In the distance, we could see the last bikes disappearing around a bend. The main body of the battalion would be following in a few minutes; it was waiting for the artillery to join the column. Silently we looked ahead and enjoyed the beautiful, sunny day. It was astonishing how quickly one forgot the murderous struggle during an operation and considered even a few quiet minutes as a godsend. The heavenly silence was interrupted by an event that was perhaps the most interesting experience of all my years at the front.

To my great surprise, I saw a Russian antitank gun in a firing position as I rode past. Behind it were the tense faces of some Russians. I wanted to cry out but I suppressed the cry and let my car roll on for about another 200 meters before I stopped it. The dispatch riders were briefed about the gun and I led 2 officers and 4 men into the forest to mop up the "forgotten" antitank gun from the rear. Like redskins on the warpath we crept through the tall heather and bilberry bushes from tree to tree. I observed the gun through the trees. The crew was no longer to be seen. Had the Russian gunners escaped? Had they abandoned their gun? My machine-pistol was gripped in my fists at the ready, my finger on the trigger. My eyes were still directed at the gun emplacement. I heard the breath of Drescher behind me. I did not dare to turn around. Step by step we moved closer to the gun.

"Stoi! Stoi! Rooki verkh!"

Grimacing Russians gazed at me. I was standing in the middle of a Russian company. The whole time my comrades and I had been sneaking through a gap between two Russian platoons. The blood threatened to stop flowing through my veins. Countless rifles were being pointed at us. Spellbound, I looked at the circle surrounding me. Force would no longer achieve anything here. With subdued voice, I called to my comrades: "Don't shoot." The muzzles sank. An athletic, good-looking officer stood ten meters in front of me. I went towards him. He also scrambled past his comrades and moved in my direction. Not a sound could be heard. The Russian soldiers and my comrades watched the encounter. We stopped two meters away from each other, took our weapons into our left hands and saluted almost simultaneously. Then we took the last step and shook hands.

Up to then, I had had no sensation. I felt neither beaten nor victorious. As we were straightening up from our slight bow, we declared each other prisoner. The Russian laughed as if I had cracked the best joke of the year. His big, blue eyes beamed at me cheerfully, while I put my hand in my pocket and held a pack of cigarettes out to him. He politely waited until I had also stuck a cigarette between my lips, then he struck a match. We both behaved as if we were alone on a broad plain and as if the war were a long forgotten event. The Russian only spoke broken German and I spoke no Russian. I called Drescher and was able to whisper to him: "We have to gain time!" Drescher and the Russian then began a long palaver about which side should lay down its weapons.

While this was going on, I moved from Russian to Russian and offered cigarettes. With a grin, the young soldiers wrangled cigarettes out of my pack and held them under their noses before shoving the butts between their lips. They enjoyed sniffing their cigarettes. I clasped the shoulders of each of the individual Russians in a comradely fashion, indicating they should lay their weapons on the forest floor. But, in the twinkling of an eye, the pack was empty. Only then did I realize that I was relatively far away from Drescher and standing alone among the Russians. I was happy as soon as I rejoined the group around Drescher. From the officer's intonation I realized his patience would soon run out. Very slowly I crept ever closer to the forest's edge to cause Drescher and the Russian to continue their negotiations outside the woods. I was waiting for the approaching battalion. It had to appear at any moment and end this nightmare.

The three of us stood at the forest's edge and I tried once more to explain to the Russian that his unit was surrounded and that our armored spearhead had already reached Kiev. He then shook his head energetically and got Drescher to tell me that he was an officer and not a dunce. At that moment, there was a bang on the road and I saw a light armored car in flames. The Russian antitank gun had hit it from a range of about 20 meters. Thick clouds of smoke rose straight into the sky. Since I knew all my vehicles always moved well dispersed in order to have good fields of fire, the next one would arrive at any moment. Its turret would be appearing over the rise any second. The Russian was vigorously demanding that I lay down my machine pistol. I asked Drescher to explain to him that I could not understand the last sentence and he should show me what he meant.

The Russian looked at me in disbelief and set his wonderful, telescopic-sighted automatic weapon down on the road. He shouldn't have done that.

Quick as a flash, I stood on the weapon and pressed my shoulder against his. We then both stood like statues between the Russians on one side and my comrades on the other side of the road.

All my soldiers had snuck over to our side of the road. From the depths of the forest came the exhortation: "Russki, Russki!" A fanatical voice was issuing a call to action. More and more Russian rifles were being pointed at me and I pressed myself even closer to the officer. Even Drescher had taken cover in a ditch. At that moment a shadow passed over me. I did not dare to look but, nevertheless, I saw that it was the wingman to the original armored car. I heard the brakes grinding. He slowly pulled up to me. Everything happened at sizzling speed. The calls of the commissar left no doubt that the fireworks would start at any moment. A last glance into my opposite number's eyes. He sensed what was coming. Calmly he returned my look. Then I roared out: "Fire!"

The 20 mm high-explosive rounds and bursts of fire from the armored car slammed into the forest. My comrades threw grenades across the road and I jumped into a ditch like a shot. The Russian commander was lying on the road. The war was over for him.

Grenades came rolling across the road as we tried to make ourselves scarce but it was impossible. A small bridge barred the way. The armored car has to advance a couple of hundred meters as the Russian antitank gun had changed its direction of fire. It was getting uncomfortable. We were expecting the Russians to attack across the road at any moment. At that moment something happened I would never forget. Our youngest dispatch rider, Heinz Schlund — later the German 1500-meter champion — jumped up, ran towards his motorcycle combination, leapt into the saddle and disappeared. I watched him drive to the armored car, shout a couple of words to its commander and then return to us. He waved at me. I jumped up and landed crossways between saddle and sidecar. We then sped off towards the battalion.

Hugo Kraas had already got the *2./SS-Aufklärungs-Abteilung 1* to dismount and was leading the attack at that point. The heavy weapons, mortars and infantry weapons were quickly deployed. The fighting was hard. The Russians fought for every tree. But it was no use. The engagement was over in 15 minutes.

I sought out the Russian commander and found him with some bullet holes in his chest. He was buried in the same grave as my fallen comrades.

*

After fighting at Marshilievsk we reached *Rollbahn Nord*. Units of the *25. Infanterie-Division (mot.)* and supply units of the *13.* and *14. Panzer-Divisionen* were moving along it. Among the Russian dead in Marshilievsk was a Commissar Neumann. We wondered whether he might have been a German. *Kradschützen-Bataillon 25* was involved in hard fighting north of

Sokolov and was asking for our help. I put a couple of assault guns at its disposal that would soon bring relief to the battalion.

Our battalion took over a covering-force sector east of Sokolov, roughly 20 kilometers wide. The traffic on the *Rollbahn* was interrupted at about 1455 hours. Enemy forces had crossed it west of Sokolov and had thus cut the entire supply route to the *III Armee-Korps*. Our battalion's sector was under harassing fire. Our headquarters was located under a magnificent oak 100 meters south of the *Rollbahn*. Machine-gun fire swept through the thick foliage of the tree so the green leaves fluttered to earth. Peter "found" a plate of rice pudding for me that I ate with great gusto in the protection of the thick trunk. Only then, hours after the incident with the Russian commander, did I find time to think about the incident. Suddenly the rice had lost its taste. I got goose bumps all over.

At dawn, heavy artillery fire ranged in on the headquarters. The companies reported enemy concentrations in the forested areas north of the *Rollbahn*. Our artillery and heavy weapons tried to eliminate the identified enemy concentrations. But the enemy was not to be discouraged. I was worried about my *Kradschützen*, since the sector was unusually large and there were no available reserves. The armored car company swept back and forth along the *Rollbahn* to keep the gaps in our defenses covered. It looked like it would be a hot night. I moved over to the sector's right wing once again to visit the two motorcycle companies. The *1./SS-Aufklärungs-Abteilung 1*, with its right wing on the Tenia bridge, was exposed to especially heavy attacks. Both companies had dug themselves in well. My *Kradschützen* had burrowed into the earth for the first time in this war.

My vehicle was under machine gun and rifle fire along the whole stretch. The rounds hammered on the armor plate and ricocheted — "singing" — into the surroundings. The battalion staff dug itself in and everyone, from cooks and clerks to the last driver, was in position and awaited the night.

At 2330 hours the expected Russian attack began. A terrific hail of steel rained down on our sector. The repulsive scream of "Urrah, Urrah!" congealed the blood in our veins. Those nerve-wracking shouts were something new to us. Flares illuminated the darkness. Tracer rounds created colorful paths in the night. The Russians had pushed through to the *Rollbahn* and the attack only collapsed as a result of flanking fire from several armored cars.

Another attack followed around midnight. Its main effort was directed against the *1./SS-Aufklärungs-Abteilung 1*. The Russians penetrated the position and eliminated two machine-gun nests but were wiped out by spades and bayonets in hand-to-hand combat. The old positions were retaken. The enemy artillery fire became more intense. A heavy rail-gun battery plastered the road junction south of Sokolov and the battalion headquarters east of the junction.

The motorcycle companies had run out of ammunition and were urgently requesting resupply, but how were we going to get ammunition to the right wing? There were no routes to the right wing behind the front line and the *Rollbahn* was the demarcation line between the Russians and us. Only the road separated the fighting elements from one another. I heard Kraas' voice in my headphones once more and could see terrific fireworks in his sector to my right. The *1./SS-Aufklärungs-Abteilung 1* and the heavy mortars were also screaming for ammunition. The shouts of "Urrah! Urrah!" came right at me as the Russians attacked our headquarters. The 20 mm antiaircraft platoon fired high-explosive rounds at the attacking Soviets and concentrated its fire especially on a line of bushes on the far side of the road. The attack collapsed with high casualties for the Russians at that location.

After the last attack was beaten back, my armored car was loaded to the brim with ammunition and *SS-Sturmführer* Grezech and Peter dashed along the contested road. Peter had removed the camouflage cover from the left-hand headlight and moved down the road with blazing lights. The vehicle was hit several times during the move but the companies were supplied with ammunition.

Without luck, I asked for reinforcements. The battalions could only be expected by noon at the earliest. They were only just disengaging from the fortified line near Miropol. Once again, heavy enemy artillery fire rained down on the right wing of the sector. The heavy rounds of the 150 mm railway guns shattered the woods in the sector of the *1./SS-Aufklärungs-Abteilung 1* and, at about 0100 hours, the third storm of Russians broke over us. A few minutes later the flank company requested help. The enemy had penetrated the position in depth and the fighting was hand-to-hand. Demented creatures were fighting for their lives. It was every man for himself.

I couldn't take it any longer. I tore the headphones off, jumped into an armored car and rushed off. Without looking to the left or right, the driver raced through the Russian attack groups crossing the *Rollbahn* between the gaps in our defenses. We were moving without lights and could therefore only see shadows bounding across the road. The brakes squealed as the car took a left-hand bend in the road and then disappeared behind a small farm building. The *Kradschützen* were just fending off the last Russian attack. It collapsed deep within their defenses. The attacking forces were completely annihilated.

The men's screams mixed with the hammering of the machine guns and the hollow bangs of grenades made any communication impossible. Kraas and I crouched behind the remains of a wall perhaps 20 meters south of the road and tried to pierce the mists of the dawning day. Heavy Russian mortar fire was falling on the whole sector. Several *Kradschützen* dragged a badly injured comrade into cover and tore open his uniform jacket to render first aid. I bounded over and saw my old comrade Grezech lying mortally wounded in front of me. His eyes were closed and his chest rose and fell almost imperceptibly. I shouted at him, called his name — I wanted to call him back to life, but all was in vain. Death had already assumed his dominion. His lips were twitching as if he wanted to entrust me with a final farewell to his wife and children. His eyelids lifted slightly, so that his eyes were visible, but they could see no more. His head slowly fell to the side. A round in his heart had ended his life.

SS-Hauptscharführer von Berg — a veteran Spieß in first the *14.* and then the *15./SS-Infanterie-Regiment 1* — was in the first foxhole. He was a platoon leader in the *1./SS-Aufklärungs-Abteilung 1*. A round to the chest had torn him from our midst. Jupp Hansen, von Berg's longstanding friend, was crying for help. His screams were choked by blood. A round to the lung got him. He was pulled into cover, lying on a shelter quarter. Jupp recognized me, wanted to speak, but couldn't manage it any more. He died some hours later.

The new day illuminated a macabre landscape. Burnt areas, craters, uprooted trees, mangled equipment and blackened farm ruins bore silent witness to an insane night. A 1-ton prime mover was in front of me. During the night it was a glowing aiming point, at this point it was a smoking wreck. Thin smoke blew across the roadside ditch. The driver was sitting upright behind the steering wheel. His uniform had been burned from his body; only black ashes concealed his charcoaled chest here and there. The blackened skull with its empty eye sockets was still facing in the direction of travel. I wanted to scream, to curse the whole insanity of war, but I tumbled into the next hole and returned the fire of a Russian who was lying behind a bush not 50 meters away on the other side of the road. Glancing at the prostrate, clawing humanity around me, the bloody fields of Verdun appeared before me in my thoughts. My comrades and Russian soldiers lay dead in the foxholes; they had killed each other. The survivors threw out the dead; they wanted to live and were looking for cover.

Daylight had arrived. There was no longer a single living Russian to be seen; the battlefield was empty. Before us were flower-strewn meadows and rippling grain fields. Not a shot disrupted the silence of the morning. My comrades sat up, initially with care and then without circumspection. The first one stood up, lit a cigarette, and took a look in the enemy's direction. Everyone looked at the standing *Kradschütze* in fascination. Nothing happened. Shouts of encouragement flew from hole to hole. Life had us in its grip again. It was asking for its share.

I shook Hugo Kraas by the hand. During the night, he was the backbone of the defense. His eyes blinked nervously. His hand shook and his words sounded bitter. Both of us had lost more comrades in the last two days than during all the previous campaigns put together.

The radio reported we would be relieved by the *III./SS-Infanterie-Regiment 1*, which was supposed to take place by noon. Just as we wanted to return to battalion headquarters, *SS-Untersturmführer* Baumhardt asked me to investigate the vegetation 150 meters north of the road. He noticed movement there only a short while before. Peter steered towards the vegetation and wanted to circle it at a distance. The circle around the vegetation became smaller and smaller, but I couldn't spot any enemy. Peter then moved really close. I was standing in the turret. As if from nowhere, a Russia officer leapt up. With one bound he was standing on the front slope of our vehicle and fired. Completely taken aback by his tremendous speed, I fired my pistol and crouched down. The vehicle stopped; it didn't stir. I roared at Peter. There was a bang, the vehicle heaved to and fro and moved on in fits and starts. Peter tried to throw an egg grenade into the vegetation but it promptly rolled under our vehicle. Peter never repeated his close-combat tactics.

On reaching the *I./SS-Infanterie-Regiment 1*, we landed in a firefight which could not shake the battalion. The grenadiers under the command of Fritz Witt did not waver. This battalion had also formed itself into a strong, unshakable fighting team. In the roadside ditch, I recognized an old comrade whom I had trained at Jüterbog in 1934. Quasowsky, a fellow more than 1.9 meters tall, had a severe wound in his leg. A grenade fragment had shattered his foot. With his pocketknife, the East Prussian made the final cut and severed the foot from his body.

The relief in place was executed without interruption. *SS-Hauptsturmführer* Hempel was killed a short time later; he had been a good friend of Grezech. We put our dead to rest beneath the thunder of Russian artillery. The eulogies were drowned by the noise of roaring engines. The *Rollbahn* was open again. The columns rolled eastwards.

We marched to Kopylovo by way of Zhitomir and fought outside the gates of Kiev together with the *13.* and *14. Panzer-Divisionen*. Our battalion was relieved by the reconnaissance battalion of an Infanterie-Division and freed up for other missions.

While *Panzergruppe Kleist* had been advancing on Kiev via Zhitomir, parts of the *6. Armee* had been attacking the Russian forces near Uman by way of Vinnitsa. The *17. Armee* crossed the Bug on a broad front and advanced further east. The divisions of the *III Armee-Korps* employed against Kiev were then replaced by units of the *6. Armee* and turned southeast.

At about 0300 hours on 30 July we paused on the southern outskirts of Zibermanowka and secured the right flank of our division, which was attacking. By 0500 hours we had reached Leschtschinowka, held by strong enemy forces. The *2./SS-Aufklärungs-Abteilung 1* bogged down outside the village and was pinned by heavy artillery fire. The *1./SS-Aufklärungs-Abteilung 1* was also under heavy fire. Around noon, Russian forces tried to break to the east. Enemy tanks sped through the *Kradschützen* and tried to ram or overrun the antitank weapons. A disengagement of the motorcycle company proved impossible due to the artillery fire.

SS-Untersturmführer Baumhardt was killed in action during the extremely bitter fighting. He was a platoon leader in the *2./SS-Aufklärungs-Abteilung*

1. The company commander, *SS-Hauptsturmführer* Kraas, was wounded. *SS-Obersturmführer* Spaeth assumed temporary command of the company. A grenade fragment tore off the lower jaw of one of our youngest comrades, *SS-Kradschütze* Husmann. His comrades evacuated him while risking their own lives.

In a surprise attack, a Russian company was taken prisoner by some armored cars. The company had orders to keep the bridge near Laschtchewoje open.

On 31 July, the battalion was placed under the operational control of the *XXXXVIII Armee-Korps* and was given the mission of reaching Nowo Arkhangelsk and closing the Uman pocket. Nowo Arkhangelsk was reached exactly at noon without a fight. The village was located on both sides of a small river which cut through the town from north to south and which could not be forded by tanks due to its steep banks.

On reaching the first houses we came under lively artillery fire from the west. Northwest of the town, I noticed a German battery, some armored cars and motorcyclists disappearing to the northeast. Those German forces must have assumed we were Russians. We tried unsuccessfully to communicate with the German unit by flares. They could not be stopped and disappeared. As we later determined, it was *Aufklärungs-Abteilung 16*. It had disengaged from the enemy in error and consequently caused us a great deal of extra effort.

In the town we came under rifle fire and had to fight our way through to the bridge. There we found an abandoned armored car of *Aufklärungs-Abteilung 16*. The bridge had been damaged by mines. In our rear, the *I./SS-Infanterie-Regiment 1* reached Nowo Arkhangelsk and cleared the southeastern part of the town. I then felt considerably more comfortable. We could rely on Fritz Witt. The *I./SS-Infanterie-Regiment 1* and *SS-Aufklärungs-Abteilung 1* had been joined in "marriage". We always got along well together.

Towards 1800 hours the town was completely in our hands. However, apart from my armored cars, there were no reserves available. As darkness fell, the Russians attacked the battalion positions on the northeastern outskirts of the town with strong infantry forces and eight tanks. Our 37 mm antitank guns opened fire at pointblank range. But what was going on? The tanks weren't bothered in the least! They continued their advance into the *Kradschützen*. Shells ricocheted all over the place. We all could have done well without that bullshit! Thank God, the *Kradschützen* were among the buildings and were thus able to evade the tanks. Russian infantry also penetrated our lines and pushed our company back. The tanks had made a deep impression on my soldiers. I hoped this wouldn't turn into a rout!

I wondered if our gun on the bridge would have better luck. I leapt onto a *Sturmgeschütz* and raced off to the bridge. There I encountered Montag's platoon, which wanted to position itself on the near side of the river. I couldn't believe my eyes! The motorcycle companies came rushing onto the bridge and wanted to get to the near bank. A couple of words were enough to restore some order to the situation. The companies turned round and threw out those Russians that had already infiltrated. Four Russian tanks fell victim to the assault gun and a 47 mm antitank gun.

At 0330 hours there was a hefty infantry attack that collapsed under the fire of the *Kradschützen*. A Russian assault force had penetrated into the town along the brook from the south. However, the bridge, which had been protected by our combat engineers, was not in danger at any time. At dawn the enemy troops that had penetrated were either wiped out or captured. All further attacks were repulsed with high enemy casualties.

The encircled Russian forces in the pocket were constricted ever more tightly. The pressure on our positions became ever greater. On 2 August a *Nebelwerfer-Abteilung* and a 21 cm mortar battery were attached to us.

From a low rise I observed heavy enemy troop movements in an easterly direction. Several columns of mixed weapons were disappearing into the woods 5 kilometers away from the positions of the *1./SS-Aufklärungs-Abteilung 1*. Obviously, the enemy wanted to risk another breakout to the east at nightfall. Harassing artillery fire fell on the entire battalion sector. More and more columns disappeared into the patch of woods. The slender trees covered thousands of men. Infantry, cavalry and artillery were seeking cover from view in the forest's shadows. Didn't they realize we could follow every movement? The Russian officers were leading their men and themselves to their deaths. Those assembled forces could not be allowed to attack; they had to be smashed first. We would not be able to stand up to these masses in the dark; they would overrun us and gush through like dammed up water through a sluice gate. The Uman pocket would have had a hole in it.

All the battalion's heavy weapons, the *Nebelwerfer-Abteilung* and the 21 cm mortar battery had the mission of laying down destructive fire on the patch of woods. I reserved the order to open fire for myself. Up to that point the *Nebelwerfer* had remained an unknown factor to us. We had no idea of the weapon's effects. The hands of the clock ticked on. I was waiting patiently for the most favorable moment. The stream of Russian soldiers had not yet dried up. There was starting to be a bulge from the inside of the pocket to the outside. The moment had come; the night announced its arrival. The outlines of the woods could only just be made out. Everybody was waiting tensely for the order to fire.

Fire! Hell opened its mouth.

Above us there was a horrible howling. Thick smoke trails arced over us and disappeared between the dark trees. Fiery rockets whizzed towards the enemy without interruption and exploded in death and destruction. The heavy mortar rounds rolled through the air like freight trains and complemented the hissing of the rockets. The forest burst asunder like an anthill beneath the feet of a prehistoric monster. Men and animals struggled for their lives but still ran to their deaths. Galloping horses, wildly running away along with carts and guns, collapsed under the fire of our armored cars' automatic weapons. The waning sun glided over a field of death. Men breathed their last or remained forever crippled. Animals awaited the coup de grace. Thousands marched into captivity.

I no longer wanted to see or hear any more. The butchery disgusted me. Russian and German doctors went to great pains over the tormented humanity. We saw Russian women in uniform for the first time. I admired their attitude; they behaved in a better manner than their male counterparts. The women worked throughout the whole night to ease the suffering.

After midnight, all hell broke loose! The cry of "Paratroopers! " echoed through the position. And, indeed, I could see big four-engined transport planes above us and hear the rustling of parachutes. The parachutes fell to earth all around us. This was an extremely grave situation. We had no reserves. We awaited the descending paratroops with our peashooters ready to fire but nothing happened. After a few minutes the first report arrived and the first parachutes were brought over. The enemy has been dropping ammunition, food and fuel for the encircled troops.

In the afternoon our sector was taken over by the *297. Infanterie-Division*. The battalion was pulled back to the center of the town. During the following night, trains, vehicles and ammunition haulers suddenly sped across the bridge. Enemy infantry had infiltrated the artillery park. The battalion immediately launched a counterattack and captured several hundred Russians.

The fighting for Uman was over. The greater parts of the Russian 6th and 12th Armies were effectively destroyed. Both army commanders, 317 tanks and 858 guns had fallen into German hands.

Panzergruppe1 was freed up again and was deployed southeast to cut the lines of communications of the Russians fleeing in front of the *11. Armee*. On 9 August, the battalion received orders to reconnoiter in the direction of Bobry. For many hours we moved through deep dust and clay to the southeast. Russian cavalry shadowed our advance at a respectful distance, without giving us the opportunity of getting rid of this vigilant company. At dawn on 10 August, I saw the cloud of an explosion on the horizon to the right of our line of advance. An armored car of *Aufklärungs-Abteilung 16* had encountered a mine.

Before long, we captured a Russian mine-laying detail. The detail was made to unearth the recently buried mines under protest and complaint. The prisoners belonged to the Russian 12th Cavalry Division that was making a fighting withdrawal to the east. I led the battalion directly east through tall grain and cornfields and then headed south. We were suddenly to the rear of the enemy's rearguard. Our Kradschützen destroyed several armored cars, trucks and antitank guns. The Russian armored cars burned like tinder. The thick clouds of smoke rose straight into the sky. In pursuit of the fleeing Russians, the advance-guard platoon encountered strong field fortifications that were excellently laid out and only spotted at the last moment. During the fighting around a small rise, one of our best noncommissioned officers was killed. The ever-cheerful "Bubi" Burose died at the head of his squad. The platoon leader, *SS-Untersturmführer* Wawrzinek, was wounded.

The objective of the attack of the *16. Panzer-Division* was the harbor of Nikolajew. We were given the mission of covering the left flank of the division against attacks from the east. *General* Hube led his division across the Ingul and then advanced along the east bank of the river to the south. By doing so, he cut off the way east for many Russian units.

On reaching the bridge at Kirjanowka our battalion came under lively artillery fire from the southwest. That meant the enemy was behind us and was trying to prevent the crossing of the Ingul using artillery fire. But that fire couldn't prevent us from executing our mission. Each vehicle dashed across the bridge individually and attempted to close up with the force. Neither personnel nor vehicles were lost.

The *16. Panzer-Division* had shot like an arrow through the retreating Russians and was positioned with its spearhead outside of Nikolajew. In ceaseless movement we also cut through the enemy's retreat and reached the extended village of Sasselje by way of Nowo Poltawka. The five-day advance led us through many a critical situation and a lot of skirmishes with withdrawing units of the Russian Army. We had not learned to fear the enemy but we were certainly beginning to respect the endless spaces.

From Sasselje To Cherson

The extended village was bordered on the west by two lakes and could be easily defended from this side. The last Russian forces had left Sasselje around midday, heading east. They belonged to the 162nd Infantry and the 5th Cavalry Divisions. The church had been used as a grain store and movie theater. The staircase in the belfry had probably gone up in smoke in the chimneys of the village elders. Red banners hung down from the ceiling. A decrepit, toothless man introduced himself as an orthodox priest and asked for permission to hold services. Tears ran down his furrowed face as the villagers entered the plain room and he gave the first sermon for many years in a shaky voice. The older people listened piously to his words. Younger people, curious but slightly embarrassed, stood on the square in front of the church.

During the night the sound of heavy fighting raged to the west. The *16. Panzer-Division* was attacking Nikolajew. At dawn I stood in the church tower and gazed out over the typical southern Russian countryside. Huge fields extended in all directions and deep, dusty tracks ran toward the village like a spider's web.

On all the tracks dust clouds could be seen and, near the neighboring village of Nowo Petrowka, enemy planes were taking off and landing. As expected, the Russians had surrounded the battalion. We no longer had to be told what would happen, should Nikolajew be held for a prolonged period and enemy forces attempt to advance deep into the flanks of the *16. Panzer-Division*. The thick dust clouds to the west and east told us enough. The heat would be on!

The motorcycle companies had moved into position on the outskirts of Sasselje and were busy with their morning wake-up call when heavy enemy forces were observed to the east. At the same time, an enemy column rolled toward Sasselje from the west. The Russians had no idea that Sasselje had been occupied since the previous day and so they moved straight into a trap set by our engineer platoon and some armored cars. A dam, which separated the two lakes, was the Russians' undoing. Without resistance, without firing even a single round, countless Russians were taken prisoner.

The tower became my command post. I was able to observe the enemy's movements and take timely measures from there. More and more Russians appeared from the direction of Shuwanka and disappeared into the huge cornfields. The corn was as tall as a man and offered superb concealment and an outstanding means of approach. Company after company disappeared into the waving corn to spread out for the intended attack on Sasselje. I watched this development calmly because only a narrow patch of corn reached as far as the village. Apart from that, we had an open field of fire for at least 400 meters. Across such a distance any attack had to choke in its own blood.

On the horizon enemy batteries were going into position. Mounted messengers galloped across the black earth. In front of us were numerous targets for our artillery, but we had to conserve our ammunition. Our lines of communication were endlessly long and ammunition had become precious. We awaited the moment when each round would successfully fulfill its bloody task. Small black dots moved toward us through the ripening fields. From time to time, the sun struck bare metal, flashing brightly through the shimmering air. The dots moved closer and closer. They no longer resembled ants but could clearly be recognized as Russian infantry. Their battle order was good. They advanced in a widely dispersed formation. Teams of men drew antitank guns; the horses remained behind. I considered the situation carefully, like a chess player, and got ready those weapons that would prove deadliest to the enemy.

The Russians advanced upright but cautiously. No German soldier could be seen on the outskirts of Sasselje. The village had to appear to be unoccupied to the Russians. Only the heavy platoons of the motorcycle companies remained in position. All the other units awaited the order to attack north and south of the village. I did not want to merely repulse the attack, but to destroy the Russian unit! The moment arrived at 1100 hours. Fire struck the Russians from all weapons and tore horrible gaps in their lines. Mortars and infantry guns attempted to eliminate the enemy guns with precision fire. The attack waves went to ground. They then jumped up and ran forward to their deaths. Commissars and officers attempted to get the attack going again, but I only saw individual Russian soldiers running forward; the majority remained on the ground as though nailed there.

The moment had come for the *Kradschützen* and armored cars waiting on the flanks. They advanced east and then turned in and pushed the Russians toward our position. By midday 650 prisoners had been taken and more than 200 dead counted. According to prisoner statements, the commander of the 962nd Rifle Regiment shot himself after shooting some of his officers. Because that regiment was said to have comprised only 900 rifles, it was rendered completely combat ineffective as a result of those losses.

During the next few hours we were attacked several times by the Red Air Force. In the afternoon intense enemy artillery fire was directed on the village. The fire came from the west. Although we had expected the attack from both directions, reality nevertheless took us by surprise. With lightning speed the front line was switched, and we occupied prepared positions west of the village. In contrast to the first attack, motorized forces carried out this one. Some amphibious tanks and armored cars formed the Russian spearhead. A direct hit struck our battery position and an ammunition truck exploded. Once more we let the Russians get close and run into our fire.

The amphibious tanks were the first victims to fly into the air. The familiar smoke clouds covered the battlefield while our armored cars and fast *Panzerjäger* moved out and shot up the wildly and chaotically scattering columns. One *Panzerjäger* came to a smoking halt. Red flames shot out of its interior and engulfed the driver before he could be saved. The survivors jumped into the glow. They pulled their comrade out of the burning hell and smothered the flames on his body. My ears were ringing with the screams of the badly burned driver. I looked away from the group and pointed my binoculars to the southwest. Thick clouds of dust announced the arrival of more columns, which were trying to force a way through to the east. Like panthers, our armored cars and *Panzerjäger* pounced on the columns and shot up the vehicles into flaming wrecks. They tried to scatter in all directions but only a few managed to flee. The majority of the attackers had to trudge down the road to captivity.

During the situation report I heard the wailing of the injured and the heartrending cries of our burned comrade. He lay on the stretcher and begged for me to give him the coup de grace. His hands lay crippled on his charred body. The hairless head, swollen lips and blackened trunk were a single, continuous wound; the lower part of his body was protected by his trousers during the fire and was therefore not that badly affected. I was too ashamed to speak words of comfort. In such a situation, comfort was a barefaced lie. My soldiers watched me keenly. Again and again the young soldier begged for relief. The doctor held out no hope; he scarcely knew where to give the painkilling injection. A helpless shrug was all the doctor could manage. A bloodcurdling scream drove me away from the field dressing station. I was unable to make my farewell. I could not even lay my hand on his brow. One last glance and I rushed away. With the departing of the day our comrade also departed this life.

We held Sasselje against all Russian attacks until 17 August and reconnoitered deep to the south, as far as Snigirewka. Snigirewka Station was reported clear of the enemy at 1900 hours. The town itself was heavily occupied by the Russians. *SS-Untersturmführer* Thede's armored car received a hit from an antitank gun. Thede was reported missing when the rest of the crew linked up with the *Kradschützen*. Thede had learned of the birth of his first son only hours before. The *16. Panzer-Division* succeeded in capturing the harbor of Nikolajew and pushing the Russians back to Cherson. In the harbor the astonished grenadiers found a 36,000-ton battleship that had not yet been run down the slips.

On 18 August our battalion received the mission of reconnoitering in the direction of Cherson, about 60 kilometers from Sasselje on the lower Dnepr. Since 1918, it had developed into a considerable industrial city. However, only a tiny, wretched village was marked on our maps. Well before sunrise we began the march south. Hours later the glowing red face of the sun pushed up over the Dnepr and warmed our stiff joints. Dust clouds followed our rapid advance. A group of *Kradschützen* moved in front, covered by two armored cars. The rest of 1st Platoon and my command vehicle followed the cars. The advance-guard leader was *SS-Hauptscharführer* Erich upon whom I could rely unhesitatingly.

After a short move we encountered a Russian column, which was looking for a river-crossing site. We disarmed it and marched it off in the direction of Sasselje. Russian trucks were captured and incorporated into our own column. The prisoners were glad that their war was over and obeyed orders willingly. On both sides of our line of advance we discovered huge vineyards and tomato and cucumber fields which all gave the impression of being well tended. It was not long before the whole battalion was devouring tomatoes. Fruit orchards covered the western slopes by the bank of the Dnepr, but the fruit was not yet edible. Our southerly progress continued unabated. My men grinned at me whenever I passed them or they brought me radio messages. As old hands they had, of course, noticed for some time that we were not reconnoitering, but heading for some "monkey business".

In *Wehrmacht* reports such actions were termed a coup de main, but this so-called coup de main was far from being either a child of impulse or the idea of some reckless commander. No! The coup de main was, in most cases, an operation planned long beforehand by a responsible commander who had situational awareness and was gripped by the intent of achieving great success through level-headed planning, bold daring and swift action.

The prerequisites for this type of warfare were, apart from military abilities, outstanding human qualities. The commander must possess the absolute trust of the soldiers under his authority and he must, in the truest sense of the word, be the leading soldier of the unit. A coup de main could not be ordered from above! Higher headquarters lacked the ability to make the "on-the-ground" factual and intuitive judgment calls. The prerequisite for a coup de main is vested solely and completely in the person of the military leader.

Often a swift action seems to be the reckless operation of a commander particularly favored with luck, but reality has a different aspect. Such a commander literally puts himself in his opponent's position. He is aware of the reversals and defeats, which strike the enemy. He knows the physical and mental strain that his opposite number is under; his strengths and weaknesses are well known to him. He does not rely on intelligence reports from higher echelons. They only provide the framework for the current operation, not a whit more. He prepares his own estimate of the situation. From a thousand bits of information he pieces together the picture of his enemy. He knows how to read the operational terrain like a book. Long submerged instincts are reawakened. He sees and smells the enemy. Prisoner faces reveal more to him than page after page of interpreter interrogations. He is not a superior officer but someone who leads by example! His will is the will of the troops. He draws his strength from his grenadiers who believe in him and would follow him to Hell.

The skyline of Cherson could be made out on the horizon. Grain silos towered over the Dnepr. The western area of the town was full of tall chimneys. Tall, shady trees beckoned in front of us. The sun had scorched us. We looked forward to water and shade in the town.

A few kilometers outside of Cherson I stood on an armored car for a long time and observed the town lying before us. Busy traffic was moving in an east-west direction on the river. Gunboats flitted back and forth. Large ferries steamed at a leisurely pace to the far bank, discharged their loads and returned to Cherson. The city seemed close enough to touch. It enticed, it touted itself, and seemed to mock my hesitation. The company officers were watching me. I could tell by the artilleryman's face he was looking for emplacements to support effectively what he thought was going to happen. At that moment my soldiers were, once again, sitting in the tomato fields, eating the magnificent fruit with relish. I envied them their lack of concern.

I smoked another cigarette, puffing the smoke into the shimmering air. I felt absolutely sure of myself and had no misgivings that the huge town might swallow us up. My decision remained firm. The town had to fall to a coup de main. The Russians were expecting the attack from the direction of Nikolajew. They had prepared their defenses on that side. It was there that the *Leibstandarte* stood ready for the attack. (One more reason for considering the reconnaissance completed and for breaking into Cherson by the "back door".) We followed a country lane alongside the Dnepr up to the city and overran a Russian company building a roadblock in the outskirts. Out of sheer fright the Soviets forgot to exchange their shovels for weapons. Modern high-rises rose in front of us. Enemy machine-gun fire ripped up the earth around us. The struggle for Cherson had begun.

SS-Hauptscharführer Erich touched the rim of his steel helmet with his forefinger, shouted "Move out!" to the lead section and then tore away from us at full throttle across the broad square. He disappeared down the wide street which led to the center of Cherson. The platoon followed its leader. Armored cars swept the fronts of the houses with 20 mm cannon rounds. The muffled bang of hand grenades indicated hand-to-hand fighting. I followed the lead platoon and, suddenly, landed back by the Dnepr. The road twisted like a snake through a very old, fortified area. From the east bank of the river lively artillery fire struck the road. Soviet sailors were fighting with the agility of wildcats. The firing forced us to dismount our vehicles and continue the fight as infantry.

A row of houses protected us from view from the east bank of the river. Forced close to their walls, we fought for each house down both sides of the street. Erich fought like a lion; the blond soldier from Schleswig-Holsteiner jumped, roaring, from doorway to doorway, setting the tempo of the attack. Machine-gun fire struck sparks from the pavement. The attack came to a standstill. The gunfire was an almost insurmountable barrier in front of the lead elements.

But there was no standstill for Erich. He knew we had to reach the harbor to prevent a planned defense by the Soviets. He was laying behind some steps, his legs drawn up to his body, holding the grip of his machine pistol tight in his strong hands. He shoved his helmet to the back of his head and called out, "First platoon: Move out; cross the road. Ready. Go!" I watched the young soldiers leap up, dash across the street, and dive to the pavement as if they had been fired from a gun. Their concerted dash outwitted the enemy machine gunner; after a short time he fired no more.

We reached a small square. Sailors were positioned among the ornamental bushes and tried to check our advance. While jumping, I suddenly saw Erich pitch onto the flagstones. The machine pistol clattered with a metallic ring over the rough paving. Erich buckled; his hands sought something to grab on to. He was scratching in the dirt of the street. Grenadiers drug their platoon leader over to the wall of a house and shouted for a medic. A round to the head had torn open his skull. I wanted to say a few words and squeeze his hand but he could hear no more; it was his last battle. A few days later he closed his eyes forever after having dictated a letter to his wife. In Erich the company had lost one of its best noncommissioned officers and I had lost one of my most loyal soldiers.

The struggle increased in ferocity. Russian guns fired down the street. Fuel dumps burned. Dense smoke clouds and explosions swirled high into the air. An entrance gate offered cover. I threw my full weight against the gate, but it did not give. It had been bolted shut. Rounds hammered on the paving and the ricochets whizzed off. I thought the jig was up. I ran down the street pursued by the fire of the sailors. A small kiosk offered some cover. Rounds split the thin wooden walls. It was being sawn to pieces by the machine-gun bursts; a chain saw couldn't have done it better. Pressed flat to the ground, I awaited the outcome of the firefight between the Soviets and the men of 1st platoon. Once again I was between the lines. In a few minutes the situation was cleared up and the advance on the harbor continued.

The Soviets withdrew into the harbor district. Two large ferries were at the jetty and taking on fleeing people. We worked our way ever closer. Whistling mortar rounds tried to hinder the pace of our move on the harbor, but we were not to be stopped. House after house fell; street after street was taken. Hob-nailed boots rang out on the pavements of this important town on the Dnepr.

Our machine guns fired on the ships quayside. They were only light weapons but their effect was devastating. Without regard for the people still streaming onto them, the hulks got underway and pitched at full power toward the eastern bank. People hung on to the ships' sides like bunches of grapes as they moved ponderously and awkwardly away from the quay. A 50 mm antitank gun engaged a motorboat. The boat drifted off to the south, burning. Ships of all conceivable types tried to reach the safety of the far shore while Russian artillery covered their withdrawal. They fired at the harbor with no consideration for the stragglers remaining on the quay. Oil and fuel drums flew into the air and burning people leapt to their deaths, disappearing in the suction of the strong current.

Russians came tearing at us out of this inferno and others sought safety in the Dnepr's torrent. A prime mover moved through the dense smoke of the burning drums with an 88 mm *Flak* gun and occupied a favorable firing position. Scarcely had the gun unlimbered when the first round burst forth from the barrel and exploded in the bowels of one of the big ferries. The gun was standing in a hail of fire in full view of the Russian artillery. Ammunition, vehicles and fuel that had been left behind burned, exploded and whirled around the gun standing alone on the quay. Horses sought an escape route and went gurgling under in the brown flood. A large barge drifted, rudderless, toward the bank and ran aground. Russian soldiers tried to swim to the safe bank. Only a few succeeded in their flight; the majority of them drifted toward the sea.

I heard *SS-Obersturmführer Dr.* Naumann yelling as he was directing the second gun of his platoon into a firing position. He roared through the raging of the conflict and rushed over to the first gun. My God! At that point I could see the danger, but the grenadiers at the gun apparently did not notice it, being occupied with the firefight. The gun slowly rolled across the quay and plunged into the Dnepr. As if by a miracle, all the soldiers were successfully rescued, but our gun has disappeared. That was a bitter loss.

The firing in the harbor slackened off; only a few rounds howled over us and exploded somewhere further to the rear in town. The link up with the *Leibstandarte* — attacking from the northwest — was achieved at 1600 hours. The fighting for Cherson was over. With the extinguishing of the flames in the harbor, the job of rebuilding and maintaining the city was already started. Rubble was cleared away; the civilians came out of hiding and the first children sought contact with German soldiers. On 22 August *Regiment Hitzfeld* relieved our battalion. The *73. Infanterie-Division* had reached the Dnepr and made preparations to cross it north of Cherson.

July 1941, Russia. Meyer during the battle of the Uman pocket. (Jess Lukens)

The first wounded in Russia.

Murdered German soldiers from the *III. Armee-Korps* outside of Rowno.

A commander's conference in Russia, summer 1941.

The scene of Meyer's personal encounter with the Russians. In the background is his knocked-out armored car.

Destroyed Russian vehicles on the *Rollbahn* (lines of communications).

Assault troop leader Bergemann fell while trying to destroy this T 34.

Russia, summer of 1941. Left to right: Kurt Meyer (commander, *SS-Aufklärungs-Abteilung 1*), Sepp Dietrich, Walter Staudinger (commander, *SS-Artillerie-Regiment 1*), Max Wünsche (Divisional Adjutant). (Jost Schneider)

Upper right: 1530 hours, 31 July 1941. Kurt Meyer receives orders to attack Laschtschewoje. From the left: *SS-Sturmbannführer* Kurt Meyer, *SS-Hauptsturmführer* Lehmann (hidden), *SS-Obergruppenfürer* Sepp Dietrich. (Rudolf Lehmann)

Lower right: The *I./Nebel-Lehrregiment* fires on the forest northwest of Nowo-Archangelsk.

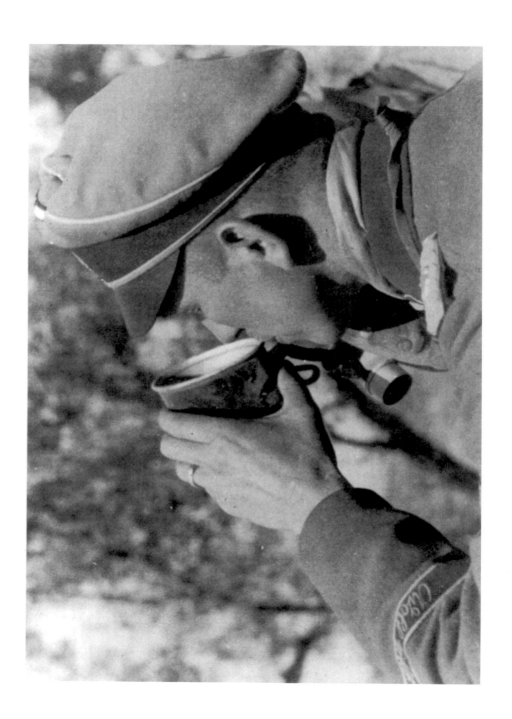

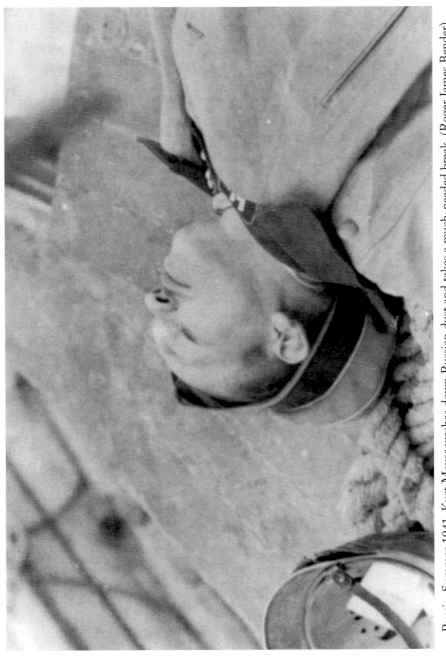

Russia, Summer 1941. Kurt Meyer washes down Russian dust and takes a much-needed break. (Roger James Bender)

Gerd Bremer leading his motorcycle company in Russia.

Staging outside of Cherson. 22 August 1941.

Moving through Cherson.

Above: Russia, summer of 1941. Meyer at a command conference. (Roger James Bender)

Right: Russia, summer of 1941, awards ceremony. Left to right: Sepp Dietrich, Max Wünsche, Kurt Meyer.

Left: Russia, 26 September 1941. Kurt Meyer with Walter Staudinger,
Commander of *SS-Artillerie-Regiment 1*, on the Perekop Peninsula. **Above**: Meyer
a few moments later. (Roger James Bender)

Above: Bubi Burose has destroyed a Russian armored car North of Cherson.

Left: *Waffen-SS* messenger battling the mud, 8.10.1941.

Below: *Batterie Fendt* at Mariupol. An 88 mm *Flak* engaging ground targets

Russia, 13 October 1941. Kurt Meyer speaking at the grave of his driver, Peter. (Jess Lukens)

Russia, 18 October 1941. Sepp Dietrich and Kurt Meyer. (Jost Schneider)

Russia, winter of 1941/42, on the Ssambek Front. Left to right: Keilhaus, Sepp Dietrich, Kurt Meyer. (Jost Schneider)

From The Dnepr To The Don

The constant fighting up to that moment had made unheard-of demands on our battalion and exacted severe casualties. The losses presented a poignant picture of the troops' need for rest and refitting. Officers and men had done their utmost to achieve their assigned missions. The battalion was already feeding off itself and was in urgent need of an overhaul. Above all, we lacked direly needed replacements of men and equipment. The *Leibstandarte* was pulled out of the frontline and was to be refitted within a week. We were glad of this break and enjoyed those few sunny rest days like a gift from God. All our wishes were satisfied. We slept late into the day and luxuriated in the peace and quiet, but concerns quickly set in.

We soon became aware there would be no question of refitting. The maintenance facilities really tried to conduct the servicing of weapons and equipment and repair the vehicles, but their efforts were only a drop in the bucket. There was a lack of repair parts. The battalion equipped itself with captured vehicles. Personnel replacement was also long in coming. Day after day slipped past without the arrival of the eagerly anticipated replacements from Germany. The result of this "refitting" was that serious problems had to be addressed among the leaders. Up to that point in time we had had the attainable objective of the Dnepr in view. We had been at full strength as we neared that objective and fought with total dedication for its attainment.

We were no longer a fully operational force. The formations could only field a fraction of their former strength. It didn't take too much effort to calculate how long it would be before our proud battalion would cease to be combat effective. What would happen when we forded the Dnepr with this battered battalion and pushed east? What were our objectives? Could we still reach it before the onset of winter? We were talking about the Don, the Volga and the Caucasus. The interminable vastness of Russia's expanses oppressed us. We began to think in Russian: "Nitchewo!"

I led my advance-guard battalion back to the Dnepr on 8 September and crossed the river on 9 September at 1630 hours. The bridgehead had been forced by the *73. Infanterie-Division*, under *General* Bieler. We slowly crossed a swaying pontoon bridge over the wide, muddy waters. Assault guns and tanks were brought over individually on ferries. The crossing-area command-er pressed an order from the *LIV. Armee-Korps* into my hand and simultane-ously informed me that I was attached to the *73. Infanterie-Division* upon crossing the river. The commander was expecting me southeast of Berislaw. Accompanied by a few dispatch riders, I moved slowly to his command post. Along the roadside were the fresh graves of fallen German and Russian sol-diers. The traces of battle were engraved in the face of the earth. In view of the expected night operation, I let my soldiers go to the river and wash up.

I found the staff of the *73. Infanterie-Division* in an orchard. I received the mission of breaking out of the bridgehead to the south — advancing on Nowaya Majatschka by way of Britany — and establishing an all around defensive position for the night. *Oberst* Hitzfeld's regiment was to be our neighbor on the left.

Right after the general's first words I gave one of my dispatch riders the signal to tell the battalion to get ready. He raced back to the battalion to start the preparations for the march. In the course of receiving the order, I gave another messenger the signal for the battalion to get underway. He dashed off and delivered the march order. A few minutes later I saw my company com-manders already waiting for me. They were chatting with *Major* Stiefvater. After a detailed briefing, *General* Bieler offered me a cup of coffee and asked: "When can you move out?" He followed my glance with astonishment as I pointed to the waiting company commanders and an approaching dust cloud and answered: "*Herr General*, the battalion is already on the move." I will never forget his startled expression.

A sunken dirt road slowed our pace. We soon left the last outposts of the *73. Infanterie-Division* behind us and were moving into the darkness. Bremer was in front. Slowly, inching carefully forward, we rolled into the night with engines throttled down. Pitch-black darkness swallowed us up. We made con-tact with the enemy at 2100 hours, 4 kilometers north of Nowaya Mojatschka. A Russian outpost was surprised without a fight. The Soviets gave the impression of total exhaustion and readily gave up information. According to their statements, Nowaya Majatschka was held in strength. After eight days of rest I felt unsure of myself. I was out of touch with my opposite number. We had been led into a situation in which we felt like strangers and, for that reason, we advanced hesitantly. I awaited the new day. The bright morning light would afford us the necessary security. A tight "hedgehog" defensive formation constituted our "fortress". I sat in the radio truck and talked with a Russian officer about the probable courses of action his commanding officers might take. The town of Perekop and the term "tar-tar ditch" cropped up again and again in the conversation. The prisoner was convinced that the "tartar ditch" was being held and defended by the Soviets.

The night was still, not a shot rang out. This calm was eerie. A couple of rounds would have revealed where the front was. As it were, we felt as though we were completely surrounded by Soviets. Moist dew clung to the dry blades of grass as the first dim gleam announced the new day. I was strained and tense as I tried to penetrate the dawn and fix my gaze on Nowaya Majatschka. The outlines of the town emerged slowly from the surrounding darkness. My *Kradschützen* perched on their vehicles and awaited the order to attack. *Regiment Hitzfeld* attacked Nowaya Majatschka from the north at 0400 hours. The peace and quiet was over. The noise of battle filled the air and drove all exhaustion from our bones. The infantrymen of the *73. Infanterie-Division* advanced in the gray early morning light. The shells that came smashing down could not stop the men as they steadily worked their way toward the town. In the disappearing morning mist we could make out the enemy fortification system. It was well integrated into the terrain west of the town and outstandingly well constructed, as always with the Russians. They were masters of the art of building field fortifications.

The attack of the *73. Infanterie-Division* was gaining ground and so the time had come to attack the fortified system. If the Soviets withdrew from our attack, they would only run into the arms of *Regiment Hitzfeld*. We prepared ourselves for the attack behind a thick hedge. The Soviets had not yet noticed us and their artillery was making every effort to thwart the infantry division's attack. We had to cover two kilometers to get to the enemy positions and the terrain offered no cover of any kind. Far and wide there was no woodland to be seen, apart from our windbreak. The Nogai Steppe lay before us — flat as a board, cracked, hard, and sporting only steppe grass.

I observed the movements in the Russian frontline with the company commander of the *1./SS-Aufklärungs-Abteilung 1* and realized that a hasty attack by my battalion would shatter the Soviets completely. Furthermore, our attack would also considerably benefit the division's advance. I considered which attack formation would cover the no-man's-land without loss and at the greatest possible speed.

The shimmering expanse of the steppe tempted me to think about the cavalry charges of days gone by. The Devil was tempting me at that moment. Why wouldn't a charge by my *Kradschützen* lead to success? I didn't dare voice my thoughts; I still considered such a charge to be madness. Yet, while common sense and instinct battled for the final decision, I could already see my *Kradschützen* roaring like the Devil across the steppe and piercing the enemy positions in my mind.

My soldiers watched me silently as my gaze roamed over the steppe again and again, estimating the distance. I let my binoculars fall and probed Bremer's eyes. How would he react, after assessing the situation, if I ordered his *Kradschützen* to charge? Did he already suspect that something extraordinary was in the offing? His gaze was frank and his face betrayed no surprise when I expressed my intention to charge the enemy with motorcycles and armored cars. My "greyhounds" received the order coolly and as a matter of fact.

Artillery and heavy infantry weapons took up position. *Kradschützen* deployed across a broad front behind the protection of the hedge. Armored cars moved forward into predetermined gaps so as to be able to give unhampered covering fire. Gripped by feverish excitement, I climbed into my vehicle and thrust my arm vertically into the air. There was no turning back! The spell was broken; the uncertainty had disappeared. Our armored car rolled slowly out from under cover. We were in the Russian's line of sight.

The first enemy rounds had to come crashing down. I perched, hunched over in our armored car, and looked ahead. My good-old Erich changed gear; we started moving faster and a dust cloud swirled into the air. We moved between the two motorcycle companies. The *Kradschützen* clung to their machines like monkeys. After moving a hundred meters I could no longer make anything out. Shadows swooped and darted in front of us. The attack became a race. Would it be a race toward death?

Russian rounds howled over us and exploded where we had been a few seconds previously. The artillery fire spurred us on even more strongly; the speed increased. We had to outmaneuver the Russian fire control system and break into the Soviet lines like the Devil incarnate.

Gripped by the intoxication of speed, stirred to the marrow of our bones by the roaring of the engines, we sought out the enemy through squinting eyes. Our objective was over there — where our artillery was fulfilling its deadly task, where the blood of Russian soldiers was already irrigating the soil. The scene of destruction spurred us on. We rushed toward death as if possessed.

Four soldiers occupied our armored car but only one was recognizable. He sat behind the wheel and drove with a sure hand. The rest of us hung at the sides like Cossacks ready to open fire or disappear into foxholes at a moment's notice. Nothing would shake Erich. East Prussian sangfroid triumphed over death and destruction. Did he actually realize that he had already been leading our proud battalion for minutes, and that it was he who was propelling it forward? His speed defined the pace of the attack and he raced toward the objective like a race driver.

The first Russians suddenly appeared. Terrified faces stared at us. They threw their weapons away and ran off to the west. We advanced through the defense system, past foxholes, past smashed soldier bodies and helpless wounded. Countless Soviets ran headlong westward and were rounded up by our combat engineers.

Russian artillery, however, still remained emplaced somewhere. We could not break off our frenzied hunt! Enemy trucks attempted to escape but burned up in the 20 mm cannon fire of our armored cars. We moved past unlimbered guns and pushed on past Nowaya Majatschka, heading in the direction of Staraya Majatschka. The tension slowly eased. There was no form of life to be seen around us any longer. The steppe lay there as if dead. On the other hand, the terrain behind us resembled an anthill. Friend and foe alike gave first aid to the wounded.

A handshake with *Oberst* Hitzfeld, a short orientation on the situation with the *73. Infanterie-Division*, and the advance east continued. Five hundred thirty four Soviets were taken prisoner in the charge. Our losses were two soldiers killed in action and a noncommissioned officer and two enlisted personnel who died of wounds. The attack was a complete success but, nevertheless, never again did I order a motorized charge.

We positioned ourselves outside of Kalantschak in the darkness, attacked the town using the element of surprise, and took it. An enemy armored car went up in flames and 221 Russians trudged their way into captivity. Reconnaissance patrols reported the area up to ten kilometers east of Kalantschak clear of the enemy.

At midnight I received an order from the infantry division to break through the Perekop Isthmus with a coup de main and await further orders south of the Ishun Straits. In the course of various campaigns, I had often received orders and missions that had nothing in common with the fundamentals of classical military leadership, but this order surpassed them all. Were my responsible superiors really of the opinion that a coup de main on the isthmus could open the gates to the Crimea? My company commanders looked at me in somewhat of a daze as I passed the order to them and briefed them on the situation.

The Crimea was separated from the mainland by the "Bad Sea". This so-called sea was a few hundred meters wide and was usually impassable. It was an insuperable obstacle even for assault boats due to the low water levels. Three approaches led to the Crimea: The neck of land from Perekop in the west, the railway crossing at Saljkoff in the center, and the narrow access route at Genitschesk in the east. The Perekop Isthmus was a few kilometers wide and was bisected across its whole width by the "tartar ditch", which was up to fifteen meters deep in places. The terrain was as flat as a pancake and was bisected by a few dry waterbeds. These steep and often deep cuttings were called "Balkas". They afforded the battalion its only cover. Hard by the northern side of the "tartar ditch" was the old, fortified town of Perekop. A railway line ran through Perekop to the south. In view of those favorable defensive opportunities and the fact we had taken prisoners from three separate enemy divisions in the last few days, nobody believed the Isthmus was going to be an easy obstacle to hurdle.

On 12 September, around 0430 hours, my battalion moved out for Perekop. We established contact with the advance-guard battalion of the *73. Infanterie-Division* under *Major* Stiefvater at 0455 hours. Stiefvater joined my battalion. The distant horizon slowly emerged from the half-light of dawn. The steppe glowed in the most magnificent colors. No other human form was to be seen; only my soldiers moving forward in leapfrog fashion. *SS-*

Untersturmführer Montag led the advance-guard platoon, *SS-Unterscharführer* Westphal, the point. I followed, nervously scanning the horizon for signs of movement. Neither man nor beast could be seen. Only the play of colors from the sun gave the plain a wondrous appearance. South of Nowo Alexondrowka I sent von Büttner's platoon to reconnoiter along the coast to Adamanij. The terrain north and south of the "tartar ditch" had to be observable from there.

I suddenly picked out several riders on the horizon who wheeled around on the spot and then disappeared in the direction of Preobrashchenko. Their fleeting appearance electrified us. Preobrashchenko was behind a low rise and only a few individual houses were visible. Keenly alert, we scanned the horizon while moving through the stillness. We were spaced at intervals of several vehicle lengths. We had the feeling that the peace might be shattered at any moment by screaming mortar rounds. The Soviets had to make use of this favorable defensive position. The eerie calm promised battle. No fleeing Russian, no madly dashing horse team, no rushing motor vehicle indicated flight or even retreat. The steppe was empty. Not a soul could be seen far and wide. This fact alone pointed to tight command and control on the part of the enemy command.

My soldiers were again hanging from the sides of the vehicles; even the riders of the lead platoon sat sideways on their machines. I stood on the running board of my armored car. My vehicle followed in the main body. My watch showed 0605 hours as Westphal's section moved slowly up to the houses of Preobrashchenko. A huge herd of sheep blocked the entrance to the town; it took off towards the steppe. A bang destroyed the silence. Sheep spun through the air. The animals ran madly for their lives, climbing over each other to get away but blowing themselves into the air. In turning away, the sheep had strayed into a minefield. The bleating of the tortured beasts filled the air, mingling with the dull thud of exploding mines. Crouching, ready to jump and trembling with excitement, we awaited the flickering flash of Soviet weapons. We ran alongside the vehicles, wanting to get inside the town and establish a firm hold there. Still not a round was fired; only the mines completed their deadly task. All that remained of the big flock of sheep was a twitching, bloody heap and a couple of animals dragging themselves laboriously away.

Then it happened! Then came the long awaited voice of the front! Rounds whizzed over us and exploded among Stiefvater's march group. They streamed toward the rear, at first individually and then in great swarms. I rushed forward, wanting to reach the first building and have a look in the direction of Perekop. After a few bounds I dove into the dust amid the hail of bursting shells. A dark monster had moved around the slight rise and was firing right into us. Just a few hundred meters ahead of us was a fire-belching dragon, sowing death and destruction among our ranks. An armored train, bristling with weapons, had positioned itself across the route of the battalion.

I gave the signal to pull back. The *Kradschützen* wheeled around on the spot and raced back on a broad front while armored cars fired at the train and withdrew behind a smokescreen. A 37 mm antitank gun hurled its rounds at the armored train and, a few seconds later, flew into the air itself. The shattered steel of the gun carriage smothered my soldiers' cries.

We were bombarded with fire from five heavy and two light batteries. Only dust clouds were to be seen behind us. I heaved a sigh of relief. There were no burning tanks or vehicles to be seen. I crept forward a few more paces and could make out deeply echeloned field fortification with trenches and barbed-wire entanglements. The armored train puffed away in the direction of Perekop.

I could see Russian infantrymen in their foxholes only 50 meters away, their machine guns rattling and forcing us even closer to the ground. We had to get clear of this or start the path into captivity. It was then the turn for our rounds to hiss over us and force the Russians into cover, although we had to press ourselves into every fold in the ground as well.

Wounded soldiers lay around me; *SS-Unterscharführer* Westphal had lost an arm; *SS-Rottenführer* Stoll was a few meters away from me, groaning in pain. Helmut Belke was uninjured. Stoll's motorcycle combination was still usable; its engine roaring amid the machine-gun clatter. Belke shouted something to Stoll, pointed to the machine and worked his way toward Stoll. I took care of *SS-Untersturmführer* Rehrl. All aid was in vain; shrapnel had torn open his back. When he gasped I could see the rise and fall of his lungs. The roar of an engine announced Stoll's rescue. Belke carried the wounded soldiers to safety in defiance of the Russians. He made the trip three times, risking his life for the sake of his comrades. Three times he returned with a groaning cargo. One more comrade was still lying wounded in the steppe grass. Like us he was in the blind spot of the slight elevation. Soldier "G" was a reservist, married and the happy father of two small boys.

He crouched in the shallow depression, his blond mop of hair smeared with blood. The words came groaning from his lips: "Get out of here. It's no use — I'm done for." I tried to console him but in vain. Several machines came roaring up to fetch the rest of the advance party. I couldn't keep my eyes off "G", his clenched fist gripped the butt of his pistol. Slowly he raised the weapon and pulled the trigger. His body slumped forward and lay still. He was thrown by the horrified *Kradschützen* onto one of the vehicles, which had pulled up in the meantime. We all reached the battalion position despite artillery and infantry fire. I shakily related the experience to Doctor Gatternig and only then heard that Comrade "G" had suffered the loss of his manhood. Rehrl died while the doctors were working to save him; the power of mere human beings no longer sufficed.

We established a position four kilometers west of Preobrashchenko with Stiefvater's detachment and awaited the arrival of the *73. Infanterie-Division.* Von Büttner's platoon reported Adamanij clear of the enemy at 0650 hours.

The terrain south of Perekop, including the "tartar ditch", could be readily observed. Von Büttner reported strong field fortifications, barbed-wire obstacles, dug-in guns and tanks. Half an hour later I verified the accuracy of his report. An advance through the isthmus could only be achieved by several divisions with heavy artillery support.

I reported by radio to the *73. Infanterie-Division* that a coup de main on the isthmus was impossible and, in addition, an orderly would deliver an after-action report and a thorough estimate of the situation.

Therefore, I was thunderstruck around midday to receive the order to continue the advance on the isthmus. I refused to lead my comrades to their certain deaths, referring to my first report and to the extremely strong fortifications in the strip. Division ordered me to report to the divisional commander in person. After a trip lasting several hours I reached the commander in a small village northeast of Kalantschak. I was expecting an unholy ass-chewing due to my refusal to carry out the order and so I was not a little surprised when *General* Bieler greeted me in an extremely friendly fashion and agreed with my estimate of the situation.

During the evening I briefed *Oberst* Hitzfeld on our battalion's sector and then led the battalion to Tschaplinka where new orders were awaiting us. Enemy dive-bombers and artillery accompanied us as we disengaged from the front.

At Tschaplinka I received orders from Sepp Dietrich to push forward into the central neck of land near Saljkoff and, if possible, take it by surprise. Meanwhile, it had already become 1600 hours and we would have to operate in the approaching night. My young soldiers were already perched on their machines as I reached the battalion and, five minutes later, we moved off into the glimmering steppe. We passed through Wladimirowka at 1750 hours and received fire from the "Rhinoceros" Peninsula as a 122 mm battery attempted to stop our advance. We roared east without loss. I wanted to exploit the daylight and put as many kilometers as possible behind us before the onset of darkness. We spent the night in Gromowka without enemy contact.

At 0430 on 15 September the *2./SS-Aufklärungs-Abteilung 1* stood ready to move out as the lead company. Hot coffee steamed in the hands of my soldiers as I weighed the situation with *SS-Obersturmführer* Späth and evaluated the most recent reconnaissance findings and prisoner statements. Air and ground reconnaissance reported well developed field fortifications, which ran in a semi-circle around Saljkoff rail station. A breakthrough, piercing the fortifications, seemed impossible. We had neither the requisite forces nor the appropriate weapons at our disposal.

Apart from that, the aerial rconnaissance photos showed guns in concrete emplacements south of Saljkoff which commanded the narrow crossing point. Thick fog lay over the steppe and visibility was less than 20 meters. That last factor gave me the idea of using the fog to advance right under the noses of the enemy battery gun barrels and into the defensive circle around Saljkoff. I was convinced that the fortifications nearest the shore were not as strong and it would never occur to anyone that a motorized unit could be so crazy as to assault a fortified defensive system two to three hundred meters in front of the muzzles of emplaced batteries. The fog would hold for another hour at the most. The last wisp would disappear by 0700 hours at the latest. It was imperative to have taken the crossing by then.

I quickly explained my intent to the lead company and shook *SS-Obersturmführer* Späth's hand. Späth was moving with the point platoon. *SS-Obersturmführer Dr.* Naumann's 88 mm guns, including the one that had been recovered from the Dnepr, were incorporated into the march column behind the lead company. He was to fire on the bunkers south of the crossing point.

The motorcycles, armored cars, prime movers and guns rolled slowly into the impenetrable wall of fog. In a few seconds the gray nothingness had swallowed them up. Pete swore aloud as he started up our armored car and pushed into the damp wall. My adjutant, the small but wiry *SS-Obersturmführer* Weiser, looked out to the right to try to identify the bank of the Siwasch. It was important that we moved close to the shore of the "Bad Sea". Although we were, at the most, fifty meters from the bank, we couldn't make it out.

After twenty minutes we stumbled upon a road junction from which a whole lot of tracks led off in all directions. A soldier approached us out of the fog and we assumed Späth had left behind a guide to make the route clearer for us. I called out to the figure, asking: "Where do we go from here?" The fellow nearly fell over backwards when he heard my voice. We had no idea where he had run off to. He disappeared in a fraction of a second, as if from the face of the earth. Only later did we determine we had passed within a 150 meters south of a Russian outpost.

We pushed further and further toward the crossing point. It didn't take too long before we could make out the causeway. The fog lightened. It was time to act! Was the fortune of war to smile upon us or would we have to pay a high price for our brazen deed? We gazed silently into the wall of fog. To our right, water was splashing against the low bank. A few hundred meters south of us, beyond the flat Siwasch, the dark Crimean coast rose out of the fog. However, what was to the north? Where was the enemy? Step by step the move continued to the east. The tracks of the prime movers ground into the sand, their engines scarcely audible. Everything was as tense as tense could be. I pointed to the south and drew the attention of the 88 mm gunners to the reported bunker installations. Everything was still quiet. Were we advancing toward death once more? Would the coup de main of Preobrashchenko be repeated?

A dull rumbling interrupted the morning's stillness. Was that going to be another armored train? Suddenly the spell was broken. As soon as it fired we recognized the sharp crack of one of our 37 mm antitank guns.

Simultaneously, we heard the bark of our 20 mm cannon and the angry clatter of German machine guns. The *2./SS-Aufklärungs-Abteilung 1* had reached the Isthmus of Saljkoff and stopped a logistics train carrying weapons and equipment. The 37 mm had knocked out the locomotive. We had penetrated the defensive system better than we had thought and could actually roll up the positions from the rear. At 0855 hours the train station was firmly in our hands. By doing that we had smashed the Russian defenders' command post and interrupted their lines of communication. The Soviets' confusion was indescribable; they were simply unable to believe we were there. The Russian artillery from the emplacements south of the strip only opened fire after being fired at by our own 88's.

The fog had disappeared and protected neither friend nor foe. Bremer advanced north and penetrated the outskirts of Nowo Alexjewka. *SS-Obersturmführer Dr.* Naumann rushed an 88 into a position from where it could knock out a few enemy guns. Thanks to the decisive action of the crew, the objective of silencing the Russian guns was attained. Unfortunately, in the process, Dr. Naumann had been severely wounded. He was brought to safety with the aid of an armored car. We had completely shattered the defensive ring north of the isthmus. However, it was impossible to cross it. Dug-in emplacements with thick barbed-wire obstacles and minefields necessitated the employment of heavy artillery and infantry formations. The battalion succeeded in taking several hundred prisoners belonging to the 871st and 876th Rifle Regiments. The 276th Rifle Division was defending the isthmus. We captured 86 new Ford trucks, 26 tracked prime movers, two 47 mm antitank guns and several ammunition cars loaded with 122 mm shells from the supply trains, which came from Melitopol, that were intent on reaching Sevastopol.

We were very happy to capture the train; it wasn't long before we were equipped with Russian Ford trucks. Our own vehicle inventory was sadly depleted; the lack of motorcycles had made itself felt. I was relieved to hear a few hours later that we had only lost one soldier, his life taken by shrapnel. The unorthodox conduct of the engagement had made our success possible.

We were relieved in place by the *II./Infanterie-Regiment 1 "Leibstandarte" (mot.)* during the night. The vehicles were topped off and checked over at all possible speed for the next operation. Our new mission: "Take Genitschesk immediately and block the route across the third neck of land."

By 0500 hours we were once more sitting on our machines, moving into the sun. A short time later we caught sight of the Sea of Azow for the first time. The surface lay before us like a mirror. One large and five smaller ships steamed toward the east and soon disappeared below the horizon.

We reached Genitschesk around 0630 hours. The outskirts of town were silent with not a soul to be seen. Was this quiet part of the Russian defender's tactics or was this harbor town simply undefended? A section of

Kradschützen cautiously approached the houses but, despite their expectations, encountered no resistance. We then dashed full steam ahead into the eastern part of the city and the harbor district. The situation there looked different. A column of trucks was attempting to escape in the direction of Melitopol. Enemy infantry ran into the closest houses in panic and, somewhat later, marched off into captivity.

An explosion in the harbor announceed the demolition of the bridge, which connected Genitschesk with the mainland. The *2./SS-Aufklärungs-Abteilung 1* saved a footbridge over the channel from destruction through its prompt seizure. During the storming of the bridge, the company commander, *SS-Obersturmführer* Späth, was killed by a round to the head. The loss of officers had taken on dangerous proportions. Nearly all the original company commanders and platoon leaders had been killed or wounded. The *2./SS-Aufklärungs-Abteilung 1* received its third leader, *SS-Obersturmführer* Böttcher.

From the steep bank at Genitschesk we could see far to the south over the spit of land and observe all movements. Consequently, I was not a little surprised when the Soviets suddenly attacked us from the south, presenting themselves like targets on a range. Company after company moved slowly but steadily toward our steep embankment and into certain death or captivity. It was a mystery to me why the Soviet commander was carrying out this attack. We allowed the enemy infantry to get within about two hundred meters of us before our machine guns reaped a bloody harvest. The result was horrific. Within minutes countless brown dots covered the sparsely grassed area whilst others staggered toward our positions with arms raised.

The Russian mortar emplacements were engaged by the superior firepower of our 88 mm *Flak*. By 0900 hours the attack had been called off and the *1./SS-Aufklärungs-Abteilung 1* moved out over the footbridge on a reconnaissance to the south. My intention was to set up a bridgehead and push on across the narrow strip as far as possible. Unfortunately, that reconnaissance came to a halt after only three kilometers. Field fortifications and coastal batteries emplaced in concrete formed an insurmountable obstacle and the fire which emanated from them, together with a few bomber attacks, caused some casualties during the night.

At about 2100 hours on 17 September the battalion was relieved by the *III./SS-Infanterie-Regiment 1 "Leibstandarte" (mot.)* and received the mission of reconnoitering to the north and seeking a link-up with von Boddin's advance-guard battalion of the *XXX. Armee-Korps*. During the night the battalion received six officers and 95 men as replacements. The first replacements in Russia made an excellent impression and integrated rapidly into the battalion. In a few days the young soldiers had become "veteran warriors". With the customary nightime aerial nuisance raid, we left Genitschesk and moved in the direction of Melitopol. We felt our way north over sunken, sandy roads and thick scrubland and soon made contact with von Boddin's battalion positioned south of Akimowka, which was occupied by strong Russian forces.

Von Boddin and I had a lot of acquaintances in common in beautiful Mecklenburg, where we had both spent some memorable years. A typical cavalryman, von Boddin had changed his uniform at the beginning of the Thirties and worked with *General* von Seeckt in China until he was recalled to service in the German Army. He was a dashing, almost impetuous officer, a born commander of an advance-guard battalion: Swift in decision, bold in venture and decisive during the attack. He was killed in January 1942 in the fighting at Eupatoria in the Crimea, murdered by treacherous partisans.

The two advance-guard battalions fought south of Melitopol until 21 September, awaiting the arrival of the *72. Infanterie-Division*. A further advance on Melitopol and to the north of the town could not be executed due to a lack of infantry. We had outdistanced the rest of the *XXX. Armee-Korps* by 200 kilometers.

On 21 September I received the order to disengage from the enemy and lead the battalion back to Kalantschak. The unit withdrew almost 200 kilometers in twelve hours and stood ready for the assault on the Crimean Peninsula on 22 September.

During the long movement across the Nogai Steppe I felt the terrifying emptiness of the open spaces for the first time. We moved west for hours without encountering even a single German soldier. Granted, the rapidly moving advance-guard battalions had pushed through the area and were hammering their way forward far to the east, but German troops did not hold sway over this region. The yawning emptiness of the steppe had a depressing effect on us. How were we to operate to the east? Which units would be employed against the Crimean forces?

We began to look beyond the context of the immediate fighting and looked for an operational goal for the Eastern Army. No one believed the formations on hand were sufficient to hold the front during the winter months. The units were battered and in urgent need of refitting.

We were attached to the *LIV. Armee-Korps* and, after the brave *73. Infanterie-Division* succeeded in taking Perekop by storm and overcoming the "tartar ditch" on 26 September, we were to cross the isthmus rapidly and drive deep into the withdrawing enemy. The battalion stood ready for the operation in the late afternoon, four kilometers north west of Perekopskij Bay.

The fighting proceeded with extreme intensity and bitterness on both sides and, only on 27 September, did a battalion of the *72. Infanterie-Regiment* enter Armyansk. Neither did the fighting on 28 September provide conditions for the use of motorized formations.

The Soviets attacked again and again with strong forces and plentiful tank support. At 0430 hours the battalion was attached to the *46. Infanterie-Division* and moved up to a position three kilometers northwest of Perekop.

The positioning of the battalion north of the "tartar ditch" was called off at 0905 hours. There were no opportunities for employment in the area of operations of the *46. Infanterie-Division* and the battalion returned to the control of the *Leibstandarte* at 1100 hours.

The Soviet defenses on the isthmus south of Perekop collapsed only after 10 days of grim fighting. The neck of land was not pierced until 28 September. That cleared the way through to the Crimea. On 29 September the pursuit of the beaten Soviets began, culminating in the heroic assault on the fortress of Sevastopol on 1 July 1942. While the *LIV. Armee-Korps* was doggedly fighting for every meter of ground south of Perekop and the *Armee-Oberkommando* was planning to set the *Leibstandarte* in pursuit following the successful breakthrough, things happened on the Eastern Front between the Sea of Azow, Melitopol and the Dnepr which demanded the immediate redeployment of the division.

The Russians had established a front based on the line indicated above and, following the arrival of two new armies, the 18th and the 9th, had begun to attack the *XXX. Armee-Korps* and the Rumanian 3rd Army with a total of twelve divisions. The attacks on the *XXX. Armee-Korps* collapsed when they met stiff resistance but, further north, in the Rumanian 3rd Army sector, the Rumanian 4th Mountain Brigade was overrun, creating a grave breach in the front.

Because of this new situation we were hurled north on 29 September and, together with the the advance-guard battalion of the *4. Gebirgs-Division*, had the mission of attacking and eliminating the enemy where he had broken through the line at Balki. Working together with units of the *Gebirgs-Korps*, the breach in the Rumanian line was completely cleared and we inflicted heavy losses on the Soviets.

Beyond a doubt, the sudden emergence of two new Russian armies near Melitopol had caused the German High Command many hours of consternation but, as a result of the way they executed their operation, the Russians had presented *Heeresgruppe Süd* with a unique opportunity. As a consequence of the Soviets' costly massed attacks, they lacked the forces to prevent a break-out by *Panzergruppe von Kleist* from the Dnepr bridgehead. *Panzergruppe von Kleist* commenced the offensive to the southeast on 1 October, threatening to cut the lines of communication of both Russian armies and, in conjunction with the *XXX. Armee-Korps* and the Rumanian 3rd Army, to eliminate them. The pursuit by the Sea of Azow had begun.

From 2 to 4 October *Abteilung von Boddin*, the *Panzerjäger-Abteilung* of the *72. Infanterie-Division* and our battalion fought together against strong enemy forces in the area of Jelissawetowka. The Soviets suffered heavy casualties in these engagements. They ran up against our superior weapons and more mobile battalions time after time without heeding the losses they incurred. The steppe naturally offered our advance-guard battalions immense advantages and led their far superior infantry forces into unpleasant situations.

On 5 October our infantry divisions attacked the deeply echeloned positions between Melitopol and the Dnepr. A deep antitank ditch extended across the whole width of the attack sector and was stubbornly defended by the Soviets. Minefields and wire entanglements hindered our forward progress.

Our battalion was assigned the mission of taking the crossing over the Molotschnaja River and keeping the bridge open for the following infantry.

Once again we were positioned behind the attacking infantry battalion and awaited the signal to commence operations. Our great infantry literally had to thread their way through the many minefields. The mines were encased in wood and could not be located with detectors. Around noon the infantry had crossed the antitank ditch and, in so doing, had cracked the enemy's main line of resistance. A crossing point through it was quickly created for my battalion.

Feverish with anticipation, Bremer's *1./SS-Aufklärungs-Abteilung 1* was positioned at the head of the column behind my armored car. The Soviets were withdrawing. We could see individual batteries changing positions to the rear. Our time had come. We had to push through the retreating enemy and create an insurmountable barrier out of the Molotschnaja Bridge.

The hunt began! Accompanied by all our best wishes, the lead platoon shot off like an unleashed grayhound. Bremer and I followed the whirling dust. Only sporadic mortar rounds attempted to slow our pace. The battalion advanced through the withdrawing enemy and, at 1230 hours, moved through Federowka, where the first limbered batteries were overtaken and several hundred prisoners captured. We could observe the Russian withdrawal for as far as the eye could see. The whole front had started to move. We encountered antitank fire from a maize field and lost our lead armored car to a direct hit. The *Kradschützen* wiped out the enemy antitank gun. The prime mover of our lead antitank gun fell victim to a mine. The pace of the hunt quickened. We moved as if on thin ice. After all, the fleeing Soviets had made abundant use of their effective mines.

Before us was the town of Terpenje, which was, bisected from north to south by the river. The land fell away to the east. Thousands of Soviets raced wildly alongside us on their horse teams and tried to reach the crossing ahead of us. The made for the river on a broad front. We reached the first houses of Terpenje. The ground fell away sharply and encouraged us to increase our tempo. Distraught Soviets leapt into the houses and sought cover. At a bend in the road, guns, vehicles and wildly thrashing horses were gathered in a confused throng. Machine-gun fire was received. I caught sight of the bridge.

Several columns of Russians accumulated on the riverbank, trying to reach the safety of the far side by crossing through the water. 20 mm rounds from the armored cars whistled into this chaos and colored the waters of the river red. We had got between the Soviets like a whiplash.

The lead platoon neared the bridge, firing to both sides. The tumult could hardly get any worse. A dense river of escaping men flooded over the bridge. To the left and right of the crossing, men and horses fought for their lives. Only another 50 meters separated us from the bridge. The crowding, heaving, shoving and rushing came to a gruesome end. Just as the lead platoon was steering for the bridge approaches with 20 mm cannon cutting a bloody swath in the living mass on the bridge, I saw men, horses and vehicles fly high into the air, hang for a moment with girders and support beams in the mushroom cloud of an explosion and then disappeared into the mud of the river. The enemy had blown up the bridge without regard for his own losses.

I stood at the site of the explosion with the bitter taste of sulfur in my mouth and desperately sought a way across the river in order to prevent the consolidation of the fleeing Soviets. Although our fire dominateed the terrain on the far side of the river, we observed a small range of hills three kilometers away where the Russians were establishing defenses. We had to get across; the fluid front could not be allowed to stabilize.

A few meters to the right of the bridge we discovered a ford. The *Kradschützen* of the *1./SS-Aufklärungs-Abteilung 1* had already reached the far bank and established a small bridgehead. A detonation cloaked an armored car in smoke and flames. It had run over a mine on the approach to the ford. Only then did we notice the Russians had mined the area around the bank and a number of Soviets had fallen victim to their own mines. I alerted Bremer to the mines and directed their elimination before any more forces crossed the river. Bremer, standing a few meters in front of me, suddenly gesticulated, pointed wildly at my feet and cried out: "Look — there, there — you're standing on one!" He was right. I was standing on one of those hellish spots and, at any moment, could trigger off an explosion with a slight shift of body weight. We cleared this inhospitable area with the utmost speed.

By 1500 hours it was possible to lead the entire battalion across a temporary bridge to the far bank and we succeeded in establishing a bridgehead three kilometers wide. We had been joined by Witt's battalion. The pursuit would be continued in the early hours of the morning.

With relief I heard we had only lost four soldiers and only one injured soldier had succumbed to his wounds. The Soviets had sustained huge losses in terms of men and equipment. The prisoners could scarcely be counted by the frontline troops. During the night the battalion's sector was hit by harassing fire, indicating by its nature a further disengagement by the Soviets.

From prisoner statements, operations so far and the results of combat reconnaissance during the night, I suspected a headlong retreat by the Soviets. The pressure of *Panzergruppe von Kleist*, which had moved out of the Dnepr bridgehead and was advancing in a southeasterly direction, was probably making itself felt within the Russian command. While it was still night, I drew back the outposts and prepared the battalion for the coming day. My soldiers knew we wanted to drive a rapier thrust deep into the enemy's retreat and, perhaps for days on end, we would be dependent on our own resources. Thrust and parry would alternate with one another including, perhaps, brutal and ruthless stabs in the back. We wanted to confuse the Soviets, wreck their plans and deal them a fatal blow.

At the first glimmer of morning, I looked at my sleeping soldiers, hunched or curled up by their motorcycles, guns and tanks, protecting their bodies from the cold night with shelter quarters. It was getting to be cold in Russia and we had no winter equipment. As always, whenever I faced a decision and held the power of life and death over my soldiers, my whole body trembled and I chain-smoked cigarette after cigarette. The time before the first exchange of fire, before first contact with the enemy, weighed heaviest on me. But when the forces clashed and I was standing in the midst of the melée it was as if this pressure had evaporated.

Our combat engineers moved forward. Heavy demolition charges hung from their arms and bowed their powerful backs. A few hundred meters in front of us was the crossing over the antitank ditch. It had been blown up during the night. We had to blow the ditch walls into the ditch itself to flatten out the steep sides and use the masses of loose earth as a way across. *Kradschützen* stood by to dash across the ditch. The mortars and artillery had been ranged in.

The hands of my watch crept inexorably on and, keeping pace, night fell back to let in the light of day. Vegetation and trees emerged from the darkness. Sporadic Russian rounds crashed into the bridgehead. The rooster crowed his morning greeting from the Akkermen collective farm. Morning had broken. I moved to my armored car with stiff strides and climbed onto the rear deck. From there I could look out over some vegetation far across the antitank ditch and watch the advancing engineers. Armored cars silenced a few enemy machine guns. The weak enemy rearguard was overcome and the crossing of the ditch was achieved. As expected, the Russians had withdrawn during the night, so we would probably cnot encounter stronger forces for at least two or three hours.

That day the *2./SS-Aufklärungs-Abteilung 1* had been placed in the lead. I had a few misgivings since the company had a new commander who, up to then, had been teaching tactics at the *Junkerschule* at Braunschweig and who looked upon my unorthodox fighting methods with skepticism. I gave *SS-Hauptsturmführer* "L" another good talking to and told him not to halt without my orders. He was to advance at the greatest possible speed in the prescribed direction. I moved with the lead company. After a short engagement, the enemy rearguard at Schiroki was overrun and we reached Astrachhanka at 0845 hours. The prisoners that had been taken up to then belonged to the

35th, 71st and 256th Regiments. They were hastily retreating in a southeasterly direction. We pushed into the area of operations of the 30th Division.

An interesting, indeed impressive, tableau was spread before us. As far as the eye could see, Russian units were running, riding and galloping to the east. A horse-artillery unit raced down a slope with its guns bouncing around. It was attempting to get into firing position in order to give us a proper welcome and frustrate our further advance. Mortar fire was already landing uncomfortably close to our battalion, but the retreat was degenerating into a wild flight and slipping away from its commanders. It would have been senseless to begin a firefight with the Russian rearguard and lose valuable time.

I gave *SS-Hauptsturmführer* "L" the order to continue the advance with unabated speed and not worry about the threat to our flank. My order and the accompanying signals to my young soldiers added fuel to the flames. The *Kradschützen* landed among the Soviets like a tropical storm and slashed the fleeing masses apart with their machine-gun fire. Heavy armored cars fired past the flanks of the *Kradschützen* and provided cover. Assault guns fired their rounds at distant targets. Enemy ammunition trucks flew into the air with a crash and artillery batteries formed confused throngs with their horse teams and guns. In the background, two old Soviet planes flew over the wide steppe like singed moths. They did not dare approach within range of our 20 mm cannon.

In a few seconds Bremer stood by me. I pointed to the enemy artillery positioning itself and the lead elements of the battalion disappearing into a dustcloud. Words were unnecessary. What we had learned through long, laborious work in the training areas and at the sandtable was coming to full fruition at that point. Widely dispersed, the company raced towards the firing guns. The flanks of the company tore into the emplacement like wolfpacks and silenced four 122 mm cannon and two 7.62 cm guns. *SS-Obersturmführer* Hess and Wolf were severely wounded in the hand-to-hand fighting. Countless Russians moved westward with their hands in the air.

As far as the eye could see the Soviets were stampeding in wild flight. Bremer rampaged like a loosed hunting dog, leading his company toward the densest throng. Guderian's words again proved true: "The engine is a weapon." Our lightning speed had literally unnerved the Soviets. I followed the lead company and stormed through approaching Russians with my dispatch riders. The main body of the battalion followed at a five-minute interval. In front of us was a collective farm. Fruit plantations and tall deciduous trees surrounded the farm buildings. Even as I approached this small group of buildings, I started to have doubts. I could see no movement at all. Neither Soviet soldiers nor civilians were to be seen even though the *2./SS-Aufklärungs-Abteilung 1* had just moved through the settlement; the dust still hung in the air. What was going to happen?

Peter had his foot on the pedal and wasn't concerned about my reservations. After we passed the first building I shouted: "Step on the gas! Move out!" The buildings to the left and right were crawling with Russians. In a farmyard I noticed a Russian radio center with its aerial still up. Evidently we had moved right through a Russian command post whose elimination the *2./SS-Aufklärungs-Abteilung 1* wanted to leave to the rest of the battalion. Bearing my orders in mind, the *2./SS-Aufklärungs-Abteilung 1* had roared straight through the collective farm.

Remembering the unpleasant experience on *"Rollbahn Nord"* I attempted to catch up with the advance-guard company with utmost speed. The main body of the battalion would have no problem with smoking out the Russian staff. We were relieved when our armored car reached the open steppe that, it should be noted, had taken on an undulating character. To our right to the southeast the Russian flight continued. To the half right behind us we could hear the sounds of fighting from Bremer's company. There was still nothing to be seen of the lead company. The new company commander had really adapted himself quickly to our style of fighting. The *Kradschützen* had to be intoxicated by the speed.

In a small hollow was the village of Inriewka. The village was extended along the road. It was as if this village had died as well; at a fork in the road we found a 20 mm *Flak* that had had minor mechanical problems. It soon got going again. The gun commander wanted to take the eastbound road even though the southbound road was actually the right one. I only succeeded in directing the gun down the right road with some difficulty. The soldiers still wanted to tell me something and kept pointing to the east. Furious over the delay, I gestured to them to move down the southbound road as ordered and at full speed. The gun commander shrugged his shoulders in resignation and followed my vehicle. The route ran parallel to a windbreak hedge that was perhaps five meters thick and consisted of vegetation and individual trees. Small rises broke up the monotony of the landscape.

In a hollow we encountered a group of armed Soviets who had been marching south and who suddenly stopped as if rooted to the spot. Angrily I indicated to the Soviets to throw down their weapons and start heading north. At the same time, I was complaining to my adjutant because the *2./SS-Aufklärungs-Abteilung 1* had not even disarmed the Russians. We had scarce-ly moved out of the hollow when I spotted more Russians cheerfully tramp-ing away to the south, carrying their rifles with them. Some officers were still carrying their map cases. At that point I really lost my temper and cursed the commander of the *2./SS-Aufklärungs-Abteilung 1*. Of course, we needed to move fast but not at the expense of not disarming prisoners. The last section needed to be given that task.

In front of us and off to the right were thousands of Soviet infantrymen and countless artillery batteries streaming away over the slopes and ridges. We had penetrated about 30 kilometers into the flood of fleeing Russians. It would soon be time for the battalion to close up and then risk the final assault on the bridge at Stanitsa Nowospasskoje. The retreating swarms would have to mass together in front of that bridge. There we could harvest the fruits of our operation. Herman Weiser, my trusty adjutant, nodded his head in agreement as I informed him of my decision.

A few hundred meters ahead of us was the Romanowka collective farm. The windbreak hedge ended and the land rose perhaps four or five meters to the left. In place of the hedge, a small stream flowed to the south and, directly in front of the farm, formed a small marsh. I couldn't make out the left-hand side of the village; I was able to observe a row of buildings to the right, extending away to the south. The farm road was at least 20 meters wide. To the west of the farm, on the ridgeline of a small mass of hills, the Soviet flight continued. I had never seen such a mass flight before.

Another 50 meters and we reached the first houses of Romanowka. In the meantime it had already approached 1445 hours and the sun was beating down on us mercilessly. A stormy atmosphere hung over the countryside and over the panic-stricken masses fleeing from us. On the long farm road the heat shimmered between the houses. As in the other villages, there was not a sign of life to be seen. The escape route skirted the village.

A shrill shout from Weiser hit me like a whip. Peter halted. A shadow flitted across the wall to our left front. I saw Weiser firing his pistol into the corner of the wall and, with his other hand, throwing an egg grenade into the same spot. Damn! I noticed a Russian antitank gun in firing position and a line of Red Army soldiers lying in the corner. I leapt out to the right, bounded across the street, dove over a dung heap and gazed into the terrified eyes of two Russians lying behind their machine-gun position; they had obviously only just woken up. We lay opposite each other and waited for the first move. I did not dare look at the road. A direct hit had torn apart the crew of the 20 mm *Flak*. The scream of a soldier segued into moaning. Peter shouted my name and looked for me. He had to be behind me and to my half left. I heard the grinding of tank tracks on the other side of the road. A second shell tore into the *Flak*.

I had to act quickly if I wanted to leave that dung heap alive. The two Russians were still gazing at me, spellbound. They must have thought we were attacking the village with a large force and their last hour had arrived. They disappeared like a pair of rabbits when I gestured to them to beat it. Almost simultaneously with the two Red Army soldiers and in one bound, I leapt past the destroyed gun and pressed myself against a small slope where I found Drescher, the messenger, and Peter. Weiser had disappeared. Drescher claimed that Weiser had leapt into the house to the left.

Only then was it clear to me why we had encountered armed Russians en route. The lead company had taken the wrong direction at Inriewka and I had promptly overtaken it, thus getting myself into this tricky situation. If only the main body of the battalion would appear soon! I yearned for *SS-*

Hauptsturmführer Fenn and his 88's. Hopefully the Russians wouldn't counterattack; otherwise it was curtains for all of us.

I almost fired at Hermann Weiser. How did he get onto the roof? He kept pointing excitedly over to the left. Did he want to draw our attention to the tank? We were exceptionally lucky that the slope was protecting us from the tank's field of fire. I edged a little higher until I spotted a small aircraft at the end of the village. The crate was parked on the collective-farm green. Had we ended up in a headquarters again?

The minutes became hours. To our right the Russians were still racing away over the hills. We pressed ourselves closer to the ground. The tank changed its position. The sound of its tracks came closer, then receded once more. A staff car shot out of the first farmyard at full speed, turning into the bend so sharply that it tipped onto two wheels. At first we were dumbstruck by the car's sudden appearance but then we fired at it with all our weapons. A great dust cloud was all that remained.

At last we heard the sound of tracks behind us. A section of 88's came racing up. Our position and the destroyed 20 mm *Flak* were orders enough for the grenadiers. With an elegant slew, the driver swung the prime mover around so the gun was immediately in position and could open fire. A few seconds later the high-explosive rounds screamed down the road. *SS-Hauptsturmführer* Fenn attempted to destroy the plane on the green, but it was not quite as easy as it had seemed. The gunners could not identify the target and the 88 was ill-suited to the task. The staff car reached the plane and, a little later, the tired bird flew off. It circled wide above the collective farm and disappeared over the horizon.

The 88 worked miracles. Our spirits returned and the paralysis of the last half hour was shaken off. How wonderfully the rounds whizzed past us! What soothing music for soldiers who considered themselves already dead and buried. The heavy tank broke out through the back gardens. We couldn't engage it as the slope was blocking our view. Infantry guns were brought into position; an antitank gun was moved up and fired into the fleeing columns together with the 88. Finally, I had a platoon of combat engineers at my disposal. I could use them to make a small advance into the village. I wanted to know what was going on there!

We entered the first house without a fight. Weiser came up excitedly and led a group of combat engineers to a cellar entrance. In amazement I saw Russian officers emerging from the dark portal. Weiser led me into the building and reported:

Suddenly, to our left, I noticed a Russian tank. The crew was busy with its lunch. Before I could even let you know about my discovery, we were confronted by the antitank gun. I didn't know what to do. I fired at the crew as I jumped and threw an egg grenade at them. I don't know what happened to the Russians: I only wanted to make myself scarce as quickly as possible, to find somewhere safe. I tore open the house door and found myself standing in front of a number of high-ranking Russian officers who were discussing things round a map table. My appearance had a terrifying effect. The Russians leapt headfirst out of the window; I vaulted up the stairs in sheer terror and saw you behind the dung heap. On the other side of the house countless Russian soldiers and officers were running away. A few senior officers climbed onto a tank and took off. We've definitely taken out a higher headquarters!

Before I inspected the spoils I took to task the commander of the *2./SS-Aufklärungs-Abteilung 1* who had appeared in the meantime. His guilt-ridden countenance spared him a thorough dressing down and he went about executing new missions eagerly. Weiser's whoop of joy called me into the building. Peter was already wearing a Russian general's litevka. There were innumerable maps laying on the table and in the next room was a complete radio installation. We had stormed into the headquarters of the Russian 9th Army but, unfortunately, the commander had escaped in the plane with an air force general, both saved by the staff car. Among the prisoners were a few staff officers and the air force general's secretary. They all behaved in a correct and soldierly manner and only at the interrogation did we discover that we had overrun the staff of the Russian 30th Infantry Division at collective farm "X". The 9th Army was rushing, panic-stricken, in the direction of Rostow.

Proud of this unusual success, I sent a corresponding report to the division and was thunderstruck when the radio operator gave me division's answer: "What's all this boasting?" A bucket of ice-cold water could not have had the same shattering effect. Somewhat sobered, I sent a staff officer and the secretary to the division.

A short while later my radio operators started to transmit orders in Russian over the air. We were trying to draw the Russians into a determined defense. We wanted to give *Korps Mackensen* the opportunity of closing the pocket. But the Soviets couldn't be stopped. Their wild flight continued.

Bremer's company was asking for assistance. They had become decisively engaged during the pursuit. We had to wait for the next day and the infantry coming up behind us. For the first time we were unable to inventory the spoils. All around us were abandoned guns, prime movers and horse teams. Countless prisoners moved westward and many Russians had been killed. We suffered 3 dead, 27 wounded and one missing soldier. Just before the onset of darkness Witt's battalion arrived and completed the clean-up of the village. Bremer reached the battalion at midnight.

With the *I./SS-Infanterie-Regiment 1 (mot.)* to our rear we continued the advance on 7 October with orders to take the harbor town of Berdjansk. We moved out past the forward outposts in the cold autumn mist. The *1/SS-Aufklärungs-Abteilung 1* was in the lead. Russian guns, armored vehicles and tanks marked the line of retreat. In Neu-Stuttgart we encountered weak enemy forces that retreated as quickly as possible to the east. In the main street of Neu-Stuttgart we found countless Russian civilians who had been shot. Some of the victims with severe bullet wounds crawled towards us, imploring us to help them. Unfortunately, I never found out why the Soviets slaughtered those civilians.

The collective farm at Andrejewka was taken after a short but energetic fight, and we continued the pursuit at full speed. Strong enemy artillery columns were racing away over the low hills ahead and on our right flank. Two cannon batteries were cut off so quickly that the Soviets did not even have time to defend themselves. By 1000 hours we could no longer count our prisoners and booty. The day was the spitting image of the one before, and we could not push the pace any faster.

We were then moving through rolling countryside. A few kilometers ahead was a small river flowing south into the Sea of Azow near Berdjansk. If the Soviets were to organize their retreat they would have to hold the west bank of the river and engage our forces so their fleeing troops could be led in an orderly manner to the Mius or even as far as the Don. They had to fight for time and space and we could not allow them either. Their hard-pressed units had to be wiped out while they were still on the move. The speed of the German units had to secure victory. They could not be allowed to enter into any more long, drawn-out engagements.

According to conventional wisdom, our units should no longer have been combat effective. None of the units were at full strength any more and could only deploy a mere fraction of their original fighting power. Even the lowliest soldier knew we had to effect a crossing at Nowospasskoje at headlong speed and, above all, prevent the destruction of the bridge. Once again it was a question of minutes.

The advance-guard platoon of the *1./SS-Aufklärungs-Abteilung 1* swept around a low hill and was able to look down on Stanitsa Nowospasskoje. The village was in a depression. The riverbed was eroded and the banks on both sides were almost vertical. A modern concrete bridge spanned the river. Infantrymen, motor vehicles, artillery and a few tanks were streaming over the bridge. Horses waded through the water and attempted to climb the far bank. The entire depression was a heaving mass. But, sharp to the right of us along the heights, shattered units were running eastward and then pushing over the bridge or trying to ford the river to relieve the pressure on the bridge. It was intended for the *1./SS-Aufklärungs-Abteilung 1* to exploit the confusion and then take the bridge with a coup de main.

From the depths of the column *Kradschützen*, armored cars and *Panzerjäger* roared down the smooth slope on a broad front into the midst of the Russians. The fleeing Soviets had no notion of putting up any defense. They streamed down the slope and into the riverbed in wild panic. Horse-drawn batteries, unit trains vehicles and infantrymen formed a jumbled throng. Horses waited patiently in the water, unable to scale the other bank. Still more men crossed the river to escape certain capture, but they were running to their deaths! From countless barrels came rounds that sowed death among the fleeing mass. The panic became greater and greater. Command and control by the Soviets had become impossible. Men raced, trampled and stampeded for their lives. The *1./SS-Aufklärungs-Abteilung 1* thrust into that confusion and, firing from all barrels, rushed up to the bridge. Several armored cars cleared the bridge by fire from the heights. Only the dead and dying remained.

Suddenly I saw the 1st platoon shoot ahead as if bitten by a tarantula. It leapt through a hedge from its still moving machines and disposed of the crew of a Russian heavy antitank gun. The gun was just going into position. A few more seconds and the crossing would have become a bloody affair for us as well.

With fire support from the *2./SS-Aufklärungs-Abteilung 1* and the armored cars up on the hill, the *1./SS-Aufklärungs-Abteilung 1* attacked with speed and was soon at the eastern outskirts of the village. I advanced down the main street with the company and wrestled with the decision of whether to continue the advance on Mariupol. Due to the fast retreat, the situation demanded we advance further on Mariupol and attempt to force larger Russian units on to the coast of the Sea of Azow. The retreat was being conducted in a southeasterly direction, that is, toward Mariupol. Why then should we have wasted valuable time with the seizure of Berdjansk? The town would inevitably fall when Mariupol fell further to the east.

I stood with several soldiers deep in the study of a map. We were beyond the last houses of the village, intending to go over to the windbreak hedge which was preventing us from observing the land ahead of us. Franz Roth, the war correspondent, was with us when Franz called out as if stung. He pulled me behind the hedge and none too gently. Franz was unable to utter a word and was literally beside himself. He wanted to start "shooting" with his camera. Not twenty paces ahead to the left was a monster of a tank that looked like it was going to start up or fire at any second. The street emptied in the twinkling of an eye.

SS-Unterscharführer Bergemann grabbed a demolition charge and worked his way under covering fire through a small orchard to knock out the tank. We followed this man's progress with baited breath. The tank didn't move. The engine was also silent. Had the monster suffered a mechanical problem?

Bergemann got ready for a sprint, took a deep breath and shot off towards the tank. Any moment and the charge would be flying onto the vehicle's rear deck; the next second destroying the engine. It had to happen any second. A round broke the tension. I saw Bergemann fall and the charge roll in the sand a few meters from the tank. A pistol round from the tank had dealt our comrade a fatal blow. The Russian tank remained undamaged after the explosion. An assault gun moved into position and hurled round after round at the steel giant from a distance of barely 25 meters. Nothing happened; the rounds didn't penetrate. The Russians appeared invincible. The gun commander, Iseke, shook his head in resignation and cursed a blue streak. His weapon had found a worthy opponent. We eventually succeeded in destroying the first T 34, whose acquaintance we had just had the pleasure to make, with burning fuel.

During the firefight with the T 34 I stood in the garden of a small cottage with a few soldiers and climbed a small rise to get a better view. The rise in the ground was apparently either a potato pit or the improvised vegetable store of the Russian housewife. A hole in the straw led into the interior of the pit. We had been standing on the rise for quite a while when, suddenly, my loyal Pat dove furiously into the hole and tugged at a Russian coat. The dog had found more than a dozen Russians in the pit. We looked at each other quite nonplused as the Russians scrambled out of their hiding place carrying hand grenades, sub-machine guns and other weapons. We discovered once more for ourselves that soldiers needed an uncommonly large amount of luck not to fall prey to the many hazards of war.

My intention to continue the pursuit to Mariupol could not be carried out at the moment as incoming radio traffic ordered the seizure of Berdjansk. Once again the *1./SS-Aufklärungs-Abteilung 1* took the lead and moved over the hills to the Sea of Azow. Several batteries were taken by surprise and disarmed without a shot being fired. Cold wind whistled around our ears as a spotter plane flew over the column and dropped a message attached to a smoke canister in front of the armored cars. It read: "Only weak forces in the city. An enemy column 10 kilometers west of Berdjansk. Several parallel enemy columns moving in the direction of Mariupol east of the town."

I received the message 8 kilometers north of Berdjansk. Cold, hunger and fatigue were forgotten. The enemy had to be wiped out while still west of Berdjansk. He could not be allowed to reach the town ahead of us! A couple of words to Peter were enough to send our light armored vehicle racing past the company like a race car. Dust clouds hung behind us but ahead we could already see the shimmering surface of the sea. Gerd Bremer and the lead elements received the signal: "Follow me!"

The town could not be identified from the north. It was directly on the coast beneath the steep cliffs. Peter slowed down as soon as we spotted an airfield from which the last aircraft was just taking off and disappearing to the east. We carefully felt our way up to the first houses. A motorcycle section was moving ahead of us. The street was totally dead, not a soul to be seen. Potholes and a rough road surface forced the *Kradschützen* to move slowly.

I gave the signal to pass them and crossed the town as quickly as possible. At that point, we were the first vehicle in line and soon had a lead of more than 100 meters. We were already moving back to the west, hot on the heels of the Russian rearguard. We moved ever deeper into the "ghost" town. Intersections attracted us like magnets. The nose of the armored car edged around a corner and then "sniffed" out the situation down the street. Our vehicle then took a swift bound and stopped at the next corner. We moved like this from street to street, setting the pace for the *Kradschützen*.

I crouched behind the turret and put a death grip on my carbine while Peter prepared to move on to the next corner. Not a sound was to be heard, not a window was open, no movement to be seen. Before the armored car disappeared around the corner I looked back to check if the *Kradschützen* were still following. The armored car suddenly jerked back and I found myself lying on my side in the street. Bullets cracked around my ears; horses reared up and galloped down the street. Wildly shooting Cossacks sprang to their feet and dashed into the nearest houses. An officer jerked his heavy Nagant up and fired. At that very moment I heard Peter's voice behind me, "*Sturmbannführer*, I just got him!" He was right. The Nagant fell to the road. Riderless horses galloped west and wounded Cossacks pressed themselves against the walls of the houses.

Our unarmed command armored car had encountered a Russian cavalry troop. Even greater haste was required if we were to surprise and eliminate the column west of town. Bremer raced off to the outskirts of town like greased lightning and waited there for further orders. The column came nearer and nearer, completely oblivious to our presence. We could already make out every man and vehicle. It had to be the remnants of a reinforced infantry regiment that had been engaged in fighting on the shore south of Melitopol and was trying to regain contact with its main body.

Meanwhile, the battalion had gone into position on both sides of the road and was awaiting my order to attack. I had time. I waited until the column had disappeared into a depression and was in the process of ascending the near slope. Minutes dragged by in expectation of things to come. The *Kradschützen* perched on their machines and enjoyed the last drags on their glowing cigarettes. At that point, the enemy lead elements were only 300 meters away and moving without a care. Somehow, I was feeling sorry for that Russian unit. They had covered their comrades' retreat and had then been left on their own. Before the decimated Russian unit knew what had happened the motorcycle companies and armored cars had raced down both sides of the column and encircled it without much of a fight. More than 2,000 Soviets were taken prisoner with their weapons and equipment. Two batteries were captured. We had one loss from 7 October: The brave *SS-Unterscharführer* Bergemann lost his life in the engagement with the tank.

After taking the enemy column prisoner we established contact with the advance guard from *Abteilung Boddien*. I went over to the battalion myself and congratulated Boddien on the award of his Knight's Cross. His battalion occupied Berdjansk while we moved back to Stanitsa Nowospasskoje and got ready for further pursuit in the direction of Mariupol. For the first time in quite a while we had the privilege of taking a break under the protection of the infantry; the *III./SS-Infanterie-Regiment 1* covered to the east. Mail from home arrived at midnight; it was brought forward by our efficient logistics officer, *SS-Sturmbannführer* Walter Ewert. He sought to use every opportunity to help the line troops and get a general idea of their needs. We owed him a debt of gratitude.

All too quickly it was daylight again. I heard the engines starting up and the clatter of kitchen utensils, but I couldn't find the energy to lift myself from where I was sleeping. Peter didn't give up trying to awaken me, however, until I was standing in the middle of the room putting a hot mug of *Muckefuck* (ersatz coffee) to my lips.

The deep roaring of the *Sturmgeschütze* and the higher-pitched sound of the motorcycles drew me out onto the village street. The engine noise was music to my ears. In order to even out the requirements on the units, we chose a leapfrog advance by the lead elements. The advance-guard platoon, reinforced by armored cars and antitank guns, roared off at top speed ahead of the main body and awaited its approach at prearranged points. The battalion followed the lead elements at a constant speed.

The terrain was very hilly and treeless; newly planted orchards could only be seen in the built-up areas. For the first 20 minutes we moved along the broad, dusty road without encountering the enemy. Neither abandoned equipment nor Russian stragglers indicated that this was the route of a military retreat.

We had been moving into the light of a new day since 0500 hours. We were expecting defensive fire at any moment, but we only encountered enemy resistance at about 0745 hours at Mangusch. The *1./SS-Aufklärungs-Abteilung 1* attacked from the line of march and advanced right through the town. It was difficult to get an idea of how Mangusch was laid out. It straddled a fold in the land across both sides of our line of advance. While advancing through the town, I spotted enemy infantry between the houses and in the gardens. But we didn't have the time to mop up the town completely. We had to get to Mariupol. A larger objective was drawing us on. Our following infantry would be charged with eliminating the remaining enemy elements. The battalion rolled on without stopping. The firefight was initiated from moving vehicles.

Two kilometers east of Mangusch the leading elements discovered well-constructed Soviet field fortifications. This was the outer defensive ring around Mariupol. It could no longer be fully garrisoned as a result of the swift advance of the German forces, but it did afford a strong rearguard the opportunity of defending itself stubbornly.

The position ran along a ridge across the line of march and commanded the approach march completely. We once again chose an otherwise unconventional form of attack. We intended to race up the ridge under the covering fire of our artillery and heavy infantry weapons. The advance-guard company had orders to initiate the engagement only from a point 500 meters to the rear of the position and then work back toward the Soviet defenses. The follow-on units, with the help of the assault guns, were to roll up the enemy position to the sides.

The order was passed around orally and through dispatch riders. I was with Bremer as he was preparing for the sprint over the ridge and into the unknown with my old fighting companions. The assault guns rumbled forward behind the company; rounds landed among the ranks of the defenders and forced them down. The position on the ridge was reached in a few seconds, penetrated by the lead elements and then rolled up on both sides by the follow-on units.

The advance-guard platoon leader, *SS-Untersturmführer* Schulz, was wounded during the advance. The Soviets recovered from their surprise fairly quickly. Bitter fighting was going on, especially to the right of the road. A young, spirited commissar was spurring his unit on again and again. It was not only his yelling that fired up his men, but also his bold example that kept them coming on. I shall never forget the last image I had of this man — pulled up to his full height, throwing his last grenades at Mahl's section. He then solemnly dropped the last one to the ground in front of him and covered it with his body. A quick lift and shudder of the body, a fall of the shattered remains — that was the end of a fanatic.

Two batteries that had not opened fire during the advance were taken at Mangusch. Unfortunately, the younger brother of our fallen comrade, Erich, was killed during this operation. I had transferred him to the trains to protect him. More than 300 prisoners were brought in from the newly won position and several batteries were taken as booty. The prisoners stated they had had orders to withdraw toward Rostow. So — onward! There was no time to lose. The Soviets had to be hit while on the move.

At 0930 hours I was standing at the highest point to the right of the road looking at Mariupol. It was a few kilometers ahead of us. The road led dead straight to the town. No one could be identified. At the edge of the town we could observe a road obstacle and the coming and going of a few armored vehicles. But what else was happening? A long column was marching into the town from the northeast. The Soviet column was several kilometers long. Enemy artillery sheered off from the column and took up position against us. We could also see a Soviet march group west of the town. Russian troops were withdrawing along the Berdjansk—Mariupol road. According to the map, both columns would have to use a single bridge in Mariupol if they wanted to continue their retreat east.

I intently observed the snaking columns for a few minutes without being able to come to a decision. The huge city with its enormous steelworks, shipyards and airfields and the soldiers that the town was ceaselessly soaking up did not fail to impress me. The eastern Soviet column stretched back to the horizon off to our left. The dark line rolled inexorably towards the town. Wasn't it presumptuous to even consider attacking such a mass with less than 1,000 soldiers?

Aircraft started up and flew off to the east. Wasn't that a direct challenge to us to risk the attack despite all our misgivings? Didn't the departure of the planes imply the abandonment of the town? As always at such moments of decision I went to the lead element and listened to the voices of my young soldiers. If a plan was "bad", if it did not look as if it promised success then my "point men" looked at me with indifference or fumbled around with their weapons. If, however, there was the slightest chance of success, I then sensed their willingness to attack and felt their unspoken trust which led me to give the order to attack.

The leader of the advance-guard elements was Sepp Mahl, a comrade-in-arms from my old *15./SS-Infanterie-Regiment 1*. Sepp had taken over the platoon from his wounded platoon leader. He nodded to me, made a dismissive gesture and puffed nervously at his cigarette. The company commander, Gerd Bremer, looked at me calmly. I could always see in his eyes that he would follow me to the ends of the earth. Nowhere did I see doubt or even the hint of being uncomfortable with what was going on.

My soldiers were waiting for the order to continue the pursuit. Their instinct and the experience gained in many engagements led them to believe they would be successful. Their confidence, their fraternal trust and belief in their own powers drove me forward. Fear of my own cowardice drove me onward!

The clock showed 0945 hours as the first Russian rounds burst uncomfortably close and the lead elements moved out. Artillery and an 88 mm *Flak* battery engaged the enemy batteries and disturbed the traffic on the airfield. During the advance I noticed field fortifications on both sides of the road right at the edge of Mariupol. Bursts of machine-gun fire clattered away over us and mortar rounds slammed into the black earth far behind us. Our advance could not be halted. The *Kradschützen* raced towards the town down both sides of the road. A road obstacle had not been completed; the defenders fell to the fire of the *Sturmgeschütze*. Only a few hundred meters away, to the left of the road, enemy aircraft started up and disappeared low over the rooftops. They were flying eastwards. None of the planes attacked us. Perhaps the Soviets did not have any time to load ammunition?

In contrast to other Soviet towns, high, multi-story buildings were on the outskirts of Mariupol. As there had not even been a single tree or a small cottage up to then, the sudden change was intimidating, even oppressive.

The lead elements halted after the initial high-rises and started to advance like infantry. I also wanted to jump from my armored car and take full cover. The high walls threatened to crush us. I was then dragged back out of my thoughts as Peter tore into a round plaza. *SS-Hauptscharführer* Fritz Bügelsack was suddenly to our left. Streetcars were moving towards us. Trucks, prime movers, horse teams and thousands of people enlivened the large circular plaza. Our armored car was suddenly in front of a fire department ladder truck that had got itself stuck in the confusion.

Explosive rounds from Bügelsack's armored car ripped the vehicles apart. The machine-gun fire from the *Kradschützen* echoed gruesomely over the square. Burning soldiers leapt from the ladder truck and ran across the broad plaza like torches. A fuel drum had been hit and the explosion had set dozens of men on fire. A panic-stricken mass of humanity stormed into the side streets, trampling anything in its way underfoot.

Out of breath, we continued to advance and block the roads. We were yelling and screaming. Rounds from our assault guns then began to land in the overcrowded streets. The columns pushing into the town were ripped apart. All semblance of order in the Soviet columns had been lost. We swarmed through the streets like locusts and tried to reach the road to Rostow.

The square was strewn with smoking wreckage. There were only wailing or dead men to be seen. The mass of the Soviets had disappeared. A corner building became the command post from which the fighting could be continued. Bremer pressed on in the direction of Taganrog. At 1310 hours the *1./SS-Aufklärungs-Abteilung 1* was in Sartana 2 kilometers east of Mariupol. A strong enemy column was fleeing down the main road towards the east.

We got the following reply from division when we announced the fall of Mariupol: "Must be a mistake. You can only mean Mangusch." It was no mistake, however. The city had been taken through a daring assault by a handful of German grenadiers whose speed had triumphed over inertia and indecision. The day's success has to be reckoned against a wounded officer, a non-commissioned officer and four men. Four soldiers were gone forever. The following infantry took over mopping-up operations in the city and established outposts far out to the east. We inspected the huge Azow Works with astonishment. It extended several kilometers along the coast and was equipped with modern manufacturing plants. The works had fallen into our hands unscathed.

The "Battle on the Sea of Azow" had ended with the fall of Mariupol. More than 100,000 Soviets had been taken prisoner; 212 armored vehicles and 672 guns had been captured.

From 10 to 12 October the battalion was engaged in fighting between Mariupol and the Mius sector, a few kilometers west of Taganrog. At 0430 hours on 12 October the battalion attempted to establish a bridgehead over the Mius and seize the existing bridge by means of a coup de main. We moved into the Russian bridgehead from the south and were caught in terrific fire from the enemy artillery shooting from the east bank of the Mius. The *1./SS-Aufklärungs-Abteilung 1* was forced to the ground 700 meters from the bridge and had to stick it out under enemy fire until dark. The main body of the battalion succeeded in disengaging itself from the Mius without loss but the lead elements took a vicious battering from the Soviets.

More than 20 men and 5 officers were wounded. Among the wounded were also two of the battalion's doctors. I took my loyal "Peter", *SS-Unterscharführer* Erich Petersilie, back in the twilight. Shrapnel had struck him down; the first driver to be killed at my side. At the time I had no idea that another seven would be destined to die at my side in the course of the war.

In the gray light of dawn of the next day we stood by the graves of our fallen comrades. We lowered four shelter-quarter-wrapped bodies into foreign soil. Silent men stood at the graveside and took their leave. The firing of Soviet heavy guns came booming across from Taganrog as their shells screamed overhead searching for our own artillery.

The dispatch rider section, Peter's closest comrades, Weiser and the rest of the staff waited for my eulogy. I was choked with emotion, unable to say a word. The tears ran down my face. A few wild flowers fell into the dark grave; I saluted Peter and then turned away. I subsequently wrote to his mother.

The battalion was engaged in clearing up the west bank of the Mius until 16 October and followed Witt's battalion. It had established a bridgehead at Koselkin. On 17 October the battalion advanced south as part of the attack on Taganrog and made its way into the harbor without much fighting. Witt's battalion was attacking on our left while Frey's battalion was storming the town further to the north. The infantry was attacking the northern outskirts of Taganrog with incredible decisiveness and penetrated into the city. Unfortunately, Frey's battalion came under fire from two Soviet armored trains. They tore open terrible gaps in the lines until they were wiped out by 88's. More than 80 soldiers died in the fire of those trains that bristled with weapons.

During the assault on Taganrog we were able to observe the Soviets' systematic destruction of a city for the first time. Factories and public buildings flew into the air, one after another. Thick clouds of smoke indicated the Soviet line of retreat. Up to then, we had only seen huge heaps of grain burning; at Taganrog we had a practical demonstration of the "scorched earth" policy. Fleeing ships were sunk in the harbor. Not one Russian thought of rescuing his fellow countrymen from drowning. The Russians only sailed out and brought the survivors to shore at Drescher's insistence. A monument to Peter the Great stood on the steep coastline and looked down on the sinking ships.

It was cold. Icy winds swept across the sea and heralded the onset of winter. We could make out the snow-capped mountains of the Caucasus off to our right. The giants glistened, majestic and undisturbed by the raging of humanity. We were freezing; our uniforms hung in rags. Winter clothing was not available. The unit's offensive elan was not yet broken, but we were searching for an objective to the advance. A vast, unoccupied country with no communications network lay behind us. The rail lines ran from north to south. We thought about defense for the first time.

In cold rain and on churned-up roads we moved in the direction of

Rostow on 20 October. The axis of advance was occupied by the *14. Panzer-Division* which, along with the *13. Panzer-Division*, the *60. Infanterie-Division (mot.)* and the *Leibstandarte*, belonged to *Korps Mackensen*. I looked indifferently at the remains of a former Russian battalion on the rearward slope at Ssambek. I could not read the map anymore. The letters swam before my eyes. Faintness and sickness tormented me to such an extent that I had to ask the divisional commander to be relieved. Four months of fighting in Russia had been enough to force me to my bed. I was no longer fit for combat. *SS-Hauptsturmführer* Kraas assumed temporary command of the battalion and led it into extremely hard fighting.

I returned to Taganrog in the dark and was admitted to the field hospital with jaundice and nauseating dysentery. An epidemic of dysentery raged then and dangerously weakened the front. There were no more German regiments and divisions. The Eastern Front was only being held by the weak remnants of once strong fighting units. The German grenadier was embarking on the hardest battle of his existence — he was bled dry and unprepared. He could see the approaching disaster with wide-open eyes but did not waver for one second in the fulfillment of his duty. He believed in the necessity of his sacrifice.

I left the field hospital some weeks later. I was rather shaky on my legs, but I reported back to the unit. I was transferred to the officer reserve by the commander without much ceremony and had to remain with the divisional staff. The *Leibstandarte* had moved into a defensive position west of Rostow. It was working closely with the *13.* and *14. Panzer-Divisionen* and was parrying every Soviet thrust. The area of operations of the *III. Armee-Korps* was a broad, desolate landscape without trees or vegetation. Icy winds swept across the bare fields, and a severe cold had frozen the ground as hard as stone. It was impossible to dig a foxhole or even an acceptable position. The weather had become our most ferocious enemy.

My battalion was integrated into the defensive lines and was fighting against the Russian 253rd Infantry Division, newly raised in the Caucasus in August and commanded by Colonel Ochatzky. It recruited its men from the Kuban Cossacks who were not friendly to the Soviets.

The battalion had become even weaker in my absence; the lack of officers was making itself particularly felt. The *2./SS-Aufklärungs-Abteilung 1* was being led by *SS-Obersturmführer* Olboeter. He had been wounded, but refused to leave his company, however. The *3./SS-Aufklärungs-Abteilung 1* was afflicted with severe cases of diphtheria that weakened the power of the battalion even further. *SS-Obersturmführer* von Büttner was wounded near Alexandrowskij on 1 November.

I visited my old comrades at Alexandrowskij on 2 November in the company of the divisional commander and was witness to the presentation of the Knight's Cross to Gerd Bremer. I watched with satisfaction the unrestrained joy of the *Kradschützen* at Bremer's award. The company had earned this recognition.

Frost was followed by rain. Water stood in the foxholes and in the deep ruts of the roads and trickled endlessly into uniforms. Motor vehicles, guns and tanks sank into the mud. The grenadiers waded knee deep through the muck. Supply had become almost impossible and could only be achieved by using horse-drawn vehicles. Motor vehicles moved at a snail's pace. The consumption of fuel and loss of equipment was disproportionate to what was achieved. An army was sinking into the mud. Losses due to illness mounted endlessly. A period of biting frost began in the middle of November. We had to hack the frozen-in vehicles out of the mud one by one and keep the engines warm through auxiliary equipment. We were a crippled army.

At the time I was witness to a very serious briefing at the command post at Lachanoff. It concerned the importance of oil. *Generaloberst* von Kleist, *General* von Mackensen, Sepp Dietrich and some oil experts were present. They were convinced that the Russian oilfields at Baku had to be captured in order to continue the war. The capture of Rostow was an absolute necessity and a prerequisite for doing that.

The lesser-ranking soldiers listened in silence while production and consumption figures were quoted and oil-production requirements were discussed. We lacked the necessary data and experience to be able to pass a useful judgment in that domain.

From the military viewpoint, however, things looked different. Everyone warned against an attack on Rostow, pointing out the high losses in the units and, additionally, the fact that the force was simply not operationally ready. The divisions had been bled dry, inadequately equipped and, for a winter campaign, inexcusably poorly supplied with winter uniforms. Fur coats and hats had been procured from Mariupol with great difficulty. German forces could not be distinguished from Russian units at 100 meters. The state of health among the troops was wretched. The unit commanders were judging the situation with complete accuracy when they said: "We will attack, we will take Rostow and chase the Soviets over the Don, but we will never be able to successfully defend the captured city."

In the middle of November 11 soldiers of the *II./SS-Infanterie-Regiment 1* were found in the slime of the septic tank of the GPU building at Taganrog. They had fallen into Russian hands in September and, according to statements from the civilian population, had been thrown into the pit alive.

The attack on Rostow was ordered on 14 November, with the *Schwerpunkt* assigned to the *Leibstandarte*. The attack would start on 16 November. The attack had to be postponed, however, because not enough armored vehicles were operationally ready as a consequence of the severe frost. My battalion attacked down the Sultan-Saly road and immediately came under very heavy defensive fire in heavily mined and fortified terrain. Every inch of ground was wrested in unprecedented hardship at a temperature of minus 30 degrees Celsius. The road to Rostow was won by sheer obstinacy.

My soldiers were fighting without me for the first time and were engaging in perhaps their toughest battle. The struggle raged with great fury along the entire front around Rostow. The strongly fortified and heavily mined terrain exacted a bitter sacrifice from the attacking divisions. Gerd Pleiß, the brave leader of the *1./SS-Aufklärungs-Abteilung 1* lost both legs and died while being evacuated to the field hospital. Fritz Witt fought alongside his grenadiers from the front. *Generaloberst* von Mackensen presented a shining example of real Prussian soldierly bearing as he proceeded to stamp his way fully erect through the heavy snowdrifts and accompanied the *Leibstandarte's* attack.

Grenadiers and generals made the assault shoulder to shoulder on the icy fields outside of Rostow. Attacking T 34's overran the light antitank guns of the *60. Infanterie-Division (mot.)* and threatened to force a breakthrough, but then stood burning in the fire of our 88's. Tears of rage ran down the faces of the gunners of the light antitank guns; they were powerless against the steel monsters. The German antitank gun had become a museum piece; its caliber was no longer sufficient to knock out the enemy heavy tanks.

The grenadiers and the tankers continued the attack relentlessly and with admirable toughness and stormed into fiercely defended Rostow on 21 November. The *1./SS-Infanterie-Regiment 1* succeeded in capturing an intact bridge over the Don and establishing a small bridgehead. The commander of that company, Heinz Springer, was wounded for the sixth time during that operation. The company numbered no more than twelve rifles.

The conclusion of the Battle of Rostow was announced in the following order:

The Commanding General
of the *III. Panzerkorps*

Corps headquarters, 21 November 1941

Corps Order of the Day.

Soldiers of the III. Panzer-Korps!

The Battle of Rostow has been won.

The corps moved out for the offensive before noon on 17 November with the mission of capturing Rostow and a bridge over the Don. The mission had been successfully executed by 20 November.

We captured more than 10,000 prisoners plus — at last count — 159 guns, 56 armored vehicles, 2 armored trains and large amounts of other equipment.

Soldiers of my corps! We can all be proud of this great, new and successful team performance in which each individual soldier added his own well-measured share.

Neither icy wind, biting frost, poor winter clothing and equipment nor the darkest moonless night, neither tank, rocket artillery, thousands of mines nor the field fortifications — constructed weeks in advance and whose huge dimensions we had all seen — nor, least of all, the Red Army soldier himself could halt our triumphant advance.

As a result of the carefully and skillfully prepared deep surprise advance to the east by the eager *Leibstandarte* — supported by the aggressive tanks of the *13. Panzer-Division* and soon accompanied by the highly skilled *14. Panzer-Division* — the enemy defense was lifted off its hinges on his northern front. The enemy was no longer successful — despite furious counterattacks, particularly against the *14. Panzer-Division* — in preventing the two courageous formations from penetrating into the northern outskirts of the large city of Rostow and pushing as far as the Don and its bridges.

The fleeing enemy remnants attempted to save themselves by crossing the Don. The aggressive *I./SS-Infanterie-Regiment 1* of the *Leibstandarte* — a unit accustomed to victory — even succeeded in taking the Rostow railway bridge intact.

At the same time, in a decisive attack executed far to the east and southeast, the *60. Infanterie-Division (mot.)* successfully covered the deep, open flanks of the corps and took Aksajskaja, while elements of the *13. Panzer-Division* pursued the retreating enemy from the west with swift resolve.

Additionally, all corps units as well as the air force units — in particular our wonderful and never-failing reconnaissance pilots — contributed a considerable share to our success! We have cut through the Russians' only effective connection with the Caucasus once and for all.

We now have to hold what has been won in order to open the door to new victories as soon as the *Führer* orders them.

We also extend our *Sieg-Heil* to him!

Signed: von Mackensen
General der Kavallerie

Victory had been won but catastrophe was already looming large. The corps was thoroughly over-extended and much too weak for a long defense of the objectives gained. Units that had been decimated and bled dry were attacked by superior Soviet forces without respite.

My battalion was fighting under the command of *SS-Hauptsturmführer* Kraas on the Donez. The Donez separated from the Don near Rostow and formed the most northerly water course in the Don Delta. The battalion's security zone was eight kilometers wide and had to be defended by slightly less than 300 soldiers. This figure included drivers, radio operators, members of the staff and all officers. There were no more trains, everyone who could fight was in the frontline.

While the Soviets assailed the German front again and again northeast of Rostow, hoping to achieve a breakthrough, they hurled newly raised divisions across the ice-covered Don in rapid succession, attempting to overrun the weak German forces. The German soldiers' performance in this hard fighting almost reached the limits of human capacity. Widely scattered sections lay on the steep, iced-over banks of the central Donez and gazed towards the south over the frozen plain and the Don. Only with difficulty and the help of demolition charges could shallow holes be ripped into the solid earth for use as cover. Protective clothing was taken from dead soldiers and even from dead Soviet soldiers to ward off the deadly frost.

The Soviets had been putting out feelers against our weak security front for three days, indicating impending major activity. My comrades viewed these things without fear and without getting worked up; they carried out their duties almost fatalistically. The few officers circled their sectors like sheep dogs taking care of the soldiers entrusted to them. I met Hugo Kraas and Herman Weiser in a small cottage where they were evaluating the statements of a deserter from the Russian 65th Cavalry Division and preparing their units for upcoming defensive engagements.

The position of *SS-Aufklärungs-Abteilung 1*, weakly held but defended by hardened soldiers, was attacked at 0520 hours on 25 November and bombarded by every caliber of Russian artillery. The losses from this barrage were nil: Where nothing stands, nothing can be destroyed. But then my soldiers' blood threatened to freeze in their veins! Out of the gray light of dawn came masses of Russian infantry who rushed the position singing and yelling. The foremost ranks had linked arms, thus forming a continuous chain which stamped across the ice in time to the wild singing. Mines tore great holes in the ice cover, forcing the Soviets to break their chain. But the mines could not stop the roused mass rushing my comrades like a machine. The Soviets were caught by our fire in the middle of the river and laid out on the ice like ripe corn under the swing of the scythe.

My soldiers lost faith in God and mankind as the succeeding Russian units came clambering over the fallen Red Army soldiers and continued the assault. The attack was being carried out by the Russian 343rd and 31st Infantry Divisions and the 70th Cavalry Division. Three newly-raised divisions on the attack against a few hundred men spread across 8,000 meters and practically alone, each left to his own devices and having to cope with this mass!

Two battalions of the Russian 1151st Rifle Regiment had penetrated the sector of the *2./SS-Aufklärungs-Abteilung 1* and were inside the position, threatening to break up the whole front. The 177th and 248th Rifle Regiments attacked the center of the battalion's sector and were close to a breakthrough there as well.

A counterattack in the sector of the *2./SS-Aufklärungs-Abteilung 1* needed to be launched at once, but there was no one available for that urgent task at that moment. The Soviets attacked the entire front with undiminished intensity, threatening to unnerve the handful of men behind their machine guns. The Russian assault seemed like a tidal wave surging out of the Caucasus, breaking against the steep banks of the Don and there losing its impetus. The first rays of sunlight that peeped from behind the thick clouds illuminated a gruesome picture. As far as the eye could see the Don and its tributaries were strewn with dark dots, some of which were moving painfully while others were being slowly covered with snow. The attack had been repulsed with heavy losses along its entire length. Thousands of Soviets lay in the terrain and waited for night. Riderless horses galloped south, their shrill neighing sounding like the call of death.

The enemy forces that had broken into the sector of the *2./SS-Auf-klärungs-Abteilung 1* were completely wiped out by an immediate counterattack. Six officers and 393 Red Army soldiers were taken prisoner. Three hundred and ten dead Russian soldiers were counted in that company's sector alone. According to prisoner statements, the attack was intended to cut Rostow off from the west.

The attacks continued unabated on 26 and 27 November and without consideration for the huge losses incurred. It was a mystery how men could allow themselves to be led like lambs to the slaughter so willingly. Despite the masses of dead lying stiff and shattered on the ice, fresh formations joined the battle and rushed to their destruction. The Russian attack on 27 November commenced at 1600 hours with a barrage from guns of all calibers — in particular from rocket artillery — and the last attack was repulsed in the area of operations of the *1./SS-Aufklärungs-Abteilung 1* at 1950 hours. Weak enemy elements penetrated the company's position and were sealed off. The counterattack was fixed for 28 November.

The battalion's losses were bitter because they cost us the core of the unit and affected the noncommissioned officers and officers. The battalion adjutant, *SS-Obersturmführer* H. Weiser, assumed command of the *2./SS-Aufklärungs-Abteilung 1*. *SS-Obersturmführer* Olboeter was wounded once again. The last attack on the *2./SS-Aufklärungs-Abteilung 1* hit the left flank in particular and was repelled with fearful losses there for the Russians. The attacking Russian unit had been raised in June in Krasnodar as the 128th Infantry Division and had been employed for the first time. The battalion that attacked the left flank had more than 450 men at the beginning of the assault. In the attack across the Don, 135 soldiers from that battalion were killed and more than 100 wounded were taken as prisoners. A further 37 prisoners fell into our hands unwounded.

The unit's immense achievement can only be gauged by one who had felt the paralyzing effect of the biting frost and the psychological burden as a result of fighting for days on end. I saw soldiers lying behind their machine guns with tears of despair running down their faces as, with hands flying, they poured full belts of ammunition into the attacking masses. Olboeter, the acting company commander, led his *Kradschützen* in an immediate counterattack without his boots. They had been cut from his feet shortly before the attack.

Both his feet had been seriously frostbitten.

The individual warrior proved victorious in this fighting. Left completely to his own resources — at the most, a second soldiers was at his side on the machine gun — he fought the hardest battle of his life without supervision, orders or others to provide an example.

The wounded were patched up as best as possible in the bitter cold and transported to Taganrog on trucks. The piercing screams of pain from our wounded soldiers were harder to bear than the most dangerous attack. We could figure out when the front line would have to collapse in on itself. The daily losses wouldn't permit a prolonged defense.

The fighting continued into the hours of darkness. The area of the breakthrough near the *1./SS-Aufklärungs-Abteilung 1* was mopped up with the help of some *Sturmgeschütze* and was effectively eliminated by 0900 hours. More than 300 dead remained behind in the positions. Prisoners dragged their wounded companions with them. Even after this costly fighting the Soviets continued their attacks. The enemy only withdrew at about 1400 hours to a position 2 to 3 kilometers away; he constantly increased his employment of artillery.

The fighting along the rest of the front had proceeded under similar conditions and had gravely weakened the defenses. The danger of a breakthrough at any point in the front around Rostow could not be denied and had already been expected for several days by higher headquarters. We were all in agreement the front line had to be shortened if we were to avoid a catastrophe which, under the circumstances, would cause the whole front to cave in. There were no reserves behind us. The steppe was empty. Only deep snowdrifts and telegraph poles broke up the monotonous snowy wastes. The best defensive options were available in the Mius sector behind us. It was only there that we could hope to stop the far superior Russian forces and prevent a breakthrough. Advance parties had been dispatched for some time to establish passage points to the rear. Any withdrawal had to stop there on the Mius; the position had to be manned to the last round. A further retreat across the snow-swept steppe would have brought about unimaginable losses in men and equipment.

At the same time the heavy fighting on the Don had died down somewhat and been repulsed with bloody losses for the Soviets, superior Soviet forces renewed the attack in the sector of the 60. *Infanterie-Division (mot.)* and broke through the weak German front to the north of Rostow. The Russians had also broken through on a broad front between the *1. Panzerarmee* and the *17. Armee*. The *17. Armee* withdrew behind the Donez. The front was wavering! Hard and bitter fighting went on up as far as Leningrad. The German Army of the East was no longer prepared to withstand such tremendous superior strength. The icy cold, the totally inadequate clothing and the terrible losses, as well as the lack of replacements of men and equipment, made successful operations simply impossible. We were fighting for our very lives!

In the afternoon the *III. Panzer-Korps* ordered the evacuation of Rostow and an incremental withdrawal to the established defensive line on the Mius. In extremely fierce fighting, the *Leibstandarte* succeeded in evacuating Rostow without great loss and, with the help of the *13. Panzer-Division*, in occupying the established defensive positions. I participated in the withdrawal with the divisional staff. We were happy the order for the evacuation of the city and the shortening of the front had been given. That decision had prevented a disaster of the first magnitude. As a result, the order from the *Führer's* Headquarters not to evacuate Rostow on any account and to defend the captured positions to the last round hit us like a bolt from the blue.

It would have been impossible to execute that order. It showed in an horrific way that the bitter gravity of the situation at the front was not appreciated. At that moment the units were stumbling through the dark night scarcely able to keep themselves on their feet because of the cold. Deep snowdrifts, a biting east wind and the feeling of boundless isolation made life a torture.

All this was a mystery to me. How could such an order have been issued? The order was disregarded and the retreat to the Mius sector proceeded. *Feldmarschall* von Rundstedt, *General* von Mackensen and others deserved the thanks of the units. They had preserved the lives of countless soldiers as a result of their decision and prevented the collapse of the Southern Front. It must also be mentioned that Sepp Dietrich condemned the "hold fast" order in the strongest terms and defended the decision of the *III. Panzer-Korps* as the only one that had been possible. I don't believe I'm wrong when I assert that he placed himself unambiguously at *Feldmarschall* von Rundstedt's side during those grave hours. As a result of his orders, *Feldmarschall* von Rundstedt was replaced by *Feldmarschall* von Reichenau.

Meanwhile, the Russian divisions had been attacking our defenses with undiminished strength. Breakthroughs were the order of the day and could only be eliminated with extreme effort.

My battalion was on the left flank of the *Leibstandarte* and maintained contact with the *60. Infanterie-Division (mot.)*. The units succeeded in holding the position through close cooperation with one another during the fighting and establishing strongpoints with the aid of Russian volunteers. Unit strength had sunk so low that commanders had proceeded to use anti-Bolshevik Russians in the frontline units. It did not surprise me therefore when I visited my soldiers that I almost came across more Russians than Germans in the positions. The volunteers hailed from either the Caucasus or the Ukraine. Their enthusiasm knew no bounds and it was for this reason that they were completely and utterly accepted by our soldiers.

In December I lost one of my best comrades during an artillery barrage. Our efficient interpreter, the brave Heinz Drescher, was one of the most capable officers in the battalion. He always lived and fought as an example to others. He found his final rest beside Gerd Pleiß on the railway embankment at Taganrog.

Shortly before Christmas I had the incomprehensible luck of being allowed to fly home. I flew from Taganrog in a *Ju 52* with a few comrades via Uman to Lemberg and boarded a train there. Within 18 hours I was standing, shabby and ragged, at the Friedrichsstraße Train Station and having my first telephone conversation with my loved ones. Sadly that happy time passed much too quickly and the moment of taking leave approached with giant strides.

On 30 December I received orders to report to Adolf Hitler on 1 January. The tickets were delivered to me from the *Reich* Chancellery. Bitter cold reigned in Germany. I took leave of my wife at the Zoo Train Station in Berlin and climbed into the ice-cold train. My travel companion was the Japanese Ambassador who was also traveling to East Prussia and who, on the basis of previous experiences on the special train, had provided himself with cognac. It did not take too long before we were trying to relieve the cold with firewater.

I was met by comrades at Korschen and taken to the *Führer's* Headquarters through the deep East Prussian forest. We were checked at several barriers and announced by telephone at the last barrier. Personnel of *Panzergrenadier-Division "Großdeutschland"* performed the security duty. The headquarters consisted of a number of bunkers and the usual wooden barracks — all outstandingly camouflaged — which disappeared under the tall trees. The billeting and messing were appropriate for the circumstances. Doubtlessly, expediency and simplicity were the force behind the planning of the camp.

SS-Hauptsturmführer Pfeifer received me and informed me of the reason for my being there. I inferred from Pfeifer's words that Adolf Hitler was concerned about the situation at the front and wanted to have reports from the frontline.

Adolf Hitler made a simple and energetic impression. In amazement, I determined he possessed an excellent knowledge of weapons and he was very accurately informed on the advantages and disadvantages of different types of armored vehicles. Above all, however, I was completely astounded that he was familiar with my battalion's operations and wanted to have questions on tactics answered. On the basis of the battalion's previous successes, it had been reinforced with a lightly armored grenadier company and a heavy infantry weapon platoon.

I did not mince my words with regard to the facts about Rostow and reported on the superhuman demands that had been made of the units. I made a special effort to point out the inadequate supply of replacements. *Generaloberst* Jodl reinforced my words and referred to reports from other formations. I got the impression from the conversation that the situation of the Army of the East preyed heavily on Adolf Hitler's mind and he was anxious to intervene with a helping hand.

I flew back to Mariupol on 3 January with *Oberst* Zeitzler in a *He 111*. In Mariupol I transferred to a *Storch* that was to take me to Taganrog. On the way we flew over the smoking wreck of a shot down *Ju 52*. To spare me the trip to divisional headquarters, the pilot landed in the vicinity of the command post and I climbed on a horse-drawn sled that was passing by. Thoroughly frozen, I landed back at my unit after a 16-day absence.

That same night I relieved Hugo Kraas as acting commander of the battalion and stumbled through the positions for the first time in the gray light of dawn. I was home again. From the start of the Russian campaign through 15 December 1941 my battalion has registered the following casualties:

Killed:

6	officers
9	noncommissioned officers
79	enlisted personnel

Wounded:

20	officers
33	noncommissioned officers
308	enlisted personnel

Missing:

1	officer
2	noncommissioned officers
7	enlisted personnel

Replacements received:

11	officers
1	noncommissioned officer
186	enlisted personnel

During the same period of time 112 officers and 10,142 men of the Russian Army were taken prisoner by *SS-Aufklärungs-Abteilung 1*.

The position went right through the village of Sambek and was on the forward slope of a long ridge. Flooded meadows extended in front of us, their covering of ice broken only by some willow bushes. The Russians were in their own positions across from us. In some places they were only 100 meters away. The front had become quiet. Apart from reconnaissance and artillery, no operations were being conducted. Given the situation, I considered it superfluous to conduct patrols and, consequently, I had no losses for weeks on end. The positions continued to be worked on energetically; in particular, deeply echeloned minefields were laid.

The earthworks were constructed with the help of the local population. The Russians were fed and given medical treatment by our forces. I refused to drive civilians from their homes and chase them into the snowy waste. The consequences would have been inevitable losses. The consideration shown the populace contributed, without question, to the good relationship between the civilians and the troops. It was thus no wonder that, in next to no time, the battalion had the best-developed positions of the sector and was visited by officers from neighboring units. It was practically domestic in our bunkers.

Early in the year I learned something that I do not want to withhold from the reader. One day my driver, Max Bornhöft, set a plate with small pieces of meat in front of me, explaining that they were pigeon thighs that he had "procured" through "good connections" in Taganrog. I had my doubts as I started to eat the delicacy. I told Max that while everything tasted excellent, none of it had ever flown through the air as a pigeon. Well, Max didn't keep me guessing for long. He said quite dryly: "No, the things didn't fly, they jumped! You've been eating frogs legs!"

Following the disappearance of the frost, a period of mud set in which made the supply of the troops almost impossible and prohibited any offensive action. How was the war to be continued? That question was of burning interest to us. A defensive solution was unthinkable, and the Army of the East lacked the forces for a large-scale offensive. The divisions that would be used for an offensive were still in their positions. They only had cadres available to reconstitute and refit themselves. We were afraid of being pulled out of our positions one day and having to go into the attack with improvised units.

After the *1. Panzer-Armee* and elements of the *17. Armee* had eliminated the Russian units which had broken through to the south of Kharkov, we were pulled out of our winter positions at the end of May and transferred to the Stalino area. It was there that the units raised during the winter were also supplied to us and the replacements integrated with the old "warriors". Live-fire exercises quickly returned my battalion to top form. It was then better equipped than in 1941 and, with the experience of the previous fighting, it had become a worthy opponent. The morale of the unit was good. Soldiers of all ranks had gained an unshakeable trust in their own powers and in their officers through their super-human performance, through offensive and defensive operations as well as through the battle against permanent enemy superiority.

In a surprise move at the beginning of June, the *Leibstandarte* was detached from the offensive forces and transferred to France to await possible Allied landings. My battalion was transferred to the Caen area and the staff billeted at Bretteville sur Laize. It was not long before Normandy no longer had any secrets for us. All possible situations were exercised with and without the troops and, as a result, a state of training was achieved that could be measured confidently against that of the best peacetime units.

In the fall we prepared for deployment in North Africa, but fate decided differently. The tragedy of the *6. Armee* in Stalingrad called us back to Russia. Only luck had saved us from the destruction in Stalingrad. Our neighboring divisions from the fighting in 1941/42 were fighting their last fight. Others expected our help. We left France with the greatest of speed and moved east once more. Our objective was the front east of Kharkov.

The Winter War: 1942/43

In November 1942, the biggest offensive that the Soviet High command had ever mounted began along the great curve of the Don. The following impressions were gained:

The enemy had staged an array of men and equipment, especially tanks, that was hitherto unseen. The offensive was operational in nature and planned according to German command and control fundamentals. The Soviet offensive was similar to the workings of a clock in its precision and phasing:

— Breakthrough on the Don at Sserafimowitch; simultaneous breakthrough at Krasnoarmeisk south of Stalingrad. As a result, two Rumanian armies were eliminated and the *6. Armee* in Stalingrad was also eliminated within a short time by total encirclement.

— That was followed by an attack to the west by the two army groups assembled to the west and northwest of Stalingrad. The Soviet "Southern Front" then moved out on both sides of the Don, attacking in the direction of Rostow and the southern Donez area. In the process, the lines of communication which ran through Rostow to the German Army of the Caucasus were severed.

— The Soviet "Southwestern Front" then launched an offensive between the Stalingrad — Morozowsk rail line and the line Kantemirowska — Starobelsk. It had the northern Donez as its objective. The Italian and Hungarian armies northwest of the first point of breakthrough on the Don were threatened on the flank and to the rear by that attack group. As a result of these threats, both armies vacated their strong positions without putting up any appreciable resistance. Their avoidance of contact with the enemy eventually resembled flight. The "Southwestern Front" crossed the Don northwest of Stalingrad.

— After the "Southern Front" reached the lower Donez and the "Southwest Front" reached the Oskol, the southern wing of the "Voronezh Front" joined the attack to the west.

— The sector assigned to the *VII.* and *XIII. Armee-Korps*, which projected far out to the east after the retreat of the flanking Hungarian army, was caught in a pincer attack on the northern and southern flanks. The two German corps were encircled after the linking up of the two Soviet attack groups at Kastornoje.

— Following that, the entire "Voronezh Front" moved west. It broke through the Tim position, which had been established in great haste by *Armeeoberkommando 2*. In the course of further attacks to the west Kursk and Rylsk were taken.

— The southern wing moved out of the Livny area against the right wing of *Panzer-Armeeoberkommando 2*, which was in the process of withdrawing. The northern wing moved out of the area northeast of Orel on Orel itself.

The strategic objective of the Russian winter offensive and its phasing were clearly discernible. The entry of individual army groups were synchronized so that the German defensive line — running generally to the northwest — was rolled up and cut off by each army group.

The operation ran according to plan from Stalingrad to the heights of Orel. The expected results occurred almost automatically, at least insofar as the Italian 8th Army and the Hungarian 2nd Army were concerned. The German frontline was torn open across a breadth of more than 500 kilometers between Slawjansk and the area north of Kursk. The armies of two Soviet army groups marched inexorably westward.

The operational objective — the collapse of the German Eastern Front — seemed to have been achieved in the southern sector.

The Russian High Command announced the Dnjepr to be the next offensive objective. It took no account of the exhausted troops, of growing supply problems, and the losses and casualties suffered during the course of the offensive. It seemed to be of little concern to the Russians that only parts of their artillery had kept up with the rapid advance and that rifle units were almost totally filled by civilians. Artillery was hardly used and unit strengths were maintained by untrained and poorly equipped civilians.

On the other hand, the Soviet Army's strong numerical superiority was established as a result of the loss of five German and allied armies. It was intended for the masses of troops to triumph over the far inferior numbers of defenders in the further course of operations.

The decisive factor, however, was that the Soviet High Command did not recognize the culminating point of its offensive. The culminating point had been reached on the Donez. Logistics and air force ground organizations had to fail over great distances in the face of the inevitable difficulties of a winter campaign; combat power had to wane after offensive fighting covering several hundred kilometers.

The superiority of German command and control, as well as that of the combat units, was consequently able to bring about a decisive result for friendly forces despite great numerical inferiority. The SS-Panzer-Korps, consisting of the *1. SS-Panzer-Grenadier-Division "Leibstandarte"*, the *2. SS-Panzer-Grenadier-Division "Das Reich"* and the *3. SS-Panzer-Division "Totenkopf"*, played a decisive part in this turning point.

The Fighting For Kharkov

By the end of January 1943 the Russians had reached a line on the Donez from Woroschilowgrad to Starobelsk to Waluiki and the upper Oskol and were closing up their units for a further advance to the west. With regard to friendly forces, the *320. Infanterie-Division* was at Sswatowo; the *298. Infanterie-Division* was reforming at Kupjansk after having been severely battered during the withdrawal; elements of *Panzer-Grenadier-Division "Großdeutschland"* were securing the area to the west of Waluiki; and, in the Korotscha area, *Korps z.b.V. Cramer* was assembling together parts of the shattered German and Hungarian units coming from the upper Don.

Great gaps yawned between the units. The German general assigned to the Italian 8th Army was in command of this area. At the time the *SS-Panzer-Korps* with its corps elements, the main body of *Panzer-Grenadier-Division "Das Reich"* and strong elements of the *1. SS-Panzer-Grenadier-Division "Leibstandarte"* had already arrived in the Kharkov area. The *Leibstandarte* established itself on both sides of Tschugujew to defend the Donez. The intention of the *OKH* to employ the *SS-Panzer-Korps* in a concentrated counterattack following its regrouping was thwarted by a swift Soviet advance. A breakthrough into the corps area had to be prevented. Elements of *Panzer-Grenadier-Division "Das Reich"* were moved forward to cover the area west of Waluiki.

Deep snow and heavy frost restricted the forward movement of the units. We had been offloaded east of Kharkov and the battalion had orders to set up a bridgehead at Tschugujew and establish contact with the *298. Infanterie-Division*. Over the course of the years we had got into the habit of not debating orders that seem impossible to execute or even care about the overwhelming superiority of the enemy. Normal standards no longer existed in the conduct of this war. The performance expected of the German soldiers was simply phenomenal. It did not surprise us when the *Leibstandarte* was expected to defend a 90-kilometer front (!) and break the offensive of the Russian 6th Army!

Thick planks creaked under the weight of armored vehicles moving carefully across the long wooden bridge over the Donez at Tschugujew. I was leading my battalion into old Russian positions from the winter fighting of 1941/42. They stretched along the Donez and spared my soldiers from digging positions in the frozen earth. Groups of decimated Italian units approached us across the snowy wastes. Isolated German trains elements with wounded German soldiers and half-starved horses approached from the direction of Kupjansk. The retreating men dragged themselves silently over the bridge. They were no longer fit for battle.

The battalion had to cover a front of about 10 kilometers and, in addition, two companies had to support the disengagement of the *298. Infanterie-Division* at Kupjansk. The *298. Infanterie-Division* was fighting a desperate action against superior Russian forces and was withdrawing to the town. The *2./SS-Aufklärungs-Abteilung 1*, under the command of *SS-Obersturmführer* Weiser, was fighting north of Kupjansk at Dwuretschnaja and, at the last moment, hammered its way through to Kupjansk. The German front line no longer existed. The worsening situation continued inexorably.

Established strongpoints were outflanked on both sides, cut off from their lines of communication and overwhelmed by superior forces. High barricades of snow lay across the deeply furrowed landscape making cross-country operations simply impossible and forcing the units onto the only usable road. The artillery bogged down on the rises; it could not be moved with either prime movers or horses. The road was as smooth as glass.

A look at the divisional situation map and an on-site visit to the units of the *298. Infanterie-Division* led me to the conclusion that the division's position had become indefensible and it would collapse within 24 hours at the latest. The Russians would attack my battalion in a few days and attempt to seize the crossing at Tschugujew. *Kompanie Knittel* remained with the *298. Infanterie-Division* It was to fight a rearguard action with its armored vehicles along the road until it could be received back into the battalion east of the Donez. I returned to the battalion at dusk and found numerous German stragglers who had reported to the battalion and were happy to have linked up with a German unit.

The situation became ever more threatening. Soviet forces were forging ahead towards the Donez, threatening to cut off units fighting east of the river from their lines of communication. I was worried about *Kompanie Knittel* which was still busy with the *298. Infanterie-Division*. Meanwhile, we had improved our position with all means available at our disposal and organized it for all-round defense. By chance, our weaponry had also undergone an appreciable improvement. A train on a siding provided us with a dozen 75 mm antitank guns and six heavy infantry guns. The shortfall in personnel strength was made good with stragglers.

The *3./SS-Aufklärungs-Abteilung 1* succeeded in disengaging from the Soviets on the Kupjansk — Tschugujew road and reached the battalion without serious casualties. The *298. Infanterie-Division* was fighting its way westward through deep snow drifts and icy winds south of the road, having lost all its artillery on trackless open ground. At the moment all contact with the division had been cut off.

I moved in the direction of Kupjansk with two escort armored vehicles to orient myself on the enemy situation. A biting snowstorm was whipping against our faces when I discovered an ox-drawn sled a few kilometers ahead. We slowly approached the rig. The two armored vehicles stopped and provided cover. Was the sled a trap?

On it was *SS-Unterscharführer* Krüger who, despite his wounded condition, had succeeded in dragging himself onto the sled and giving the Russians the slip. I heard from Krüger that there were more stragglers from the *298. Infanterie-Division* in the surrounding area. Within half an hour we had found about 20 members of the division in the huge snowfields on both sides of the road. I had seldom seen such grateful men as these. They had already given up their lives. The endless shroud of snow would have covered them forever during the night.

We observed the shadows of Russian tanks slowly working their way west to the left and right of the road. The tanks avoided the road, snaking across the deeply furrowed landscape, obviously trying to take our bridgehead in a pincer movement and crush it in their armored jaws.

As a result of these observations the antitank defense was appropriately echeloned and the *Panzerjäger* got ready. The clash would have to happen in the next 24 hours. Would we be able to stand fast against the storm or would we have to give way to superior force?

My soldiers were filled in on the situation down to the last little detail and also familiarized with the intended conduct of the fighting. It was my intention to deal a crushing blow to the Soviets without putting ourselves at risk. These Red Army soldiers — drunk with victory — were going to deliver their own death sentence. While the antitank guns occupied positions on both sides of the road, the swift *Panzerjäger* and the *Sturmgeschütze* were positioned on the flanks of the battalion. I was in contact with all the units, either by radio or telephone. The units could respond in fractions of a second and were convinced of their power — they had not been gripped by fear of the Russians.

In the dawn of the new day the white expanse of snow lay glistening in front of our position and the sun heralded a magnificent winter day. The land sloped uphill in front of us for about 1,500 meters, offering an attacker neither cover nor concealment. Woe be to the attacker if he acted as I expected and led his infantry down both sides of the road. In that event the long slope would become the deathbed for countless Russian soldiers. We were positioned in the shadows of tall trees and could only be spotted at the last moment. To ensure complete success, I gave the order to hold fire, opening fire only on my command. Well prepared, we awaited the fight.

From the observation post high above the west bank of the Donez, the Russian vanguard was already being reported. The artillery also remained quiet. It was intended for the Soviets to believe they had eliminated the last German combat unit — the *298. Infanterie-Division* — east of the Donez and that only weak forces from that division were at the bridge.

Around noon the Russians to the left of the road moved against the bridgehead. At first they waited on the ridge, looking down on the Donez and the wooded bank; they then got going again and advanced in attack formation. On the extreme Soviet right flank I could see two KV II's which, if they didn't change course, would run straight into a nest of antitank weapons. Their fate was already sealed.

The spearhead of the attack hesitatingly approached our positions. Not a single round had been fired up to that point. Everything was quiet. I was unable to observe Bremer's sector but was kept constantly informed by the company. More and more Russians came over the ridge and stamped down the slope. The entire slope was swarming with small dark dots. Now and then the spearhead stopped and listened intently. It heard nothing, discerned no movement and then stamped on towards the west.

And what was our situation? My soldiers were crouching in their holes and awaited the order that would get them going: "Fire!" They were freezing. They had been exposed to ice, snow and frost for days on end and held their weapons close to them with numb fingers. They would throw them down on the encrusted snow in the next few seconds and join battle with the Soviets.

I heard the range being counted off from the left flank in my handset. The phone was never silent. The artillery sent its coordinates. Bohr called out on the wire: "Another 500 meters!" A few minutes later it was only 200 meters separating the Russians from the *1./SS-Aufklärungs-Abteilung 1*. The company requested permission to fire. I refused to give the order. Both Russian tanks were moving down the slope to reach the spearhead of the attack.

The voice of Bohr, who was the acting commander of the *1./SS-Aufklärungs-Abteilung 1* in place of Bremer, came over the receiver: "Another 100 meters!" The voice became anxious because I was still not reacting at 75 meters. The tanks had moved to within about 150 meters of the position when, on the word "Fire", death and destruction smashed into the Soviet ranks and one of the tanks was knocked out by an assault gun. The harvest of death was grisly. The rear elements of the Russian units soon ceased any movement as well. They had fallen into a deadly trap; the slope was their downfall.

What was the purpose of our successful defense to the east of Tschugujew, however, when the front was wavering along several hundred kilometers! On 8 February a crisis was developing on both flanks of the defensive front outside of Kharkov. Two Russian armies were enveloping the flanks. To the south, the *320. Infanterie-Division* had been ordered to pull

162

back too late and had been fighting its way slowly back since 5 February. It was out of contact and moving over deep, snow-covered tracks.

The Russians were probing the *Leibstandarte's* front and had found the southern flank at Smijew. Between our flank and that of the *320. Infanterie-Division* was a 40 kilometer gap. The advance there threatened the southern defense of Kharkov at Merefa. A small *Kampfgruppe* with tanks from the *Leibstandarte* was rushed towards Merefa and given the mission of blocking the road to Kharkov.

Our northern flank was also threatened. A weak corps was fighting northeast of Belgorod and had already been outflanked. The operational encirclement of Kharkov had begun and could not be prevented by the forces available. At the same time the Soviet High Command was preparing an advance into the northern flank of the Donez basin. Apart from the attack on the front from Belgorod to Kharkov, our opponent's intent was to cut the German lines of communication in the Donez basin between Slawjansk and the Sea of Azow in a coup de main. The Russians then hoped to eliminate the German forces there.

The fatal blow was to move through Lozowaja and Pawlograd to Dnepropetrowsk and Saporoschje. Five armored and three rifle corps were positioned north of Slawjansk for this operation. The Soviet 1st Guards Army would flood into this area following the evacuation of Izyum. It would move southwest without meeting any resistance; the Soviet 6th Guards Army would join the attack on the right flank after the *320. Infanterie-Division* had been cut off. If this operation were to succeed, then *Heeresgruppe Süd* would be cut off from its lines of communication, the Dnjepr up for grabs and the road to the Western Ukraine open.

After an advance, contact was successfully made with the remnants of the *298. Infanterie-Division* and the survivors were ferried over the Donez. A gloomy atmosphere reigned over the units at the bridgehead. It was obvious that our position had already been threatened deep on each flank and the units had to be pulled back beyond the Donez.

The enemy was positioned in front of the entire Donez front. It was moving between the right flank of the *Leibstandarte* and the *320. Infanterie-Division* with such strong forces that desperate measures were needed. The situation was forcing us to either attack the enemy forces assigned to the southern envelopment of the city — which would result in the evacuation of Kharkov — or the tight concentration of all our forces around the city for an all-round defense — which would be tantamount to an encirclement.

On 9 February the battalion received orders to disengage from the Donez and prepare to attack south of Kharkov at Merefa. The disengagement from the enemy proceeded without loss or difficulty. We were happy to be on the move again and hear the sounds of our motors. We pushed west over laboriously cleared forest tracks through deep snow and across dilapidated bridges, reaching the area of Merefa at midnight.

It was only there that I learned of the gravity of the situation and that the *SS-Panzer-Korps* was facing complete envelopment if it carried out the order to defend Kharkov to the last round. Execution of the order would mean annihilation of the corps and, above and beyond that, would let the Russians get to the Dnjepr. There was no other formation behind the *SS-Panzer-Korps* that could be committed against the Russian onslaught. The corps commanding general, *General* Hausser, decided to move south with three *Kampfgruppen* to eliminate the threat to the right flank and prevent the encirclement of Kharkov.

Deep snow made the approach march for the attack difficult but, in the small hours of the morning, everything was ready to go. My *Kampfgruppe* was positioned on the right flank of the attack group and had orders to drive towards Alexejewka. That mission meant we had to push through about 70 kilometers of enemy-infested territory and strike one of the Russian force's main routes of advance. I took my leave of the divisional staff, moved over to our lead elements and quickly briefed the battalion on its mission.

Every soldier knew what was expected of us and that difficult days lay ahead. My soldiers listened carefully to every word as I outlined the extremely critical situation and told them about the dangerous advance we were going to make. I expected to see anxious faces, but no one seemed amazed at the mission. My young soldiers stood in front of me with red faces and hands buried deep in their pockets. I had known all the officers, including those of the attached units, for a long time, most of them for years. The noncommissioned officers and the junior enlisted personnel were devoted to me and the young replacements formed a close community. I could dare such a "ride" through the Russian hordes without hesitation with such a group of men. Our strongest weapon was the comradeship and absolute loyalty which bound us together and made us self-confident.

The *Kampfgruppe* was positioned to attack on the snow-covered road south of Merefa. The road sloped gently downward in front of us and disappeared after a few hundred meters between the houses of the settlement. Two knocked-out *Schwimmwagen* of the *2. SS-Panzer-Grenadier-Division "Das Reich"* were in front of the village. They had probably belonged to a destroyed patrol of the division. There was no movement to be seen. A strip of woodland extended along the road off to the right behind the village; not a soul was to be seen there either. Next to me was the advance-guard leader, *SS-Obersturmführer* Schulz, who had participated in the drive on Rostow with me. *SS-Obersturmführer* von Ribbentrop was the commander of the first armored vehicle.

We did not dare to advance in a dispersed formation. The deep snow made all cross-country maneuvering almost impossible and a time-consuming business. Fuel consumption would have climbed to absurd proportions.

We had to stay on the road to maintain our tempo and exploit the element of surprise. Schulz had orders to move through the village under the covering fire of the armored vehicles and await the battalion in the small patch of woods. On no account was he to halt in the village or engage in a firefight. I wanted to confuse the enemy with a rapid advance by the advance-guard platoon and lead the *Kampfgruppe* south at breakneck speed.

Schulz climbed onto the back seat of a motorcycle combination and thrust his arm into the air. He waved to me and then shouted, "Let's go!" to the driver. Within a few seconds the first section had disappeared between the houses and the rest of the platoon was racing behind the lead element.

Max Wertinger, my new driver, had heard about our past operations and tore into the village. The assembled *Kampfgruppe* followed directly behind us. At that point, things in the village started to get lively! The Soviets came rushing out of the houses and got mixed up in our march column but only a few started to fight. The bulk of the Red Army soldiers tried to reach the patch of woods. To our left was an antitank gun with full crew ready to fire. It fell to the fire of the column. Beyond the next bend was *SS-Obersturmführer* Schulz in the snow. Despite my warning, he had jumped from the motorcycle and returned the Soviet's fire. A bullet to the chest ended his life. *SS-Oberscharführer* Sander moved back and picked up his fallen platoon leader. A grave was later blown into the frozen earth for him.

Dive-bombers circled above us and flew westwards; they had left the airfield in Kharkov. They waggled their wings as a sign of their solidarity and attacked the enemy columns with their on-board weapons. We learned from prisoner statements that we had cut through the vanguard of the VI Guard Cavalry Corps and had moved straight across the corps sector. We thrust into the westward advancing columns like a dagger. A severe snowstorm set in on the afternoon of 11 February presenting us with deep snowdrifts and making forward movement all but impossible. We had to clear the way with shovel and spade. The blizzard held us in its grip with frightening strength. Our vehicles were nose to tail. It was impossible to overtake. The "road" had become a deep trench in the snow. The armored vehicles pushed through the snow like ploughs. We chewed our way through the glistening white wall meter by meter. The enemy still only appeared as shadows. Both sides were battling the omnipotent weather.

In the twilight we were in front of a wide hollow. I was considering whether we should risk leading the *Kampfgruppe* into the snow-bank filled valley. According to the map the valley had to be 1,000 meters wide and perhaps 50 meters deeper than the surrounding countryside. A village was on the other side of the obstacle; we had to reach it if we did not want to get completely snowed in. Some men on skis were sent out to reconnoiter. Wünsche and I were with the outposts awaiting the return of the patrol. When it returned we would know whether we could move through the snowscape.

We were crouching behind a snowdrift when a sentry pointed straight ahead excitedly and whispered, "Tank!" He was right! We then heard the deep roaring of the engine. The tank had to be climbing the slope a few hundred meters in front of us. Our lead tank was quickly warned. The gunner was sitting at the trigger ready to fire. We awaited the Soviet tank in silence. There it was; I could see it! It came slowly up the slope. Max Wünsche whispered, "Jesus, he's turning his turret right towards us! Can't you see the gun?"

Suddenly a sentry laughed out loud. Before us was a huge Siberian ox whose head we had turned into a tank turret and whose horns we had taken for a gun barrel. The thick snow flurries had played a funny joke on us. We laughed heartily despite the bitter cold.

An hour later we had chased the Red Army soldiers out of their warm hootches and occupied the village. It was only 1800 hours but pitch-dark night surrounded us and allowed us to identify neither friend nor foe. The vehicles rolled in slowly. The darkness prohibited any type of orientation. We were sitting in the middle of the Russian VI Guard Cavalry Corps. Our artillery and the combat trains were cut off from the main body of the *Kampfgruppe*. An enemy thrust had split the column of march into two halves. The artillery had set up an all round defense. I learned by radio that the enemy's attack elements were already 25 kilometers to our west and engaged in attacking Krasnograd. Krasnograd was being covered by *SS-Panzer-Grenadier-Regiment "Thule"* of the *3. SS-Panzer-Division "Totenkopf"*. It had been thrown quickly into the breach to thwart the enemy advance. The main body of that division was still entrained between France and the Dnjepr!

Two reinforced regimental *Kampfgruppe* were fighting desperately east of Kharkov. The German defenders of Smijew were no match for the Russian mass attacks, which were strongly supported by armored forces. Our own thin lines at Rogan held out against extremely heavy attacks pressed by new forces that were continuously being introduced to the battlefield. There was no longer anything human about the fighting. It was brutal and accompanied by crazy methods. The Soviets had perpetrated horrific acts of violence against captured soldiers at the Rogan airfield. Fifty murdered soldiers were found there after an immediate counterattack. Ten had their eyes put out and one had his genitals cut off. With few exceptions they showed severe burns. Ten men were almost completely burned to a cinder.

Following the seizure of Belgorod, an enemy army also advanced deep into the area northwest of Kharkov. By 13 February the Kharkov defensive front's left flank was extended from Russkije Tischky — north of Russkoje — Jemzow Rail Station —Feski. The *SPW-Bataillon* of the *Leibstandarte*, under the command of Jochen Peiper, succeeded in establishing contact with the *320. Infanterie-Divisi*on east of Smijew and in eliminating enemy forces south of Wodjanoje. The remnants of the *320. Infanterie-Division* were totally exhausted and gave a wretched impression. More than 1,500 wounded had survived the march of misery through the dreadful snowstorms and were immediately transported by the corps to the rear and taken care of. The starving division was fed by the *Leibstandarte*.

The Soviet flood continued further west and neared Dnjepropetrowsk. The entire southern front was in danger. As a result of this development, I received orders to push forward in the direction of Alexejewka and block the line of advance to the west.

The blizzards were still raging on the Kharkov front, whipping the snow into the optics of our tanks as we advanced east. Russian forces and my *Kampfgruppe* moved past each other. At midday a reconnaissance plane circled us and dropped a messages attached to a smoke canister. We had been surrounded by the advancing Soviets. We reached Alexejewka 24 hours later and went over to an all-round defense. At that point we were the easternmost formation on the Kharkov Front.

Would the *Kampfgruppe* be able to complete its mission? It was all by itself — no artillery, tanks or combat trains. The town was large; it extended some distance along both sides of the road. While on terrain reconnaissance we encountered a Russian patrol and opened fire at a distance of barely five meters. The blizzard had prevented us from seeing anything. *SS-Obersturmführer* von Ribbentrop collapsed a few steps to my right. A round through the lung had thrown him to the ground. I was happy with von Ribbentrop's situation the next morning. He refused to allow himself to be evacuated to safety in a *Storch* as long as a single wounded soldier was to be found in the pocket. We were completely surrounded. The Soviets flooded past us on both sides of the town.

I was happy to hear Max Wünsche's situation report. He was forcing his way through to us bringing, in addition to his armor, the artillery and combat trains. Hopefully he would arrive in time; we needed fuel and ammunition urgently.

On the morning of 13 February the *SS-Panzer-Korps* received the order from the *Führer* through the *Armee-Abteilung* to hold Kharkov at all costs. Following that, a further shortening of the defensive front around Kharkov was carried out during the night of 13 February in order to pull reserves out of the front line. The new line ran through Lisogubowka — Bolschaja Danilowka. However, by the evening of 13 February, the corps had already reported that the new line could only be held until 14 February, as the city was already surrounded. At midnight the *Armee-Abteilung* ordered all depots to be blown up as well as military installations and those useful to the war economy.

In the morning the Soviets succeeded in breaking through the thin line of strongpoints north of Satischje. An enemy armor attack consisting of forty tanks at Rogan also led to a breakthrough. A thrust on the tractor works at Lossewo was also feared. We lost Olschany as well. That allowed the Russians to keep the Poltawa — Kharkov main supply route under fire.

The Soviets incessantly attacked the southern- and eastern-most points of the Kharkov defense line and threatened to overwhelm us in Alexejewka. We knocked out several antitank guns and inflicted grave losses on the enemy infantry in the course of an immediate counterattack in the direction of Bereka, but our ranks were also thinned. *SS-Hauptsturmführer* Knittel, commander of the *SPW-Kompanie*, received his fourth wound on this occasion. Night attacks were especially dangerous since we could not see the encroaching enemy and had to be sparing with the ammunition.

The enemy penetrated the village during the night of 13/14 February and pushed us back to the middle of the village. The fighting had reached its climax. The soldiers were fighting with the courage born of desperation, but it was not long before there were no longer any withdrawal options. In the middle of this hopeless situation, however, events changed with great rapidity. Assembled close together, the armored cars developed an enormous firepower as they fired explosive rounds into the attacking Soviets. The straw-thatched houses went up in flames. We were positioned in the middle of a wreath of fire and firing out of the darkness into the brightly lit ranks of the Russians. The momentum of the Russian assault had been broken. Our immediate counterattack threw them right out of Alexejewka and returned us to our old positions.

At around this time *Regiment Witt* attempted to force a breakthrough from the north towards us and establish contact with the *Kampfgruppe*. However, the regiment encountered such strong Soviet formations north of Bereka that it could not force its way through in the direction of Alexejewka. We could see new Red Army preparations to the east and west of the village as the day dawned. If both attacks had been unleashed at the same time our fate would have been sealed.

I walked the position and spoke to nearly every soldier. Everyone was crouching together to form a strongpoint. A ring of machine guns surrounded the antitank guns. I was met by gallows humor as I greeted my young soldiers. In no way did we feel defeated. The Russian superiority hardly bothered us, but the lack of fuel, our immobility and the impending lack of ammunition brought us close to desperation.

We had been promised a supply drop for the previous 48 hours but had not yet seen a single plane. The weather conditions made such supply impossible. I sent a situation report once more and requested ammunition urgently. At the same time *Abteilung Wünsche* was gnawing its way ever closer to Alexejewka. Would Wünsche reach us in time?

I was shaken as I stood in the schoolroom among my wounded soldiers. They knew what was happening and begged me not to allow them to fall into Russian hands. My eyes sought *Dr.* Gatternig. We knew the fate of wounded German soldiers who fell into the hands of the Russians. We remembered the grisly end of the German field hospital at Feodosia in the Crimea. It had fallen into Russian hands for a time. The wounded had been thrown naked out of the windows and then had water poured over them. More than 300 corpses had been found frozen solid in the hospital courtyard after the immediate counterattack.

Dr. Gatternig shrugged his shoulders, shook his head and turned away. The voices of my soldiers were tearing my heart from my body. What was I to do? The young men looked at me with relief when I gave the order to provide the wounded with pistols. I would rather stand in the middle of a hail of fire than have to hold another conversation like that.

The clouds hung low in the sky. We heard the sound of engines. From the howling of the engines we could assume that we were being searched for. Suddenly we spotted the shadow of a *He 111*. Would it bring our salvation? A few minutes later the shadow was directly above us again; it was flying directly over Alexejewka. Supply containers fell from the sky but, unfortunately, only a few remained intact. The majority burst under the impact of the fall.

At that point I abandoned all hope. The available fuel was quickly apportioned among a few assault guns and armored cars. If we were to perish then we wanted to storm across the steppe in our armor and leave a bad taste in the mouths of the Russians. They wouldn't take us without a fight.

After finishing a final situation report in which I reported the impending demise of the Kampfgruppe, I took leave of the soldiers who were following our fate with maps in hand and listening to the humming of the radio. I looked into the faces of my soldiers in astonishment. Their expressions seem relaxed, almost curious. There was not one face that showed the distorted features of fanaticism. They followed my words solemnly. I identified the attack objective and climbed into my armored vehicle. Would this be out last attack together?

We pushed slowly out from the center of the village — past ruins and our comrades' graves — to the outskirts of the village. A few hundred meters in front of us Red Army soldiers ran back and forth. They were able to allow themselves this freedom of movement as we had no artillery at our disposal and were also low on ammunition. The Soviets did not appear to believe a counterattack was possible. In the meantime, it had become noon. The snowstorm has eased off; a few rays of sunshine slipped shyly past overhead. What would the Russians to our rear do when we attacked to the east?

Our armor was positioned on the road that led right through the middle of the assembled Russians. I intended to roar down that stretch of road at full speed and hit the Soviets, leading our armor right into the heart of the enemy position. We could only be successful if we moved into the Soviets like a thunderbolt and won ourselves at least 24 hours respite. I hoped to be able to deal with the Soviets to the west of Alexejewka during those 24 hours and thought that Wünsche would be able to break through by then. My little Cossack, a

Russian volunteer who had accompanied me since Rostow and was loyally devoted to me, pointed to a pack of Soviets in the background. The dark dots could be seen everywhere. We were sitting right in the shit!

Only a few seconds then separated us from our start into the great unknown. Our driver shifted the clutch, fiddled with the accelerator and the engine's noise grew deeper. The armored vehicle slowly set off. The *Sturmgeschütze* pushed forward along both sides of the road and left the ruins of the village behind. Our movement accelerated. *Schützenpanzerwagen* and armored cars raced ahead of the assault guns, which were providing covering fire for the more lightly armored vehicles. We were the lead vehicle. Speed was our weapon against the Russians. Bursts of machine-gun fire hammered against the armor. I could only see the unending road ahead and tried to increase speed. Tracks whirled clouds of snow into the air looking like the wake of a destroyer plunging through the waves. We cut through the Russian attack waves in a wedge formation and burrowed deep into their ranks. Heavy mortar fire was being laid down on the road ahead of us. Straight on through! There was no stopping at that point! We had to destroy the attack position or we would all meet the devil.

A particularly hard blow against the armor plate tensed every muscle in my body. The smell of burning filled my nostrils. A second blow crashed into the armored vehicle with incredible force and brought it to a standstill. Our driver, *SS-Rottenführer* Nebelung, started screaming wildly. Flames climbed around my body. I flew out of the turret and lay with Michel, the Cossack, in the deep tracks left by the vehicle. The screams from the car drove me crazy. I worked my way along the rut intending to help our driver. He must have caught his thick winter clothing on something inside the vehicle as his hatch was open. I was suddenly held firmly by the leg. Michel tugged me back and cries: "Back! Commander more important for unit! Back! I get comrade!" The Cossack sprang onto the burning vehicle, pulled the driver out and rolled him in the snow. Mortar and machine-gun fire landed around us. We crawled back down the track pressed close to the ground. We were picked up by advancing soldiers.

It was only then that I determined we had totally smashed the Soviet attack position and Soviet infantry was fleeing in all directions. Unfortunately we were not able to exploit that success as our fuel was running low and another Soviet assault group was positioned for attack to our rear.

After we returned to our jumping-off position we ascertained that Michel had a load of shrapnel in the back of his neck and that our driver had not suffered any serious injury apart from minor burns. The sound of fighting to our rear, that is, to the west of Alexejewka, transformed our mood into one of happy excitement. The sound could only mean Max Wünsche's advance had been successful. And so it was. The *Panzerabteilung* had gnawed its way through heavy enemy forces, bringing us ample ammunition and fuel. We were completely operational once again and, on the orders of division, fought our way back the next morning.

During the fighting back to the west we got to know a new phase of this inhuman war. It was impossible to distinguish Soviet soldiers from harmless civilians. For the first time soldiers were ambushed in towns and in the countryside without being able to identify enemy units. We became nervous. The locals did not dare to betray the concealed Red Army soldiers. The Soviets' enthusiasm and the attitude of the population demanded special watchfulness on our part. My old comrade, Fritz Montag, who had been given acting command of the headquarters company, drove into a minefield and lost both legs above the knee. He was brought to me fully conscious in a motorcycle side-car. A few days later he was buried in Poltawa at the side of *General* von Briesen. The fighting had taken on a treacherous character.

In the meantime, the situation around Kharkov had become catastrophic. Despite all common sense, the town was to be held. Since the requests of the *SS-Panzer-Korps* to abandon the town had been refused — with reference being made to the *Führer's* order of 13 February — the commanding general was determined to issue the order himself to withdraw the units in order to prevent their encirclement and free them for the necessary counter-offensive.

Enemy units that had broken through the eastern section of the front from the southeast pushed into the suburb of Ossnowa during the evening of 14 February. *SPW-Bataillon Peiper*, which had been employed for the immediate counterattack, grappled hard with the Russians in night fighting without being able to clear the area of the constantly reinforced enemy. In the town itself civilians commenced armed encounters. In-transit columns came under fire from the houses.

In this situation the *Armeeabteilung* ordered the attack group of the *SS-Panzer-Korps* to halt its attack south on 14 February and hold the captured terrain. The corps was to release troops for the defense of the city and send an armored formation to Walki to retake enemy-occupied Olschany. It was not feasible for this order to be executed since the moving up of the forces necessary for these missions would take two days with the road conditions as they were.

The Commanding General once more briefed the situation that evening to obtain the order for the evacuation of Kharkov. During the night of 14/15 February the enemy penetrated the rear of our formations in the northwestern and southeastern parts of the city. An armor battalion from the *2. SS-Panzer-Grenadier-Division "Das Reich"* succeeded in inflicting heavy losses during an immediate counterattack on the enemy in the northwest. The enemy's forward advance was halted temporarily. The *SS-Panzer-Korps* once more reported the seriousness of the situation to the army. No decision had been made by noon of 15 February.

It was at that last possible moment that, at 1250 hours on 15 February, the commanding general gave the order for the *2. SS-Panzer-Grenadier-*

Division "Das Reich" to evacuate its position and fight its way through to the Udy sector in order to prevent the encirclement of one and a half divisions. With armor support, the formations succeeded in pulling back through Kharkov and southeast of the city in the nick of time.

This decision was reported to the *Armeeabteilung* at 1300 hours. The commanding general joined the fighting units. An army order arrived at 1630 hours once again demanding defense at all costs. The commanding general's answer: "It's settled. Kharkov is being evacuated!"

General Hausser's decision had saved thousands of lives or spared them many years of captivity. Furthermore, it allowed the formation of a shorter main line of resistance, for which the available forces were quite sufficient. Further enemy penetration at its present pace could be halted by a deliberate defense. The rearguard of the *2. SS-Panzer-Grenadier-Division "Das Reich"* fought its way back through the city on 16 February.

The Counterattack

The operation could have been considered a success just by the fact that the encirclement of one and a half divisions had been successfully prevented through the abandonment of Kharkov and we could then conduct our defense along a considerably shortened front line. The decisive significance of this decision lies in the freeing of the majority of the *SS-Panzer-Korps* for the resumption of the attack to the south to link up with *Heeresgruppe Süd* to which the *Armeeabteilung* had been attached.

The situation on the northern edge of the Donez Basin had developed in the following manner: The enemy had outflanked *Heeresgruppe Süd* at Slawjansk with General Popov's massed armored and infantry units and was steadily advancing on the Dnjepr via Pawlograd and Nowomoskowsk. Enemy reconnaissance units were already pushing as far as Dnjepropetrowsk and Saporoschje with their left flank on Krassnoarmeiskoje. We had hardly any fighting formations with combat power at our disposal on the Dnjepr. *Gruppe Steinbauer*, assembled from personnel on leave and remnants of units, was dislodged from Dnjepropetrowsk to Nowomoskowsk and secured the western outskirts of the latter. The enemy already occupied the eastern part. The *15. Infanterie-Division* was unloading in Dnjepropetrowsk and had moved a regimental *Kampfgruppe* forward to cover Sinelnikowo.

The left wing of the Soviet 6th Army, which had strong elements positioned opposite the *Leibstandarte's* front line, had already begun a southerly envelopment of the *SS-Panzer-Korps* and has crossed the Krasnograd — Nowomoskowsk road heading west with the leading units of several divisions. Elements of the force had already turned northwest. Further forces were aiming for Dnjeprroserschinsk. Immediate countermeasures were a matter of life and death for *Heeresgruppe Süd*.

Following the evacuation of Kharkov the two *SS* divisions could be redeployed. The *2. SS-Panzer-Grenadier-Division "Das Reich"* assembled on the corps' right wing in the Krasnograd area and moved out to the northwest on 19 February to oppose the enemy pressing in from the east. During the gradual withdrawal of its left flank, the *1. SS-Panzer-Grenadier-Division "Leibstandarte"* remained on the previously established front line in support of *Korps Rauß*. It supported *"Das Reich"* in its advance with local counterattacks.

Our units could breathe again. The days of retreat were over at last and the hour of the counterattack had finally arrived. The seriousness of the situation and the significance of the upcoming fighting was clear to every man.

My *Kampfgruppe* received orders to relieve the *SPW-Bataillon* of Peiper at Jeremejewka and to disrupt any further Soviet advance. We advanced through the positions held by *Bataillon Kraas*. It was occupying widely dispersed strongpoints along the front line. The battalion had a front of at least five kilometers to hold. We solved the problem by eliminating Soviet forces that had broken through the strongpoints during the day and establishing fire-spewing hedgehog defenses in the evening.

I linked up with Jochen Peiper ten kilometers in front of the actual front line. He had taken the village of Jeremejewka after a fierce fight and was to hand over the resulting forward bastion to me. Several destroyed T 34's served as windbreaks against the cold snowstorm blasting constantly from the east for our forward outposts. Jochen Peiper briefed me on the situation and drew my attention to the heavy troop movements east of Jeremejewka. It seemed the Soviets were preparing for the attack.

We turned Jeremejewka into a strongpoint in the icy cold. I intended to lead the armored elements of the *Kampfgruppe* in lightning raids from there against the advancing Soviets. We spotted strong Russian formations moving to the south of our position in the direction of Krasnograd and only leaving weak outposts against us. We could observe the line of march in great detail from our forward observation posts and were able to plot their exact positions. A reinforced Russian regiment pushed past us and was feeling its way towards Krasnograd. It presented only a weakly guarded flank to us. It was a direct challenge to us to attack.

The *Kampfgruppe* rolled south at first light the next day; it left the strong-point in the hands of the artillery, trains and the *Panzerjäger*. We departed the strongpoint undetected by the enemy. The misty weather favored out intentions. The artillery was laying down harassing fire on known attack positions east of Jeremejewka to deceive our opponent about our true intentions.

Our column could barely be identified among the masses of snow. Each vehicle was covered with white camouflage, and the soldiers were wearing snow parkas or white winter uniforms. We snaked our way quickly through the undulating terrain.

The *Kampfgruppe* halted behind a low rise. Enemy columns were moving incessantly westwards. A village that extended along the road took in the Soviet column and hid it from our view. About 1,000 meters still separated us from the Soviets. Should we chance it and storm down the gently sloping road? The Red Army soldiers had been marching westwards for nearly 24 hours. Would their superiority be too great? Would we run into a screen of antitank weapons? I stood at the head of the *Kampfgruppe* with Wünsche, the commander of the tank battalion, and the company commanders of *SS-Aufklärungs-Abteilung 1*, searching for a suitable method of attack. I considered speed to be our best weapon, just as I had at other places. It was my intention to advance into the middle of the Soviets with a company from *SS-Aufklärungs-Abteilung 1*. Some tanks would cover the company. It would cut the march column in half and roll it up towards the west.

A section in *Schwimmwagen* volunteered to act as the lead element. The young soldiers knew what they were up against. It was almost certain mines covered the flanks of the enemy march column.

Everything was ready to go in a few minutes. The wheels spun in the snow and slowly began to grip. They got faster, reached the highest point and then roared up to the outskirts of the village at full speed. The crossing of the open ground had to be accomplished at such speed that the enemy had no time to take countermeasures. The vehicles raced down the slope like a raging storm. Tanks moved left and right of the road and hurled their rounds at the Soviets. Heavy mortar fire reinforced the effect of the tanks.

I found myself with the lead company, hanging sideways from a *Kübelwagen* as the first *Schwimmwagen* flew into the air and my soldiers were left lying with shattered limbs. The second car took the lead without a second's hesitation or braking. It, too, was immediately torn apart. The company flew across the locations of the detonations like an arrow. Our comrades had paved the way for us in the truest sense of the word, and we had broken through the mine barrier. The torn-off limbs of both drivers were lying in the snow, as were the less wounded riflemen. Their squad leader had lost both his legs. We couldn't help them, but the company that followed looked after them.

The Soviets abandoned the village street in great haste and rushed into the houses or sought salvation in flight towards the south. Our machine guns felled them on the white snow-covered fields. Their accompanying artillery was either overrun by our tanks, pushed aside or shunted together into a tangle. The destruction we had wrought was indescribable. A few tanks fired round after round into the eastward-marching column. They forced it into wild flight that was further accelerated by the advancing tanks.

The column was smashed by the charging tanks as if by a giant fist. Once again, speed revealed itself to be a terrible power. There were hardly any Russian antitank guns that succeeded in unlimbering and taking up position. The grinding tracks and the weight of the tanks crushed most of them. We had pushed right through the extended village within the space of a few minutes and the route of the enemy advance had been turned into a road of misery. Shattered steel mingled with the flesh of Siberian oxen that were serving as draft animals for the antitank guns.

The hunt to the west continued from the outskirts of the village. The Soviets had been taken completely by surprise. They did not understand how death could reach out to them from behind. Almost without resistance, the column fell victim to the onslaught.

175

A damaged tank was located between the last few houses of the village. It had been knocked out by a Russian antitank gun that was positioned no more than 150 meters away in a fruit garden. Infantry were already in the process of eliminating the gun when a burst of machine-gun fire landed between us. In a flash we took cover on the far side of the tank. Franz Roth, the ever-ready war correspondent, did not make it, however. He had received a round in the chest. We pulled him to cover and then took him to a small house where *Dr. Gatternig* immediately attended to him. Roth died a few days later in a field hospital. He had been one of our best photo correspondents.

The next village went up in flames from the tracer ammunition. The Red Army soldiers ran for their lives and died from the machine-gun fire. The pursuit continued as far as the next village and sent the Soviets into wild flight. Equipment and weapons were left behind in a mountain of debris. The danger to our southern flank had been eliminated temporarily. At the onset of darkness we returned, tired, to our strongpoint. The elimination of the Russian march column had cost us two dead and several severely wounded soldiers. The speed of the operation, the exploitation of the ability to maneuver and the employment of firepower had brought us success.

Additional enemy forces had occupied attack positions east of Jeremejewka according to battlefield observation reports. We had to believe they would be attacking very soon.

In the meantime, the *2. SS-Panzer-Grenadier-Division "Das Reich"* had attacked to the south with three attack formations. It destroyed strong enemy concentrations south of Krasnograd. The attack won ground and was continued during the night of 20 February. The lead elements, composed of armor, advanced during the night to the south. Round after round impacted into the flanks of the enemy columns, which were crossing the road to the west. Pursuit formations relieved one another until the leading elements of the division reached Nowomoskowsk on 20 February at 1400 hours. There they established contact with the outposts of *Gruppe Steinbauer*.

The *Luftwaffe* supported the attack groups with *Stuka* sorties on the enemy pockets of resistance and caused the massed formations heavy casualties. The enemy formations, which were already west of the Krasnograd — Nowomoskowsk road, started streaming back. Further to the south, however, large enemy formations were following the German lead elements that had halted outside of Nowomoskowsk.

Pawlograd was designated as the next attack objective for the *2. SS-Panzer-Grenadier-Division "Das Reich"*. The reason was the advance of strong Soviet forces on the bend of the Dnjepr south of Dnjepropetrowsk via Sinelnikowo. After hard fighting, the division was able to link up with the *XXXXVIII. Panzerkorps* east of Pawlograd.

Independent of what was happening south of Jeremejewka, it was time to advance into the Soviet assembly areas at Nischij Orel and launch a pre-emptive strike against them. I had reservations when I briefed the unit leaders on my decision. Max Wünsche was on fire, however. I intended to move the tank battalion with a *SPW-Kompanie* and two companies of mounted infantry far to the north. Those forces would then turn to the east and penetrate into the assembly areas from the rear. The reconnaissance platoon had already scouted and marked the route. The movements were to be orchestrated in such a manner that we would appear in the enemy's rear at dawn. At the same time, all of the trains drivers and anything else that had legs — supported by artillery — would launch a feint and divert the Russian's attention to the west.

In the deep of night vehicle after vehicle was positioned nose to tail and waited for the lead elements to move out. We moved out into the darkness slowly. Armored vehicles were located at the turning points and showed us the direction while, at the same time, providing cover. The rumble of the tanks could scarcely be heard in the high snow. We snaked our way forward to our objective as if on 1,000 cat paws.

We had moved too quickly. We waited for the right time between two villages (we had bypassed all villages). The tanks drew up. Close to one another we waited for the first shimmer of daylight. Were we in the right position? Did we make a mistake somewhere and get lost? Had the enemy already seen us? A lot of "What if's" occupied my time. Eventually I thought I could make out the outline of a tank behind me. It was about 100 meters behind the lead elements. This meant there was enough light to launch a surprise attack. The moment had come. Radio traffic started. I gave the artillery in Jeremejewka their fire mission and everyone tensely awaited the first rounds. Their impact would indicate whether we had approached the right place.

Yes, we had! Our howitzers hammered the Russian positions off to our right. Dazzling flashes flickered across the snow. The 20 mm tracer rounds from our armored cars identified the objective. Machine-gun and rifle fire rattled along the entire front and mortar rounds thudded in the village. The forward artillery observer concentrated the fire and trained it on our intended penetration point. We were to the side of the impacting rounds and able to observe their effect precisely. We could recognize the enemy artillery by its muzzle flashes. The batteries were not quite 400 meters away. The Soviets had not spotted us yet.

The moment had arrived! The tanks pushed into our opponent's deep flanks along a broad front and opened fire at very short range. The enemy antitank guns did not manage to fire. They were deeply echeloned and oriented towards Jeremejewka. What was the point of emplacing them that way? The Russian antitank officer had given no thought to his rear and flanks. The infantry dismounted and leapt into the houses, fetching out the surprised Soviets. Direct hits from tanks stuck several trucks with "Stalin organs" on them. A hazardous fireworks display rose heavenwards. The trucks literally disintegrated. Tiny bits of them came crashing down on us.

A tank company reconnoitered to the east and encountered an enemy artillery battalion in the process. Combat engineers blew the guns up. The street and house-to-house fighting was short and painless. It was as if paralysis had hit the Soviets. They had not expected our advance. The Russian divisional commander died as he fled. We found his remains in an fruit garden. We worked our way forward from house to house. *SS-Obersturmführer* Bohr, Bremer's executive officer, collapsed a few meters ahead of me. A round to the stomach threw him to the ground.

While breaking into a larger building a soldier warned me about the rooftop snipers who fired at us through the straw roofs. As he dove through the door the good soldier collapsed, shot through the head. The house went up in flames. A staff officer ran straight into our arms. He was the Russian division's chief of staff. Within half an hour the village was ours. Our artillery performed magnificently. The barrage slammed down in front of us like an all-shattering fist. No wonder, the forward observer was right with us and, as a result, also in the middle of the enemy.

We rolled up about two kilometers of the enemy positions, completely scattering the Russians. Black dots fled wildly across the vast snow fields. Enemy antitank guns were crushed under the weight of our tanks. All the guns were oriented west, but the deathblow came from the east.

Dense, choking smoke lay over the village as the last rounds whistled through the morning and ended the fighting. Field ambulances rolled westwards. Our fallen comrades lay before me on shelter halves as we took our leave. They were later distributed among the tanks. We left no one behind. Their peace would not be disturbed. In the past we discovered that if ever we gave up a sector, the Russians plundered and destroyed the graves.

The Russian staff officer made a good impression, showing an exemplary attitude. We had to leave our cottage in a hurry as the straw roof had caught fire and burned like tinder. The lieutenant colonel readily answered all questions which were not directly connected with the current operation. He had been transferred to frontline duty only a few days previously and had just ended a course at the Frunze Academy in Moscow. Before sending him over to the division we took leave of each other and he said: "We will win the war against Germany with America's help. You are losing now — but, one day, we shall all be friends. We will continue the struggle together and achieve the final victory."

At about 1500 hours the last tank wound up back at Jeremerewka. Deeply moved, I took my leave of *SS-Obersturmführer* Bohr. He already bore the mark of death as he departed from the battalion: "May I return to the battalion?" He left us forever on the trip to the field hospital.

The fearsome cold crammed us together in the few remaining buildings and only the most indispensable sentries had to endure the conditions outside. My soldiers suddenly give vent to a cry of jubilation and mobbed me like savages. My hand ached under the pressure of theirs. Completely astonished, I heard from the battalion I had been awarded the Oak Leaves to the Knight's Cross.

After the initial surprise I left the cottage and sought my fallen comrades. There was not a sound to be heard. The front was silent. Only in the distance were there bright flashes. I could hardly make out the gravesite. There was neither cross nor stone to mark the last resting places. The snow had been trampled down hard and scarcely stood out from the surroundings. Our comrades were not to be disturbed. We wanted to protect them from grave desecrators. Snowflakes fell from the sky and covered the grave that looked like a deep wound. The gloomy site disappeared. The good Lord covered the scars. I could not feel happy about the decoration. Beneath me rested the soldier who had warned me about the treacherous rooftop snipers that morning. Without his warning I would, perhaps, have been lying by his side.

I was ordered to the *Führer's* headquarters and, 24 hours later, I flew from Poltawa to Winiza. The headquarters was distinguished by its simplicity. The first thing I did was request a telephone call to my wife in Berlin. The call went through in a few minutes and I experienced the great joy of being able to speak to my wife and our children.

After I finished the call I was led to Adolf Hitler who greeted me heartily, presented me with the decoration and asked me to take a seat. For more than an hour I heard of the efforts being made on the home front and the battlefield. The tragedy of Stalingrad seemed to weigh heavily on him as his thoughts kept returning to the *6. Armee*, but I found it revealing that he did not censure any officer for his conduct in Stalingrad. Hitler was seriously concerned about the continuous air raids on Germany, and I had the feeling that the population's suffering was a particularly grave burden. Hitler made a good physical impression; his voice was calm and his comments on the situation at the front were realistic and most pertinent. He did not put forward any prognoses, but he knew that the war would last a long time. He saw Churchill as his worst enemy.

We remained together undisturbed for an hour, and I had the opportunity of giving him an unvarnished report from the front. At the same time I pointed out the shortages in arms and equipment. Adolf Hitler did not interrupt me. He listened to it all patiently, making the occasional note. After the meal together I sat with *General* Stief and some others discussing the course of the war and its further prosecution. (Some weeks later Stief requested me to visit him, apparently to discuss a few questions. I was unable to accede to his request since, by that time, I had been detailed to the School of Armored Warfare. *General* Stief was later hanged in connection with the 20 July Plot.)

I was back in Poltawa a mere 48 hours later and climbed into a *Storch* which took me to divisional headquarters. In the meantime, the conduct of the fighting had taken a more favorable turn.

As a result of the advance of the *2. SS-Panzer-Grenadier-Division "Das Reich"* to the south from Krasnograd, we had defeated strong enemy forces and broken up their spearhead formations. Considerable forces remained in position east of the Krasnograd — Nowomoskowsk road, however. We needed fresh forces to eliminate that enemy threat and establish contact with the *Leibstandarte* northeast of Krasnograd. To that end the urgently awaited *3. SS-Panzer-Division "Totenkopf"* — which had detrained in the Poltava area in the meantime — was attached to the *SS Panzer-Korps*. It had been assembled in the Pereschtschepino area.

The *3. SS-Panzer-Division "Totenkopf"* moved out to the southeast to attack on 22 February. It advanced in three attack groups in the area between the Ssamara and the Orelab sectors. The enemy advance-guard elements positioned there were eliminated, but the enemy was still capable of further attacks. The main body of the Soviet 1st Guards Army was just starting its approach march; our opponent still seemed to be of the opinion that defenders who had suddenly switched to the offensive would soon run out of steam. The Russians were also bringing up fresh forces to the area in front of the *Leibstandarte*. Elements of the Popov Group had already been cut off by the *1. Armee* operating to our right. However, five enemy armored corps were still advancing west in front of the *4. Panzerarmee*.

The milder weather which set in around 20 February favored offensive operations. Most roads were free of snow and this considerably increased the mobility of motorized units. It was important not to let those enemy forces that were escaping to the northeast off the hook. Instead, they had to be fixed and defeated.

The attacking German divisions made advances on a narrow front and established strong flank protection along feeder roads. The enemy was thrown from the villages he was clinging to in rapid thrusts. These offensive strikes were well supported by the *Luftwaffe*. His line of march — always directed to the southwest — was cut.

Our units took Losowaja at 1400 hours. Despite that, our lines of communication and, correspondingly, the army's left flank, remained under the threat of those battered enemy forces that remained cut off from their own lines of supply in the Samara and Orel sectors. These enemy units had been given orders to retreat and regroup around Orelka, Losowaja and Panjutina and were breaking through to the east and northeast in small, armor-reinforced groups. Other enemy units positioned off the major avenue of approach for the enemy moved north out of the area south of Pawlograd the following day. One of those formations — supported by armor — struck the corps command post at Jurjewka on 28 February. Another strong formation attacked the command post of the *15. Infanterie-Division* at Orelka shortly thereafter.

We achieved our first objective on 27 February. Assault Group Popov was robbed of its attacking power, squeezed out of its bridgehead by and large and prevented from attaining its own objectives.

Meanwhile, the *Leibstandarte* had solved its defensive mission by going on the offensive. Despite the large width of the division's sector, its assault groups had inflicted heavy losses on the enemy by constantly attacking first one flank and then the other. By doing so, it had stopped the enemy advance on Poltawa.

The enemy had been regrouping in front of the *Leibstandarte's* right flank since 28 February. The Soviets had pulled two armored corps and three rifle divisions of the Soviet 3rd Tank Army out of the area between Ljubotin and Walki in order to throw them against the *SS-Panzer-Korps*, but we had been unable to discover their assembly area.

We began a new phase of our own attack. The avenue of advance was switched to the northwest; the first objective was the line of high ground between Bereka and Jefremowka. The terrain in front of the *SS-Panzer-Korps* was well known as a result of the fighting in February. It was intended for the right wing of the army to reach the Donez, while the *SS-Panzer-Korps* was to take the high ground at Jefremowka. The Leibstandarte would link up with *Armee-Abteilung Kempf* on the east flank of the front's salient.

The *2. SS-Panzer-Grenadier-Division "Das Reich"* advanced on 1 March along the Krasnograd — Oktjabrskij road. It continued the advance on 2 March toward the high ground northeast of Paraskoweja. It's objective: The heights at Starowerowka. On 2 March Bereka was taken by the *XXXXVIII. Panzer-Korps.*

The *3. SS-Panzer-Division "Totenkopf"* pushed north in the region of Orel, though severely impeded by the condition of the road network. It took Lissowinowka on the evening of 1 March. It then turned northwest on 2 March to eliminate enemy units reported to be in the Nischnij Orel — Jeremejewka area. *Kampfgruppe Baum* encountered fierce resistance east of Nischnij Orel. From the reports on 2 and 3 March it could be determined that the division's left flank east of Jeremejewka had encountered enemy forces moved into the area from the north. This made it clear that the enemy had not succeeded in determining the direction of our advance. He had marched his regrouped units right between the elements of the attacking *SS-Panzer-Korps* and the *Leibstandarte's* defensive front. As a result, the right wing of the *3. SS-Panzer-Division "Totenkopf"* was turned inward. The division slammed into the enemy while he was still assembling.

The enemy tried to evade the pincer movement by making strong counterattacks to the southeast and northeast and later attempted to improve his position by dispersing his forces into smaller formations. That was futile. The main body of the enemy was eliminated by the *3. SS-Panzer-Division "Totenkopf"*, the southern wing of the *Leibstandarte* attacking to the east and elements of the *2. SS-Panzer-Grenadier-Division "Das Reich"*. The three divisions were engaged in concentric attacks during three days of hard fighting. *Stukas* supporting the attack scored tremendous successes against the encircled enemy.

Individual columns which had taken flight were routed during the pursuit. They made our rear areas unsafe for a few days but were completely annihilated in independent actions. The commanding general of the Soviet XV Guards Armored Corps was found dead only a few hundred meters from the command post of the *SS Panzer-Korps*.

Ochotschaje fell to the *2. SS-Panzer-Grenadier-Division "Das Reich"* on the evening of 4 March after hard fighting. The *Leibstandarte* had gone into the attack from the northeastern sector of its position at Starowerowka and linked up with the *2. SS-Panzer-Grenadier-Division "Das Reich"*. The *3. SS-Panzer-Division "Totenkopf"* completed its destruction of the encircled enemy on 5 March, achieving its greatest victory. Enemy personnel losses were high and the pocket was crammed with immense quantities of weapons and vehicles. The bulk of two armored corps and three rifle divisions could be considered as destroyed. The Soviet 3rd Tank Army was decisively weakened as a result of the Battle of Jeremejewka.

The *1. SS-Panzer-grenadier-Division "Leibstandarte"*, which was once again attached to the *SS-Panzer-Korps*, closed up on the line it had reached after the capture of Stanitschij and reorganized for the attack. By 5 March it was in position next to the *2. SS-Panzer-Grenadier-Division "Das Reich"* at Karawanskoje. It was positioned along the line Stanitschij — Winnikoff — Nikolskoje — Krut Balka. On the same day the *3. SS-Panzer-Division "Totenkopf"* was also freed up and reattached to the corps.

The road conditions favorable to an attack had worsened considerably, and the layer of snow in the northern part of the area we had reached was still deep, slowing operations. Was the attack to continue? Should Kharkov be retaken? Or should the elimination of the enemy forces in front of *Armeeabteilung Kempf* be continued?

In any event, neither course of action was taken. Instead, the next objective was to be the Mscha sector. The *SS-Panzer-Korps* moved out against the Mscha sector on 6 March with the *2. SS-Panzer-Grenadier-Division "Das Reich"* on the right, the *1. SS-Panzer-Grenadier-Division "Leibstandarte"* on the left and the *3. SS-Panzer-Division "Totenkopf"* behind the left wing. The *2. SS-Panzer-Grenadier-Division "Das Reich"* expelled the enemy from Nowaja-Wodalaga after hard fighting. The *SS-Panzer-Grenadier-Division "Leibstandarte"* penetrated the line Moskalzowa — Ljashowa — Gawrilowka, and one of its battalions established the first bridgehead at Bridok. The difficult terrain delayed our neighbor to the right who, nevertheless, established his right wing outside of Taranowka and captured Borki.

The *2. SS-Panzer-Grenadier-Division "Das Reich"* reached the Mscha sector during the night of 7 March, and the *SS-Panzer-Grenadier-Division*

"Leibstandarte" expanded its bridgehead. The weather warmed up, and the night frosts were no longer hard enough to keep the ground frozen. The state of the road network — fluctuating between snow and mud — became ever more critical. Men and equipment were pushed to the limit by the demands made upon them.

On the other hand, the Russians revealed clear signs of weakness. The fighting between the Donez and Dnjepr had inflicted heavy losses upon them. They took pains to throw fresh troops at the *SS-Panzer-Korps* but the forces were insufficient.

The same quandary remained: Attack northwest and roll up the enemy units in front of *Armeeabteilung Kempf* or attack Kharkov? Once again the problem would be resolved through continuation of the attack to the north. On the morning of 6 March I was on the left flank of the *Leibstandarte* with my *Kampfgruppe*. I had received orders to attack northeast and simultaneously protect the division's left flank. Deep snow made any advance difficult. The road was invisible under the masses of snow, and we could only guess where it was as we crept slowly up to a low rise from which we could guarantee a good observation point to the northeast.

A broad field of snow was in front of us. A small, defended village was 500 meters off to the right with Soviets going about their business casually with no inkling of our location. We could make out the low buildings of another village in the background, and it was this village that I decided was the initial objective of the attack. The distance was about 10 kilometers. I ordered a company of tanks from Wünsche's battalion to eliminate the first village. Mounted infantry accompanied it. The tanks would roar up to the village under an artillery barrage and remove the threat to the flank of the *Kampfgruppe*. I wanted to attack the far village with the main body of my forces, thus pushing through and into the Soviet rear zone.

The unit commanders informed the men about the situation and our intentions. The artillery and mortar battalions reported they were ready to open fire. My soldiers scoured the horizon for enemy movement. I leaned on the radiator and warmed my hands on the engine. The hands of my watch moved slowly forward. We stubbed out our cigarettes. Tank hatches slammed shut and reports of "Ready" rang in my ears.

Only seconds separated us from the drum roll of the artillery. I raised my arm and looked once again at the companies, then brought my arm down as the first rounds were fired. The tank company pushed slowly, creakingly and ponderously off towards the village and moved to engage the enemy under the howling of our *Stukas*. Plane after plane struck pre-determined targets from low altitude. The enemy was forced to the ground under the rain of bombs and on-board weapons. As the last bomb fell our first tank was already entering the village.

Meanwhile, the *Kampfgruppe* had moved out and was trying to reach the objective at full throttle. The snow was showing itself to be a far greater obstacle than the defenders, however. High snowdrifts towered up in front of the front slopes of the tanks. Over time it formed a hard and strong wall. The tanks pressed forward ponderously. After a short time I realized we were without infantry support as the *Schützenpanzerwagen* and the *Schwimmwagen* were stuck in the snow. The packed snow was lifting the vehicles out of contact with the ground and threatening to destroy our plan of attack. My armored car was wedged between two *Tigers* and we were only burrowing forward slowly. We had to halt for a while in a small hollow to let the infantry catch up. The soldiers climbed onto the tanks as the light reconnaissance vehicles were not up to coping with the drifts. We then took off again! To our left rear the Russians defending the first village were fleeing and falling victim to our weapons.

Antitank fire struck our lead elements as we approached to within range of the second village, but the *Tigers* disposed of the enemy antitank weapons. The right-hand tank company under Jürgensen moved forward swiftly, exploiting the cover of some fruit gardens and was about to outflank the village.

Our armored car received a hit from a 47 mm antitank gun which, however, did not cause serious damage. Unfortunately, we could not identify the firing position. We were then within 200 meters of the village and were looking for a point to enter. Machine-gun fire hammered on our armor plating. The lead *Tiger* encountered a mined obstacle and remained on the spot with a thrown track. T 34's emerged and joined the fight. We had to get into the village! Suddenly, there was a tremendous explosion in the armored car and I found myself lying in a rut looking at my driver sitting, headless, at the wheel. A direct hit had torn a massive hole in the armor plating. *SS-Unterscharführer* Albert Andres staggered, dazed into cover. With horror I noticed that he only had one arm left. I couldn't even see a stump among the remains of cloth and splintered bone.

Kompanie Bremer entered the village and fought its way down the street. Unexpectedly, we ran past a Russian tank. It was knocked out with a satchel charge. It was only after a few minutes that I noticed I had appropriated *SS-Oberscharführer* Sander's machine pistol. I had left my own weapon behind in the armored car. Sanders gratefully took back his weapon and I ran around with a Russian rifle.

The village was ours within the hour and we immediately set up a hedgehog perimeter. The shortness of the days made it seem advisable to spend the night in the village. We lay my driver, Ernst Nebelung, to rest in the twilight.

Liaison officers from *Panzer-Grenadier-Division "Großdeutschland"* reported the operations of their division to our left. I was happy we had a veteran formation to our north.

An amazing thing happened during the night to *SS-Oberscharführer*

Bügelsack. Good old Fritz felt the call of nature and accordingly sought out an appropriate place. Happy to find a corner of a building protected from the wind, he began the important business. "But the best laid plans of mice and men...." Fritz was not alone. Opposite him sat a Russian lieutenant who trained a submachine gun on him without a word and watched the "affairs of state" for some time. We suddenly heard the cry of a desperate man and, in the beams of our flashlights, there was Fritz with his trousers around his ankles and pointing excitedly but speechlessly at his adversary. We had rarely had such a good laugh. It was also possible the Russian had never had a cigarette that tasted as good as the one Fritz Bügelsack handed him.

The following morning found us attacking Walki which was some ten kilometers distant. Russian tanks and antitank guns tried to slow our advance but we outflanked those pockets of resistance and eliminated them by attacking from the rear. During the attack on Walki's last strongpoint, *SS-Oberscharführer* Reimling's vehicle received a direct hit. Reimling had been decorated with the Knight's Cross only a few days previously. Once again we had lost a brave comrade.

After some hard fighting I reached the Mscha River in Walki with *Kompanie Weiser*. The bridge was intact, but I didn't trust the Russians and ordered the attack to proceed across the ice on the river. The bridge had to have been mined. The company positioned itself behind small houses and sheds and hugged the river bank while preparing for the attack. From time to time I spotted a Russian head over on the far bank. Our tanks waited in the background and were supposed to cover our sprint across the ice. *Kompanie Bremer* had gotten hung up further to the rear. I feverishly considered how best to get to the other side and take Walki without employing artillery and mortars. The sound of tank tracks rattled through the town. The Russians were moving their T 34's.

My young soldiers looked at me as if to say: "Look, buddy, you got us into this crap. Now you'd better think about getting us out again!" It seemed a great joke to them that I was lying there like a chained dog, unable to reach the bone on the other side and vainly licking my chops. But then I knew what to do! The company raced across the ice and occupied the other bank as if shot from a gun. I dashed over the ice along with the company headquarters section. Crossing the ice was almost effortless. The Russians didn't fire a round but sat petrified behind their weapons and gave up. What had inspired the mad dash? Well, it went something like this: "Listen up! The first one to get to the other side gets three weeks home leave. Move out!" I had never seen such a concerted move before or since.

Then things began to move by leaps and bounds. Our tanks crossed the bridge that had been secured in the meantime. They advanced along the streets and, along with substantial help from the infantry, ejected the remaining Russians. We overwhelmed an enemy artillery battalion while it was still emplaced and took in several hundred prisoners. A few kilometers east of

Walki we encountered Peiper's *SPW-Bataillon* advancing on Walki from Bridok. Combat reconnaissance found the mutilated bodies of four of our comrades. They had been laid side by side and deliberately crushed under the tracks of a tank.

While my battalion pushed on to the north, Jochen Peiper reached the railway junction of Schljach where we linked up once more on 8 March. On the same day the *Leibstandarte* pushed forward as far as the western outskirts of Kharkov. Despite extensive antitank defenses and enemy counterattacks, our attack could no longer be halted. We wanted Kharkov back.

The *3. SS-Panzer-Division "Totenkopf"*, which was echeloned to our left, took Stary Mertschik and its reconnaissance element reached Olschany. The advance of the *2. SS-Panzer-Grenadier-Division "Das Reich"* was seriously hampered by the difficult terrain on the right. Additionally, the threat to the flank from a strong enemy presence east of Rakitnoje — Ljubotin resulted in the commitment of strong divisional forces to the east.

On 9 March we reached the Udy sector and took Olschany. *Armeeabteilung Kempf* was also advancing rapidly on its right wing. The final decision was dictated by the development of the situation. There could be only one objective: Kharkov. That evening the *Leibstandarte's* advance-guard had already reached Peressetschnaja and Polewaja.

The *SS-Panzer-Korps* decided to attack the city on 10 March. The orders went out on the evening of 9 March. The attack was to proceed down three attack corridors with the *Leibstandarte* attacking from the north and northeast and a simultaneous strike by the *2. SS-Panzer-Grenadier-Division "Das Reich"* from the west. It fell to the *Leibstandarte* to block the road to Tschugujew; the *3. SS-Panzer-Division "Totenkopf"* had the mission of covering to the northwest and north against the enemy in front of *Armeeabteilung Kempf* and any other enemy forces brought up.

In the process of the attack, Witt's regiment reached the major Kharkov — Belgorod road and pushed on towards the northern entrance to Kharkov, where it encountered fierce resistance at the airfield. The enemy had taken the opportunity to construct defensive works with the help of the civilian population. I linked up with Fritz Witt on the road and heard that he intended to attack the airfield and then move on to Red Square. On Witt's right was *SS-Infanterie-Regiment 2* of the *Leibstandarte*, commanded by *SS-Standartenführer* Wisch. It was also making good progress. In coordination with Witt I proposed to lead the *Kampfgruppe* through the woods north of Kharkov and block the Kharkov — Liptzy road.

Kompanie Bremer was once again in the lead. We moved a few kilometers in the direction of Belgorod and then turned east into the woods, which were covered in deep snow. The path ended at a collective farm and we spotted a Russian patrol disappearing to the east. There was no question of turning back. I wanted to negotiate the woods and thus penetrate into the eastern part of the city by surprise.

A footpath led into the tall spruce woods. It ran past a small lake and then turned due east. A patrol soon discovered a number of sleds with harnessed draft oxen pulling antitank and other guns. We no longer had any misgivings. Wherever sled teams could go so could tanks and other vehicles. I ordered Bremer to move off to the east and await further orders at the edge of the woods.

The lead section, led by *SS-Unterscharführer* Stoll, disappeared between the tall spruces and left behind a veil of snow spray. Two assault guns followed and it was not long before they were bogged down on a slope. They slipped sideways and threatened to slide onto the ice of a frozen lake. With the help of the company, they inched past that dangerous spot. To eliminate further risk, the path had to be improved with utmost speed. In a few minutes hundreds of *Kradschützen* and tankers were there to construct a negotiable detour. The rock-hard frozen earth was attacked with shovels, picks and axes. We made progress! In a short time the march column was moving again.

Bremer had already followed the lead section. Enemy riders observed us from snowed-over sections of the woods. The path grew increasingly smaller. The vehicles created a path over young spruces and birches. An 8-wheeled armored car followed us. The further we advanced into the woods, the more doubts plagued me. Had I maneuvered the *Kampfgruppe* into a hopeless situation once again? We could only move east. It would have been impossible to turn around. Not a single vehicle was capable of turning. Thick stands of spruce extended far into the woods on wither side of us. I thought about Greece and the crossing to Patras, about the assaults in the southern sector of the Eastern Front and about the hard fighting we had waged in the last few weeks. Despite hopeless situations we had continued to fight, only to win in the end in spite of it all. And that's the way it was on that day. No one would seriously consider that a motorized formation was advancing through snow-encrusted woods. A cadet at the academy in peacetime who had proposed such a solution would have been sent back to his unit. The decision seemed crazy. Despite that, I believed we would be victorious. As it turned out, I was able to grab the Soviets by the throat when they were completely unprepared. Frederick the Great said in such situations: "The more tricks and subterfuge you use, the more advantage you will have over the enemy."

Max Wertinger was only able to advance our *Kübelwagen* slowly through the narrow corridor. The snow drifted into the vehicle even after only slight contact with the branches. It was an unpleasant trip. It started to get lighter in front of us. We had advanced through the woods and had reached a cleared area that allowed us to leave the narrow passage.

In amazement, I saw that Bremer had turned his vehicles around and had taken cover. The section was in position. I sneaked up to Bremer and pressed myself instinctively into the ground as soon as I was able to cast a glance down the slope in front of us. Infantry, artillery and a few tanks were moving in the direction of Belgorod. I was not looking at a unit motivated by panic; on the contrary, this was a well-disciplined formation which was executing its movement in a tactical manner.

That morsel was too big for us. It would take hours before our *Kampfgruppe* would have crossed the woods and be ready for operations. We had to be content that we would not be discovered at this place. A Soviet attack would cause our *Kampfgruppe* to be caught in a bag. Our superior offensive strength and firepower could not be brought to bear. As already mentioned, however, turning around was out of the question. The unit had to close up and wait for a favorable moment to be employed. Perhaps the next day would offer a better opportunity. It would be night in an hour anyway.

At that point we consisted of four *Schwimmwagen*, a *Kübelwagen* and an 8-wheeled armored car; in all we had 23 soldiers with four machine guns and individual pistols and rifles. This group of German soldiers observed a Russian march column from a distance of approximately 800 meters that consisted of thousands of Soviets and which had all types of weaponry with it. The terrain sloped gently down to the road and then climbed gently up on the other side. While this side of the slope was covered with stands of trees, the far side opened up to an expanse of snow to the east which offered no cover. We did not stir in our positions. Observation posts would warn of approaching vehicles.

Suddenly we heard the rousing sound of *Stukas* behind us. We still could not see the aircraft, but they were coming from the west which meant they had a full load of bombs. Would they attack the Russian column moving in front of us? Off to our left was the village of Bolschaja Danilowka.

The thundering motors were then above us. Shadows whisked across the expanse of snow. We left our protective cover and stood like people attending the theater who were being offered an especially interesting presentation.

The Stukas flew over the column, described a great curve to gain height and came flying back from the south. Their bombs and cannon rained death and destruction down on the Soviets. Sleds raced up the slope and tanks were ripped apart by the bombs. All trace of order vanished in a few seconds. The horse-drawn vehicles careened away into the open countryside and the far slope was strewn with countless black dots as the infantry fled for its life. The unit was no longer under the command and control of its leaders.

I stared at that jumbled mass of humanity as if electrified. I grabbed a signal pistol from my vehicle and fired a red flare into the air. Bremer understood immediately. Stoll's section leapt into its vehicles and raced down the slope. The signals armored car hammered its machine-gun fire into the Soviets and provided covering fire for us. We tore down the path shouting and yelling — in contravention of all conventional rules of warfare — our horns and sirens making a hellish din. We were attacking the Soviets! Red flares were still climbing high into the air. The *Stukas* had recognized us; they rocked their wings and stormed into the fleeing mass, sweeping the road clear with their guns.

We reached the road. The Soviets threw up their arms. *Stukas* rushed past a few meters above our heads and flew round us in an endless chain. They provided covering fire and protected us. They howled along the road again and again preventing the Russians from bringing in troops. Our first armored vehicle came down from the edge of the woods and its rounds whistled away to the north. Stoll's section moved out with the first tank. Three additional tanks and Max Wünsche reached the road. We were advancing in both directions at that point. The Russians had to have the impression that this was a planned and well-thought-out attack. We couldn't allow them to come to their senses. Hundreds of captured Soviets gathered in an orchard.

We could not stop at that point; instead, we had to exploit the effects of the *Stuka* attack and continue the advance in the direction of Kharkov. Stoll's section, the signals armored car and a few dispatch riders roared off to the south towards the Soviet units. The advance was covered by two tanks, one on each side of the road. Our friends, the *Stukas*, took their leave; they had no more ammunition. The full consequences of our wild ride were then apparent. The sky was quiet again and the nerve-shattering howling was no longer above us. We had torn through the Soviet column in a couple of laughable vehicles.

Tank rounds whistled over us and exploded further south. Russians who tried to reassemble on the road after the bombers left fell to our machine-gun fire. Soviets ran for their lives once again. An enemy radio station was left of the road and the operators fell to our gunfire. Officers ran into the cover of a farmhouse; we destroyed their signals vehicle with hand grenades. Fire and dense smoke showed our way. Onwards, ever onwards! I was afraid to stop. Our only strength was in movement. Our tremendous speed, the cutting machine guns, the grenades thrown during our move and the bark of our tanks' guns had seen to it that the Soviets had cleared the road in great haste.

Our advance came to a halt in a brickworks just to the north of Kharkov. Just in the nick of time I noticed a good half-dozen enemy tanks in the gardens on both sides of the road. To our left a tank crew was busily engaged in removing the camouflage covering from a T 34. Machine-gun fire drove them back. The firing brought the remaining tank crews out of the houses. No one had counted on a German advance reaching that point. Despite that, it was starting to get dangerous for us. Stoll was just able to jump into another vehicle as his own had stopped. I saw the driver disappear into a haystack. We had to go back. The first tanks moved into firing positions.

We had to get out of there right away or we would come under fire from the Soviet tanks. We had advanced more than seven kilometers to the south and increased the Soviets' uncertainty. A Russian major with a stomach wound sat behind me. He really wanted to return with us. I admired the man; during the whole return trip I didn't hear a word from him about his pain. *Dr.* Gatternig put the first dressing on his wound.

When we returned we found a mass of prisoners at Bolschaja Danilowka guarded by just a few soldiers. They were happy with their lot. Not a single one attempted to escape.

By midnight a considerable part of the *Kampfgruppe* was still missing but, during the hours of darkness, they closed up in dribs and drabs. The whole unit had assembled by 0500 hours and the entire *Kampfgruppe* was ready for operations.

As soon as the first gray light of the new day appeared we advanced once again in the direction of Kharkov. This time, however, it was more slowly. We rolled south, carefully scanning the terrain round us. Far to the right we could see attacking Soviets employed against the airfield. They were attacking Witt's regiment. In front of us we spotted attacking Soviet infantry that was laying as if nailed to the ground by machine-gun fire. We soon arrived at the brickworks again and found Stoll's driver uninjured. Bruno Preger had spent the night sleeping in the haystack.

The enemy tanks were still in firing positions. Five T 34's fell victim to our tanks and were soon ablaze. A *Panzer IV* received a direct hit and burst completely asunder. The same enemy tank that had destroyed it also scored a direct hit on my own vehicle from a range of less than 50 meters. It immediately killed my driver, Max Wertinger. The leader of our signals platoon, *SS-Obersturmführer* Heinz Westphal, also fell to the round; Helmut Behlke was wounded and I lay unhurt beneath Max Wertinger's body. The Russian tank succeeded in escaping.

We fought our way forward, house to house. An enemy antitank crew was killed by a falling lamp post. Our tanks dominated the battlefield. Late in the afternoon of 11 March we were standing in the eastern part of Kharkov, having reached the road to Staryj.

At the moment of our victory a dangerous crisis surfaced. Our tanks had only a small amount of fuel left and could no longer be employed. They were assembled in a large graveyard and formed a "hedgehog" defensive position, creating a safe bulwark in the middle of Kharkov. From there we sent our feelers out along the Kharkov — Tschugujew road and attempted to block the Soviet's main line of retreat.

I had not had a report from the *2./SS-Aufklärungs-Abteilung 1* for some hours; it had been cutoff at the Kharkov Creek by enemy forces. *Kompanie Bremer* was fighting for its life and Olboetter was repelling enemy counterattacks from the east. In the cemetery we were having to defend ourselves against Soviets trying to break out. By the onset of darkness *SS-Hauptscharführer* Bruckmann had succeeded in bringing up fuel vehicles but, at the same time, he reported the road had been sealed off by enemy forces. (A few days later they were eliminated by elements of the *3. SS-Panzer-Division "Totenkopf".*)

Witt's regiment had broken into the town with a surprise attack from the north; it punched through to Red Square in heavy street fighting and had set up defensive positions for the night.

On 12 March the *Kampfgruppe* advanced several blocks and then blocked the road to Tschugujew once and for all. It was then the Soviet's turn to attack us. They wanted to overwhelm us. We were pressed together in a small area. Two platoons of *Kompanie Weiser* were cut off on the first floor of a school and defended themselves desperately against the Russian assault troops who had forced their way into the ground floor. An immediate counterattack under the command of Wünsche contributed to the elimination of the Russian assault troops. Once again the entire *Kampfgruppe* had been surrounded and was struggling in desperate fighting. A circle of burning buildings pinpointed our position in that sector of the city.

By the onset of night I no longer had much hope that we could hold out until the following morning. The enemy was within hand-grenade range. While moving through our position, we suddenly spotted a tank that had pulled up right against the school building. We were less than 20 meters away from it when the tank commander leaned out of the turret trying to establish contact with soldiers on the ground. He died from Weiser's pistol round. The tank pulled away on rattling tracks with the top half of its dead commander's body hanging out of the turret.

On the night of 12 March the *2. SS-Panzer-Grenadier-Division "Das Reich"* broke through the antitank ditch on the western outskirts of Kharkov and thus opened the way through to the city. The division arrived at the main train station on 12 March.

The enemy tried to break out of the encirclement en masse. He managed a stubborn resistance and dispatched new forces from northeast of the city in a relief attack. Jochen Peiper beat his way through to us with two *SPW*, thus establishing contact with the remainder of the division. His escort *SPW* was knocked out by a T 34, but he succeeded in bringing the men out to safety. We fought grimly and determinedly for each house until 14 March. By about 1800 hours we had captured the last two sectors of the city in the east and southeast. The tractor works fell on 15 March.

That same morning the *3. SS-Panzer-Division "Totenkopf"* reached and blocked the narrows at Tschugujew after successful armor engagements to the north of Rogan. This blocking position had to be held over the next few days against strong enemy attempts to break out as well as counterattacks from the east. We were successful in either eliminating or capturing the bulk of the enclosed enemy forces and capturing all of his equipment.

191

With that, the decisive counterattack against the Russian winter offensive was completed, contact reestablished between the sectors of *Heeresgruppe Süd*, a considerable part of the Russian offensive strength destroyed and the rest badly beaten. In the pursuit against the enemy withdrawing to the east and north in the following days, the banks of the Donez were taken and, rounding out the victories of the *SS-Panzer-Korps*, Jochen Peiper captured Belgorod on 18 March. It was there that the link-up was established with *Panzer-Grenadier-Division "Großdeutschland"*. *"Großdeutschland"* had been advancing from the west. In the past few days it had destroyed 150 Soviet tanks in heavy armor fighting.

The battle of Kharkov had been concluded victoriously despite considerable losses. In the great battle between the Donez and the Dnjepr the German grenadier had emerged victorious over the eastern hordes.

Shortly before the summer offensive, I had to permanently take leave of the faithful grenadiers whom I had led for many years. I will never forget the departure from my comrades. I was ordered to report to the Armor School and then transferred to the *12. SS-Panzer-Division "Hitlerjugend"*.

Right: Russian campaign, 1942. Left to right: Gerd Pleiß, Theodor Wisch, Sepp Dietrich, Kurt Meyer.

193

Russia, 15 June 1942. Kurt Meyer and officers of the Rumanian Cavalry Corps. (Roger James Bender)

Russia, 1942. A Dining-in among senior *Waffen-SS* officers. Left to right: Paul Hausser, Sepp Dietrich, H. Gille, Kurt Meyer. Right foreground: Werner Ostendorf. (Roger James Bender)

The winter fighting of 1943.

The winter fighting of 1943. **Below**: A briefing just prior to operations. Left to right: *SS-Obersturmbannführer* Meyer (commander, SS-Aufklärungs-Abteilung 1; a platoon leader in *SS-Panzer-Regiment 1*; *SS-Standartenführer Dr.* Besuden (division surgeon); *SS-Obersturmbannführer* Witt (commander, *SS-Panzer-Grenadier-Regiment 1*); and *SS-Sturmbannführer* Wünsche (commander, *I./SS-Panzer-Regiment 1*)

Russia, February 1943. *Panzermeyer* and Sepp Dietrich in Kharkov. (Jost Schneider)

Kharkov, February 1943. Just before the award of the Oak Leaves to the Knight's Cross. (Jess Lukens)

Right: Vinitsa, 25 February 1943. Kurt Meyer receives congratulations from Adolf Hitler on receiving the Oak Leaves to the Knight's Cross during his one-day visit to the *Führerhauptquartier*. (Roger James Bender)

Kharkov, March 1943. A soldier of the *Leibstandarte* renames the central square in honor of his division. (Jost Schneider)

Right: Berlin, March 1943. A reception for *Waffen-SS* Eastern Front fighters. Left to right: *Dr.* Joseph Goebbels, Kurt Meyer (*Leibstandarte*), Hugo Kraas (*Leibstandarte*), Hermann Buchner (*Totenkopf*), Heinz Macher (*Das Reich*). (Jess Lukens)

At the March 1943 reception, Max Wunsche is promoted. (Roger James Bender)

Below: At Kharkov, end of March 1943. Kurt Meyer presents Hermann Weiser (commander of the 2./SS-*Aufklärungs–Abteilung 1*) with the Knight's Cross. To the right of Meyer is SS-*Sturmbannführer Dr.* Hermann Besuden (Division Surgeon). (Jess Lukens)

Kharkov, March 1943. Kurt Meyer, shortly after the award of the Oak Leaves to the Knight's Cross (195th recipient). (Jost Schneider)

Right: Russia, spring 1943. Kurt Meyer and Max Wünsche. (Jess Lukens)

March 1943. Kurt Meyer, shortly after receiving his Oak Leaves. His rank was that of *SS-Obersturmbannführer*. (Left: Jost Schneider. Right: Roger James Bender)

Kharkov, Russia, 28 May 1943. A group photo of *LAH* officers attending the celebration of Sepp Dietrich's birthday. Left to right: Kurt Meyer, Hugo Kraas, Sepp Dietrich, Albert Frey, Hermann Weiser, Rudi Sandig, Bernhard Krause, Georg Schönberger. Also: Alfred Günther (directly behind Dietrich), Hubert Meyer (behind Sandig). (Roger James Bender)

Two more photos from the 28 May 1943 birthday celebration. (Roger James Bender)

Before leaving for his assignment to the *12. SS-Panzer-Division "Hitlerjugend"*, Meyer reviews his *SS-Aufklärungs-Abteilung 1* for the last time. (Jost Schneider)

Left: Kurt Meyer in a playful mood at the birthday festivities. Left to right: Wilhelm Mohnke, Walter Ewert, *Dr.* H. Besuden, Teddy Wisch. (Roger James Bender)

SS-Obersturmbannführer Kurt Meyer, spring 1943, official photo after his award of the Oak Leaves. (Roger James Bender)

The 12. SS-Panzer-Division "Hitlerjugend"

It had to be a unique occurrence in the history of warfare that a division, especially one with as complicated an organization as a modern armored division, should consist entirely of young men aged 17 and 18, except for officers and senior noncommissioned officers.

Anyone in Germany who understood anything about military matters, the rearing of young people or leading them was of the opinion that the employment of such a formation could only lead to catastrophe in the first few days. The young soldiers would not be able to withstand the physical and mental pressure of modern attrition warfare. That opinion was shared even more decisively by our opponents of the time. And it really was not just enemy wartime propaganda when leaflets and radio broadcasts spoke of a "Baby Division" whose insignia was purported to be a baby bottle.

The deeds of these young men in action and the performance of the *12. SS-Panzer-Division "Hitlerjugend"* gave lie to the critics.

It therefore appears important to me that I briefly discuss the establishment of the division, to satisfy both general and historical interest.

When "total war" was declared after the catastrophe of Stalingrad a plan was put forward to raise a volunteer division of young men who were fit for military service. They were to be a symbol of the readiness of German youth to make sacrifices and an expression of their will to persevere. These young men of 17 and 18 were fit for service as a result of accelerated paramilitary training. By proving its worth, such a volunteer division would also promote the inclusion of youths in other German divisions to compensate for the huge losses of manpower caused by the Russian campaign and also increase German military strength significantly.

Youth leaders believed that the usual methods employed in training soldiers could not be applied to these young men. They therefore wanted to allow the testing of new methods in this special division. That would take place under their supervision.

Following a discussion between *Reichsjugendführer* Axmann and Adolf

Hitler, the corresponding orders were issued in June 1943 by Hitler. The *Hitlerjugend* was to call for volunteers and prepare them in pre-military training camps. They were then to be transferred to the newly raised division of the *Waffen-SS*. The *1. SS Panzer-Division "Leibstandarte"* was tasked to provide the cadre for the division. In conjunction with the *"Leibstandarte"*, the new division, to be known as *Panzer-Grenadier Division "Hitlerjugend"*, was to form the *I. SS-Panzer-Korps*. The activation was to commence immediately.

While the *Hitlerjugend* began its recruitment and training, the selection of the cadre from the *"Leibstandarte"* took place. That division had suffered heavy casualties in the fighting retreat from Kharkov and the subsequent retaking of that city. It was preparing for Operation *"Zitadelle"*, the elimination of the Russian salient near Kursk.

The commander of *SS Panzer-Grenadier-Regiment 1*, the 35-year-old recipient of the Oak Leaves to the Knights' Cross, *SS-Standartenführer* Witt, was given command of the division. He took with him a few officers and a portion of the noncommissioned officers and the technical experts from his regiment. Their place in the regiment was filled by personnel levies against *SS-Panzer-Grenadier-Regiment 2*. The other formations of the division had to provide cadre personnel in a similar manner.

This transferred nucleus formed only an incomplete skeleton. It lacked company commanders, platoon leaders and squad leaders to an extraordinary degree. In many cases, young platoon leaders had to be given company command. Later, about 50 regular army officers who had at one time been *Hitlerjugend* leaders were transferred to the division. In order to get the necessary squad leaders, selected youth were sent for training at the noncommissioned officer school at Lauenburg as soon as they were graduated from the military training camps. A few weeks after the beginning of basic training, additional suitable young men were chosen for a three-month noncommissioned officer course within the division.

When the first 10,000 youths arrived at Beverloo camp in Belgium during the course of a few weeks in July and August, preparations for them had not yet been completed. They could not be given uniforms right away. Nevertheless, their training began at once. Gradually, the individual units were formed according to the tables of organization and equipment. These assignments were completed during September. As a result of much effort, the formation was converted into a *Panzer-Division*.

Up to that point *SS-Panzer-Regiment 12*, whose formation had taken place at Mailly-le-Camp at Reims, had 4 *Panzer IV's* and 4 *Panthers* for training purposes. Half of these had been "procured" clandestinely in Russia. *SS-Panzer-Artillerie-Regiment 12* had no more than a few light field howitzers at its disposal. Almost all transport vehicles were still missing. Captured Italian vehicles were issued during November and December. Vehicle strength then reached almost 80% of authorized levels. Simultaneously, the first prime movers and armored vehicles arrived.

The command and control relationships of the division were complicated. In matters of training, the division was under the purview of *General der Panzertruppen West*, Geyr von Schweppenburg. Tactically it reported to the *15. Armee*.

After basic training, for the most part, was complete, unit training began at the start of 1944. Following the transfer of *SS-Panzer-Regiment 12* to the area of Hasselt in Belgium, larger exercises with tanks took place. The emphasis was on combined-arms operations. In February the *I./SS-Panzer-Grenadier-Regiment 25* took part in a live-fire exercise in the presence of the Inspector General of Armor Troops, *Generaloberst* Guderian. In March, the Commander-in-Chief West, *Generalfeldmarschall* von Rundstedt, was present at a combined-arms exercise. In both cases the level of training was commended.

Staff coordination was tested in numerous radio exercises. One of these took place under the aegis of the corps in the area of Dieppe. During the exercise so many difficulties arose with the completely unsuitable captured Italian vehicles that orders finally came from the highest authority to replace them with German military vehicles.

A portion of the young men had already reported to other branches of the service and other divisions; a few had been more or less "persuaded" by requests for volunteers. The majority, however, had come to the division with all the enthusiasm of youth. They were burning to prove themselves in action. This enthusiasm and esprit had to be maintained as a bedrock value or, where it was missing, it had to be awakened. As the young men were still developing, the principles and forms of training had to be somewhat different from those which a unit used to train and educate older recruits. Many established principles of military training were replaced with new ones which, when all was said and done, had their origin in the German youth movement which came into being at the turn of the century.

There was no obvious superior-subordinate relationship recognizing only orders and unconditional obedience. The relationship between officers, non-commissioned officers and other ranks was that between those who were older and little more experienced and those who were new. The officers' authority existed in the fact that they were role models and mentors to the young soldiers. They strove to emulate the close relationship of a family inasmuch as that was possible in the circumstances of the war.

The young men were trained to accept responsibility, have a sense of community, be prepared for self-sacrifice, not be afraid to make decisions, show self-discipline and be a team player. If they had already shown those qualities, they were further developed. The leadership of the division was convinced the young soldiers would achieve more if they understood and supported the purpose of their mission and their role in it. It was therefore standard operating procedure to develop all orders based on a detailed assessment of the situation.

During their training, drill and ceremonies were avoided. Everything focused on combat training and this took place under the most realistic battle conditions possible. Physical toughening was achieved through sport; forced marches were disapproved of and considered unnecessary and harmful. Based on input from *General der Panzertruppen* Geyr von Schweppenburg, advanced techniques of marksmanship were developed. These took place exclusively in the field. There were no marksmanship exercises conducted on traditional garrison ranges.

On order of the Inspector General of Armored Troops, a detachment from the division worked alongside officers of the School of Armored Warfare at Bergen in developing a new gunnery manual *Panzergrenadiere*, which appeared in the spring of 1944. The Inspectorate of Infantry turned it down. Based on input by *General der Panzertruppen* von Schweppenburg, special emphasis was placed on visual camouflage and noise discipline as well as signals security, maintaining secrecy and live-fire day and night close-combat training. The division received a signals intelligence platoon for monitoring enemy radio traffic. It later performed very well for the division. Based on guidance from *General der Panzertruppen* von Schweppenburg, more and more situations were practiced during tactical leadership training which concerned offensive operations against air-landed troops.

Because the young men were still growing and undernourished at home, they received additional rations from the *Feldersatzheer* above and beyond the normal allocations. They developed well physically. No cigarettes were issued to those under 18; they received sweets instead.

There can be no reason to believe that the division had any priority in the issuance of weapons and equipment; that should be clear from the above. Everything had to be fought for the hard way. The organization of the division, like all *Panzer-Divisionen* of the *Waffen-SS*, differed from the *Panzer-Divisionen* of the army only inasmuch as there were three battalions in the *Panzer-Grenadier-Regimenter* instead of two. In contrast to the *Panzer-Lehr-Division*, it only had one *SPW-Bataillon*.

On the basis of the training and instruction which was given in accordance with the fundamentals outlined here — of course, not all of that was given equally well — the soldiers entered the fray animated by the thought that their employment would be decisive for the defense of Germany and for it's final victory. They were imbued with a belief in the rightness and justice of the German cause. The young soldiers went to war superbly trained. There were few divisions which had been trained as well. As a result, their employment was fully justified.

The Invasion

At about 0700 hours on the morning of 6 June 1944, the division received employment orders from the *I. SS-Panzer-Korps*. It was placed under the operational control of Rommel's *Heeresgruppe B* and was ordered to assemble in the area around Lisieux. It was to report directly to the *LXXXI. Armee-Korps* in Rouen. This order had a calamitous effect. Previously prepared approach routes were not used. The only thing that mattered was getting the division close to the coast. It was not clear how the division was to be employed.

This involved a lot of lost time in comparison to the pre-planned deployment straight to the combat zone from assembly areas. The division was unsuccessful in attempting to change the new orders with the original corps to which it had reported. There was no telephonic contact with *Heeresgruppe B*.

The march order (including the assembly areas) was prepared immediately and arrived at the formations between 0930 and 1000 hours. The reinforced *SS-Panzer-Grenadier-Regiment 25* march serials moved out around 1000 hours; the reinforced *SS-Panzer-Grenadier-Regiment 26* around 1100 hours. Both *SS-Panzer-Grenadier-Regiment 25*, which was collocated with the *II./SS-Panzer-Regiment 12*, and *SS-Panzer-Grenadier-Regiment 26*, which was collocated with the *I./SS-Panzer-Regiment 12*, were to relocate to the area east of Lisieux. For the time being the divisional staff stayed east of Tillieres, where it had radio communications. Only a reporting point was established at Lisieux.

At about 1500 hours, the division received a telephonic order from *Heeresgruppe B* via the *I. SS-Panzer-Korps* to assemble in the area east of Caen and prepare for a counterattack. The division was first put under the operational control of the *LXXXIV. Armee-Korps* in St. Lo; later it came under the *I. SS-Panzer-Korps*.

At 1600 hours, sixteen hours after the first enemy report, the reinforced *SS-Panzer-Grenadier-Regiment 25* was given employment orders. The regiment was to attack the area from the western outskirts of Carpiquet — Verson — Louvigny. On its left flank, the reinforced *SS-Panzer-Grenadier-Regiment 26* was to assemble in the area from St. Mauvieu — Cristot — Fonteney le Pesnel — Cheux. SS-Panzer-Pionier-Bataillon 12 was to attack in the area around Esqay; *SS-Aufklärungs-Abteilung 12* at Tilly-sur-Seulles. The support units were to stay east of the Orne around the Foret de Grimbosq and in the Foret de Cuiybis, moving west only after nightfall. Divisional headquarters was moved to the northern edge of the Foret de Grimbosq.

The time had come! The soldiers mounted their vehicles. Dispatch riders roared down the streets on their motorcycles; the combat vehicles' engines were bellowing. How often had we experienced the moment of moving out? In Poland, in the West, in the Balkans, in Russia, and now again in the West. We, the old soldiers, faced the future with anxiety. We knew what was in front of us. In comparison, the magnificent young soldiers looked at us with laughter in their eyes. They had no fear. They were confident; they had faith in their strength and the will to fight.

How would these young men turn out? Enemy fighter planes were above us. They were diving on the march column, tearing flourishing life to pieces. The tanks were racing across the fiendish road junctions in leaps and bounds. Von Büttner's reconnaissance company was far in front. I was waiting for reports from the front. If only we had a clear picture of the enemy; up until then, everything had been shrouded in fog.

My experienced driver moved forward, careful as usual. Dark clouds were rising in the west. Caen, the town from which William the Conqueror started his victorious journey across the Channel, had been destroyed. More than 10,000 men and women were beneath the smoking rubble. The town had become a vast cemetery.

On the Caen — Falaise road we encountered French refugees; a bus was ablaze. Heartrending cries greeted towards us. We could not help; the door was jammed and barred the way to freedom. Mangled bodies hung out of the broken windows barring the way. What horror! Why those burning civilians? But we could not allow ourselves to bunch up! We could not stop! We had to press ever onwards to gain ground. The woods attracted us like magnets; more and more fighter planes were above us. We were hunted relentlessly but could not afford to take cover. The march had to go on!

A string of Spitfires was attacking the last platoon of the *15./SS-Panzer-Grenadier-Regiment 25*. Rockets and other weapons were reaping a grizzly harvest. The platoon was moving down a defile; evasion was impossible. An old French woman came running towards us shouting: "Murder! Murder!" A soldier was lying on the road, a jet of blood shooting from his throat. An artery had been cut; he died in our arms. The ammunition in an *Schwimmwagen* exploded with a loud bang; the blast shot flames high into the sky, and the vehicle was torn to pieces. In a couple of minutes the wreakage was pushed aside — there was no stopping, we had to continue!

Darkness arrived. The *15./SS-Panzer-Grenadier-Regiment 25* had crossed the Caen — Villers-Bocage road. I was waiting impatiently for the *I./SS-Panzer-Grenadier-Regiment 25*. The constant air attacks had slowed the pace dramatically. Finally Waldmüller reported the arrival of the battalion, and I was informed the air attacks had not caused excessive losses. At about 2300 hours a liaison officer from the *21. Panzer-Division* reached me. That division was fighting near Troarn and north of Caen. The divisional commander, *Generalleutnant* Feuchtinger, was expecting me at the command post of the *716. Infanterie-Division*. I left immediately. Low-flying German bombers were flying across the road. They were met with intense defensive fire as soon as they reached the area of the invasion fleet. A few trucks were burning on the road. It was a hellish trip.

Caen was a sea of flames. Agitated people were wandering through the rubble; streets were blocked and burning smoke was rolling through the town. Beautiful churches were converted into heaps of rubble, the work of generations was transformed into a sea of ashes and ruin.

All of that had occurred despite the fact there wasn't a single combat unit in the city. Allied bombers had killed French civilians and destroyed cultural facilities forever. Seen from a military point of view, the destruction of Caen was an egregious error.

The bunker was situated in a quarry, dug deep into the earth. Wounded soldiers of the *716. Infanterie-Division* and the *21. Panzer-Division* were in the passageways, groaning in pain. Doctors and medics were working feverishly; ambulances were being loaded to transfer the wounded to the rear.

At 2400 hours I was standing before the commander of the *716. Infanterie-Division, Generalleutnant* Richter. This division had experienced the full firestorm of the Allied attack and, after 24 hours, it had virtually ceased to exist as a fighting unit. It was still conducting a defense from strongpoints, but communications between the regimental and battalion and divisional command posts no longer existed. Nothing was known about which positions had been overrun.

The commander briefed the situation to me. The silence was broken by a ringing telephone. One of the regimental commanders, *Oberst* Krug, reported from his bunker asking for further orders. He stated: "The enemy is standing on the top of the bunker. I have no means of engaging him, nor any contact with my units. What should I do?." An icy silence settled in the bunker. Everybody looked at the divisional commander. The tone of the man's speech was shocking: "I cannot give you any further orders; do what you have to do. Goodbye!"

The *716. Infanterie-Division* had been destroyed in the truest sense of the word. It did not exist anymore. It had fought bravely, but the enemy's superiority in men and equipment was too great. The *716. Infanterie-Division* had been attacked during the night of the 5/6 June from 2300 hours until dawn by the Royal Air Force. Following that, the entire 8th U.S. Air Force in England was employed in the sector of the division. More than 1,000 American aircraft attacked the coastal defenses in the half hour preceding the Allied landings.

After the Royal Air Force had stopped its bombing, the naval forces started their bombardment: 5 battleships, 2 monitors, 19 cruisers and 77 destroyers and 2 cannon boats let loose with everything they had. Once the invasion forces had landed, their own artillery joined in. Finally, naval rocket ships fired their salvos into this inferno. (The rockets ships had been specially built for the invasion.)

Despite this enormous destructive fire of the naval guns and the bombing, the bunkers that remained held out until late in the afternoon. But the human spirit was powerless against such masses of steel; the soldier had to bow before the weight of the onslaught. The sector of the *716. Infanterie-Division* had been transformed into a moonscape; only the occasional defender had escaped the firestorm.

After the dramatic events in the bunker of the *716. Infanterie-Division*, the commander of the *21. Panzer-Division* gave a run down on its deployment. He stated:

I did not learn about the invasion until I heard a report just after midnight on 6 June that airborne and air-landed troops had been employed in the vicinity of Troarn. Since I had received orders not to move, I could not do anything for the time being other than order the division to stand-to. I waited impatiently for orders all night, but not a single order was received from higher headquarters. Since I was aware that my armor division was the closest to the enemy area of operations, I finally became convinced at 0630 hours that I had to do something. I ordered my tanks to attack the British 6th Airborne Division, which had dug itself into a bridgehead across the Orne. I thought it was the most immediate threat to the German positions.

Right after I had made my decision, I was informed by *Heeresgruppe B* that I had been attached to the *7. Armee*. At 0930 I was informed that I would start receiving my orders from the *IV. Armee-Korps* and, finally at 1000 hours, I received my first operation orders. I had to stop my tank attack against the British 6th Airborne Division and turn westwards to support the forces defending Caen.

After I had crossed the Orne, I moved north towards the coast. By that time the enemy — consisting of three British and three Canadian divisions — had made good progress and gained control of a strip of land about 10 kilometers deep. The fire of the Allied antitank guns put 11 tanks out of action just after I had started. My *Kampfgruppe* managed to get past those guns and reach the coast at Lion sur Mer at 1900 hours.

The *21. Panzer-Division*, the only armored formation immediately available which could have decisively influenced the course of the invasion, was robbed of its combat power in the initial phase of the fighting. Instead of moving like lightning into the massed concentration of landed enemy forces, the division was condemned to be burnt out in dribs and drabs. It had remained inactive around Caen until 0630 hours, only attacking the air-landed division at Troarn. The enemy main effort was not at Troarn, however. It was north of Caen. Elements of the division were not employed north of Caen until the afternoon.

In the hands of an experienced tank commander — like Rommel in 1940 and many commanders after him — the *21. Panzer-Division* would have been able to create a difficult situation for the Allies, provided the commander had led the attack from the front, advancing in his own tank. Guderian's proven rule — *Klotzen nicht kleckern!* — had been flagrantly disregarded.

The command post of the *21. Panzer-Division* was still at St. Pierre sur Dives, about 30 kilometers away from the coast. During the night of 6/7 June the divisional commander was out of communications with his units. Shortly after midnight, he informed us the Allies might have already reached the airfield at Carpiquet. However, he didn't have any specific information.

SS-Panzer-Grenadier-Regiment 25 immediately launched combat reconnaissance. At 1300 hours Carpiquet, Rots and Buron were reported clear of the enemy. Buron was held by scattered elements of the *716. Infanterie-Division*. Les Buissons had been occupied by Allied troops. The left flank of the *21. Panzer-Division* was on the railway line as far as Epron; there were no German units west of Epron at the moment. The western outskirts of Caen and the airfield at Carpiquet were not defended. The unit which was supposed to defend the airfield had left its well-constructed positions in panic on 6 June; it had consisted of *Luftwaffe* ground personnel.

The reinforced *SS-Panzer-Grenadier-Regiment 25* was moving along the Villers Bocage — Caen road. It was briefed on the situation by liaison officers. The battalion commanders were ordered to report to the temporary regimental command post just west of Caen at the rail and road junction.

There was a pessimistic mood in the divisional bunker. It was time to leave. Shortly before I left the bunker, I was summoned to the telephone. Our divisional commander, *Generalmajor der Waffen-SS* Witt was at the command post of the *21. Panzer-Division* at St. Pierre and asked for a situation report. I explained the situation the way I was told by the commanders of the *716. Infanterie-Division* and the *21. Panzer-Division*. The commander did not interrupt me and said at the end:

The situation necessitates speedy action. First of all, the enemy must be denied Caen and the Carpiquet airfield. It can be assumed that the enemy has already assembled his units and they are prepared to defend insofar as they have not deployed for further attacks. Therefore, it would be wrong to commit our divisional units as soon as they arrive. The only possible option is a coordinated attack with the *21. Panzer-Division*.

Therefore, the division will attack the enemy in conjunction with the *21. Panzer-Division* and throw him into the sea. The attack will start tomorrow at 1200 hours.

I quickly took my leave of the two commanders and left the bunker with those accompanying me. There were still some wounded in the passageways; the majority had already been evacuated. The streets of Caen were devoid of people, neither soldiers nor civilians could be seen. Only combat engineers could be seen at places where rubble was blocking the roads. There was no sign of life other than the combat engineers. Caen was a dead city. The revolting smell of fire hung heavily in the streets. Smoldering beams and houses were showing us the way. We were quiet, nobody said a word — we were thinking of the fires burning back in Germany. Individual aircraft were flying over us; the light bulbs for the aerial photographs bathed the deserted streets with flickering light. Where was our *Luftwaffe*, for God's sake?

Our command post was beside the main road; it was in a little country house surrounded by taller, older trees that provided aerial concealment. That was essential if you wanted to live to see the next day. It was a mess inside the house. It must have been abandoned within the space of a couple of minutes by *Luftwaffe* troops. They had probably belonged to the airfield defense unit that had abandoned the area so quickly.

The commander of the *I./SS-Panzer-Grenadier-Regiment 25* reported; he was quickly briefed on the situation. A short handshake told all. We were about to start a difficult journey.

The battalion dismounted and its vehicles disappeared in the dark. No vehicles moved through the city; all of them turned south.

Battalion after battalion arrived. In the meantime, it had become day. The sky started to become busy again. There was no sense in looking for concealment from the aircraft which were constantly overhead.

The soldiers waved at me. They were moving forward to their baptism of fire in a calm manner. They showed no self-pity. They were determined to prove themselves. Relentless attacks by fighter-bombers and naval guns hit the approach routes. Nevertheless, the assembly areas were reached in time.

I set out for our forward command post at the Ardenne monastery. Erich, my driver, had already traded in our *Kfz 15* for a smaller *Volkswagen Kübelwagen* in order not to provide too obvious a target, but this precautionary measure was not of much use. We would hardly start moving when we found ourselves in a ditch. Machine-gun bursts from the fighter-bombers tore up the earth around us. Back into the vehicle — and after a couple of hundred meters an elegant leap back into the ditch! It almost drove us crazy, but Erich soon learned how to handle it. He set off like greased lightning and, as soon as a fighter-bomber started its dive, he stepped on the brakes so hard that the car almost overturned. In that way he got me to the monastery. I was happy to have solid walls around me. The monastery was an old ruin with a large orchard, surrounded by high walls of rough stone. Two church towers dominated the countryside, affording excellent observation.

One of the towers had already become an observation post for the heavy artillery battalion. Bartling reported the artillery was ready to open fire. The *Panzergrenadiere* had reached their deployment positions. The *II./SS-Panzer-Grenadier-Regiment 25* was right in front of me. I saw the combat reconnaissance element disappearing into the bushes. The heavy infantry weapons had taken up their firing positions and, in the meantime, the machine guns and 20 mm *Flak* were firing at the fighter-bombers.

Everything was ready, but where were the tanks? Could they reach the front? Was it not madness to expect tanks in the face of these fighter-bomber sorties? Instead of the tanks so desperately longed for, however, the commander of a mortar battalion reported. I was happy to hear that he had plenty of ammunition. The battalion established positions on the northern outskirts of Caen.

In the meantime it had become 1000 hours, and the first tanks appeared. The relentless fighter-bomber attacks had considerably hindered their approach march. The commander of the battalion reported 50 *Panzer IV's* were operational. The rest were somewhere along the approach route and would arrive during the course of the night. I then felt considerably better. Our attack would have inevitably been condemned to failure without tank support.

If only we could eliminate that damned naval gunfire. The heavy shells were roaring above our heads like express trains, digging themselves into the rubble of the town. The fighter-bombers hardly bothered us anymore; we knew we would continue to have that plague above us constantly from then on.

I climbed a tower, so as to have a look at the terrain. I thought I might be able to see the coast. What a surprise! The terrain as far as the coast was spread before me like a sandtable. There was intense activity on the coast. Ship after ship bobbed on the water; countless barrage balloons protected this armada from air attack. The latter measure was unnecessary; the *Luftwaffe* appeared not to exist anymore.

Enemy tank formations were forming up west of Douvres. The whole expanse looked like an anthill. And what was going on behind us — smoking rubble, empty roads and burning vehicles. The Caen — Falaise road was straight as an arrow for kilometers on end, but there wasn't a single indicator of German combat power. It was waiting under cover somewhere; it could only dare to move forward during the night.

Fighter-bombers attacked the monastery without causing any damage. The soldiers swore at the fighter-bomber plague.

But what at that moment? Was I seeing clearly? An enemy tank was pushing through the orchards of Contest! It then stopped. The commander opened his hatch and observed the terrain. Was he blind? Didn't he realize he was only 200 meters from the *Panzergrenadiere* of the *II./SS-Panzer-Grenadier-Regiment 25* and the barrels of our antitank guns were directed at him? Obviously not. He calmly lit a cigarette and looked at its smoke. Not a single round was fired. The battalion maintained excellent fire discipline.

I then saw what was happening! It had become clear. The tank had been sent forward to provide flank cover. Enemy tanks were rolling towards Authie from Buron. My God! What an opportunity! The tanks were moving right across the front of the *II./SS-Panzer-Grenadier-Regiment 25*! The enemy formation was showing us its unprotected flank. I issued orders to all battalions, the artillery and the tanks: "Do not fire! Fire on my command only!".

The commander of the *SS-Panzer-Regiment 12* had positioned his command vehicle in the garden of the monastery. Wire was quickly laid to the tank and the enemy situation relayed from the tower to all the tanks. One company was in the monastery grounds and another on the reverse slope south of Franqueville.

The enemy commander only seemed concerned with the airfield; it was right in front of him. He already controlled it with his weapons. He did not realize that destruction awaited him on the reverse slope. As soon as his tanks crossed the Caen — Bayeux road he would run into the tank company of the *II./SS-Panzer-Regiment 12*. Only a few meters separated the iron monsters from each other.

We were staring spellbound at this spectacle! Wünsche, commander of *SS-Panzer-Regiment 12*, quietly transmitted the enemy tank movements. Nobody dared raise his voice.

I was thinking of Guderian's principle — *Klotzen nicht kleckern!* — and the divisional plan of attack but, in this situation, I had to use my own initiative. *SS-Panzer-Grenadier-Regiment 26* was still east of the Orne and the *I./SS-Panzer-Regiment 12* could not move because of the lack of fuel. It was 30 kilometers east of the Orne. Fuel could not be supplied because of the intense air activity. Decision: As soon as the leading enemy tanks passed Franqueville, the *III./SS-Panzer-Grenadier-Regiment 25* would attack with the tank company positioned on the reverse slope. Once the battalion reached Authie, the other battalions would then join the fight. Objective: The coast.

The commander of the *21. Panzer-Division* was briefed on the situation and asked for support. An unbearable pressure rested on us at that point. It would have to happen soon. The enemy spearhead pushed past Franqueville and started across the road. I gave the signal for the attack to Wünsche. The last thing I heard was: *"Achtung! Panzer Marsch!"* The tension faded away.

There was the report of guns and muzzle flashes at Franqueville. The lead enemy tank was ablaze, and I watched the crew bailing out. More tanks were torn to pieces with loud explosions. Suddenly, a *Panzer IV* started to burn; a blast of flame shot out of the hatches. Canadian infantry tried to reach Authie and continue the battle from there, but it was in vain. The *Panzergrenadiere* of the *III./SS-Panzer-Grenadier-Regiment 25* were keen not to allow the tanks to outdo them. The wanted to enter to Authie. They had just reached it, when the *I.* and *II./SS-Panzer-Grenadier-Regiment 25* began to attack. The enemy had been struck deep in his flank at that point. We took Franqueville and Authie through energetic offensive action. Contest and Buron had to follow.

The enemy forces seemed to be totally surprised. Neither side's artillery had fired up to that point.

The attack proceeded quickly. Prisoners were collected and marched to the rear with their hands in the air. The *III./SS-Panzer-Grenadier-Regiment 25* advanced on Buron; the *II./SS-Panzer-Grenadier-Regiment 25* had already pushed through Contest and was engaged with enemy tanks.

I jumped on a motorcycle and went over to the *III./SS-Panzer-Grenadier-Regiment 25*. I encountered the first wounded. They were going to the aid station in the monastery. About 50 Canadians were standing with their hands up, guarded by some *Panzergrenadiere* in the orchard at Cussy. I let them lower their arms and ordered their immediate evacuation to the monastery. The village of Cussy had the pallor of death about it, but just beyond the last farm it was more lively than I would have liked. I had scarcely reached open ground when Canadian "greetings" were flying around my ears. The tanks at the southern outskirts of Buron were trying to fire at me, but it was not so easy — I dashed across the fields like lightning.

At that point they got me! I no longer know how I got to be lying next to a Canadian soldier. Smoke and explosions surrounded me without interruption. The Canadian and I were in a bomb crater. We watched each other nonplussed. We kept close to the crater's edge, not letting each other out of sight. We were in the middle of Canadian artillery fire and ducked especially low when the heavy rounds from the naval guns came roaring over. My motorcycle was laying on the path through the field; it was just a heap of wreckage.

I don't remember how long I laid in that damned hole. I could see, however, that the *Panzergrenadiere* were just about to enter Buron. Tanks were burning on both sides of the village.

I could not stay in the hole forever — I took off, jumping in leaps and bounds towards the *III./SS-Panzer-Grenadier-Regiment 25*. The Canadian disappeared in the direction of Cussy. Fire covered Buron at that point. A dispatch rider came roaring down the path; he recognized me, stopped, and we both raced off.

I encountered Milius between Buron and Authie. He was proud to report his battalion had high morale. His losses had been slight up to that point. Extremely heavy fire was then falling on Buron. One could no longer identify the village. Smoke, explosions and flames marked its position. The enemy artillery had concentrated all its strength on Buron and smashed it with enormous masses of steel. I had never experienced such concentrated artillery fire before. I had to think of Verdun. One company of the *III./SS-Panzer-Grenadier-Regiment 25* was in the midst of the artillery concentration. The rest of the companies were advancing towards Buissons. Milius followed his lead company.

I moved to the *II./SS-Panzer-Grenadier-Regiment 25* on another motorcycle. There was only light fire falling on Contest. The battalion had left the artillery fire behind and continued to attack in a northerly direction. Scapinie, the battalion commander, was killed at the head of his men. A direct hit put an end to his soldier's life.

While with the *I./SS-Panzer-Grenadier-Regiment 25*, I noticed with some trepidation that the *21. Panzer-Division* was not supporting the attack and that its tanks were stationary at Couvre Chef. As a result, the regiment's right flank was open, and the enemy tanks were probing the flank of the *I./SS-Panzer-Grenadier-Regiment 25*. That battalion was in the midst of a dangerous crisis.

Hesitating for the moment, our *Panzergrenadiere* remained prone. Suddenly I noticed one, then several, turn and run back to Malon. Giving orders wouldn't help the situation; something else had to be done. I ran towards the young soldiers and pointed in the direction of the enemy. They stopped, watched me and then returned to their former positions. A *Panzer IV* moved into an antitank ditch and became stuck. The enemy tanks were then coming into range of the antitank platoon. The first Sherman halted, smoking; the next turned in circles and was ripped apart in a few seconds. Our own tanks appeared and finished things up.

Back to the regimental command post! There were 150 prisoners standing around in the courtyard of the monastery. They belonged to the Canadian 9th Brigade, which was part of the Canadian 3rd Infantry Division. The prisoners were either from the North Nova Scotia Highlanders or crews of the 27th Tank Regiment (Sherbrooke Fusiliers). I had a conversation with some of the officers and then climbed up into the church tower again.

The forward observers were continuously calling new fire missions and directing battery fire. I could see no movement on the battalion's left flank. Only the reconnaissance company of *SS-Panzer-Grenadier-Regiment 26* had arrived in the combat zone. The battalions had been so delayed by air attacks that we could not count on the regiment being employed.

SS-Aufklärungs-Abteilung 12 still had no contact with the enemy on the division's left flank; it reconnoitered in the direction of Bayeux.

While the division's chief ordnance officer, *SS-Oberssturmführer* Meitzel, was briefing me, we noticed a lot of enemy movement west of the Muc. Armored forces were pushing ahead towards Bretteville. The Muc was a little brook with vegetation on its banks; at Rots the banks become higher, thus creating a natural antitank obstacle for an attack from west to east or vice versa. An 88 mm *Flak* battery in the area of Franqueville was covering the sector.

Excitedly, I observed the dust clouds west of the Muc. Tank after tank was rolling over the high ground towards Bretteville. There were no German troops in that sector that could stop the enemy's advance. Enemy tank forces were thus rolling right into the approach-march area of *SS-Panzer-Grenadier-Regiment 26* if they continued their advance along the Caen — Bayeux road. There were only a few scattered infantrymen of the *716. Infanterie-Division* in Bretteville itself. The way was open deep into our flanks.

As we later determined, the observed units were the Canadian 7th Brigade with units of the Regina Rifles and the Canadian Scottish Regiments. Under those circumstances, the attack of the *SS-Panzer-Grenadier-Regiment 25* had to be stopped immediately. It would have been irresponsible to continue the attack with open flanks and against the unbelievable field and naval artillery fire. The attack was called off, and the *I./SS-Panzer-Grenadier-Regiment 25* was withdrawn as far as the *21. Panzer-Division*. The division ordered that the recaptured territory be defended until reinforcements arrived.

In the monastery orchard our wounded comrades were being cared for; the young soldiers were laying side by side and cheered each other up. Canadians were next to German soldiers. The doctors and medics didn't look at the uniforms. There was nothing that separated them at this point. The only thing that mattered was saving lives.

It had become impossible to evacuate the wounded. The fighter-bombers were attacking every ambulance. The regimental doctors reported to me that the medical company attached to the regiment was no longer fully operational. Fighter-bombers had attacked the medical company while on the march and destroyed some of the vehicles carrying medical equipment. Unfortunately, some of the medics were killed.

The Red Cross no longer offered any protection. To distinguish the medical vehicles even more, all ambulances were painted snow white, but that measure also proved to be useless. A temporary military hospital was established with the assistance of French doctors. The French took special care of the severely wounded Canadians. Under the protective cover of night the wounded and prisoners were moved to the rear.

The first day's combat losses were painful. Besides the commander of the *II./SS-Panzer-Grenadier-Regiment 25*, some of the company commanders were killed or injured. The *III./SS-Panzer-Grenadier-Regiment 25* suffered heavy casualties at Buron under the heavy artillery fire. The tank battalion had lost six *Panzer IV's*, two of them totally destroyed.

The enemy losses were considerably higher. According to Lieutenant Colonel Mel Gordon, the Canadian 27th Tank Regiment had lost 28 Shermans and the North Nova Scotia Highlanders lost 245 men in all: Killed, wounded and captured. The attack of *SS-Panzer-Grenadier-Regiment 25* had halted the enemy's operations against Caen. Unfortunately, it was not possible to continue the attack on 7 June.

A deliberate attack by the *12. SS-Panzer-Division* and the *21. Panzer-*

Division on 7 June would have been successful and the bridgehead north of Caen would have been crushed, had it been left in the hands of the respective unit commanders. However, the divisions were held back and could only be employed after being released by the *Oberkommando der Wehrmacht*. Both of the divisions could have operated together north of Caen on 6 June. During the course of the night, *SS-Panzer-Grenadier-Regiment 26* arrived in its sector. The threat to the flanks had finally been ended.

The *I./SS-Panzer-Grenadier-Regiment 26* had been stopped during its attack on Norrey and could not advance any further. The attack of the *1./SS-Panzer-Grenadier-Regiment 26* on the right bank of the Muc, which was supposed to establish contact with *SS-Panzer-Grenadier-Regiment 25*, broke down in the face of flanking fire from Norrey.

An attack by the *II./SS-Panzer-Grenadier-Regiment 26* from the direction of Le Mesnil and Patry led to the encirclement and elimination of three companies of the Royal Winnipeg Rifles at Putot en Bessin. The Canadian 9th Brigade ordered the Canadian Scottish Regiment to counterattack; suffering heavy losses, it pushed the *II./SS-Panzer-Grenadier-Regiment 26* back to a position just south of Putot. The *III./SS-Panzer-Grenadier-Regiment 26* was in defensive positions along the railway embankment at Bronay.

The *Panzer-Lehr-Division*, under the command of *Generalleutnant* Bayerlin, reached the front on 8 June at Tilly. The division had already suffered severe losses on its approach march. More than 40 armored vehicles carrying fuel and 90 trucks had been destroyed. Five tanks, four prime-movers and four self-propelled guns had also been destroyed. Those were serious losses for a unit which had not even seen action.

Shortly after midnight, *SS-Obersturmführer* von Ribbentrop reported to me at the command post. Von Ribbentrop had been injured in the shoulder two weeks before the invasion by fire from a low-flying aircraft; he stood in front of me with his arm in a sling. He had taken off from the military hospital and was looking for his tank company. Since I already knew von Ribbentrop from earlier operations, I did nothing to send him back to the hospital.

I visited the battalions during the night and inspected the individual positions of the companies. I was struck dumb by the positive attitude and spirit of the soldiers. We, the old soldiers, had been deeply impressed by the events of the day. The artillery fire and enemy air attacks had affected all of us. Not so the young soldiers. For them, it was the baptism of fire they had expected. They knew that many hard days and weeks lay ahead of them. Their attitude deserved respect.

Important enemy documents were found during a search of the battlefield at Authie and Franqueville. All the signals instructions were recovered from the first enemy tank that had been knocked out.

Towards evening on 8 June, Canadian carriers coming from Putot moved over the mined bridge at the outskirts of Bronay directly in front of the barrel of an antitank gun positioned there. One of the vehicles burned up with its passengers and crew; the other one was undamaged. A first lieutenant and his driver were dead. In that vehicle a valuable map was found; it detailed all enemy positions on both sides of the Orne with exact details of weapons, down to mortars and machine guns. Instead of the actual names of places and terrain features, animal names were used which began with the same letter. For example, the Orne was called the "Orinoco". Terrain unsuitable for tanks was highlighted.

A notebook was found on a dead captain with notes concerning the operations order for the invasion. It also contained information concerning the conduct of operations and the rules of engagement concerning civilians.

The codenames and signals instructions were used by the enemy for two more days, allowing our own signals intelligence platoon to monitor and evaluate the enemy signals traffic.

On the afternoon of 8 June I made a trip around the regimental sector with the divisional commander, visiting the *I./SS-Panzer-Grenadier-Regiment 26* afterwards. Low-flying aircraft were on our tail. I was glad to bring my commander back to my command post in one piece. *SS-Panzer-Grenadier-Regiment 25* then received orders to relieve the pressure on the *I./SS-Panzer-Grenadier-Regiment 26* by attacking Bretteville-l'Orgueilleuse from the east. It would be joined in the attack by the recently arrived *Panther* company of the *I./SS-Panzer-Regiment 12* and the *15. (Aufklärung)/SS-Panzer-Grenadier-Regiment 25*. The attack was planned for the following night. Day attacks had become impossible given the Allied air supremacy.

The *Panthers* advanced from Caen towards Franqueville and the front shortly before nightfall. That made their way slowly toward the front. The tank commanders had been briefed in detail; the company commanders and platoon leaders had reconned the ground during the afternoon and knew every fold. The tank company was positioned to move in a wedge formation. The *Panzeraufklärer* climbed onto their vehicles. I moved from vehicle to vehicle, saying a few words of encouragement to the young soldiers.

The company commander, von Büttner, my adjutant for many years, suddenly reminded me of a promise I made to the *15. (Aufklärung)/SS-Panzer-Grenadier-Regiment 25* during combat training at Beverloo in Belgium. At that time, I told the company: "Boys, the reconnaissance company is always the spearhead of the regiment. You bear a lot of responsibility. I promise you that I will be in your ranks to witness your baptism of fire." The time had come; the company was going to its baptism of fire. Consequently, I had to accompany them.

My old friend and comrade, Helmut Belke, arrived with a motorcycle combination. He had constantly been by my side as a dispatch rider, a squad leader and section leader since 1939. He has accompanied me across all the battlefields. In the sidecar was *Dr.* Stift. I jumped on the rear seat and directed Helmut to the Caen — Bayeux road. On our right tank engines were rumbling. The grenadiers had mounted up; they were taking cover behind the turrets. The young soldiers were waving to me. They were slapping each other on the back and probably remembered my promise. They pointed to my motorcycle and shook their heads. My "ride" appeared to cause them some concern.

The commander of *SS-Panzer-Regiment 12*, Max Wünsche, wanted to accompany the *Panther* company. He had also been fighting at my side since 1939. We knew each other; there was no need for discussion. A look, a signal, and the tanks were rolling into the night.

To the left of the road was an 88 mm battery in firing position. After a few minutes we had left the last outposts behind. The tanks were rolling ahead at full speed off the right-hand side of the road. There were no obstacles and the tank drivers could put the pedal to the metal. A motorcycle section and the artillery forward observer's vehicle were following a few hundred meters behind me.

The engine was our strongest weapon. It was time to move out! The speed increased, only the outline of the tanks could be seen. I wanted to be through Rots before nightfall. That was what had been arranged with Wünsche. The first buildings in Rots appeared. The tanks were behind us, to the right. It was like sitting on a volcano, but Helmut continued to drive on. indefatigably. We were barely clinging to the bike so as to be able to get off the road as quickly as possible. We waited for the tanks at the outskirts of the village. They arrived in a few minutes. The first section of the reconnaissance company dismounted and moved ahead on the ground. The village was clear of the enemy and we pushed quickly through.

The *Panthers* had to move one behind the other through the village. As soon as they had the village behind them, they resumed the wedge formation. Two *Panthers* went roaring down the road towards Bretteville. The rest pushed ahead on both sides of the road. In the darkness, I was only able to see the red-hot exhaust stacks of the tanks. Norrey was already just behind us to the left. We would have to encounter the Canadian outposts in the next few seconds. Bretteville was only a few hundred meters in front of us.

Crack! Crack! The report and muzzle flash of guns could be seen on the road. The two lead *Panthers* were firing round after round from their guns. They cleared the road with their fire and roared into the village at top speed. That was the way we had fought in the east, but would these surprise tactics achieve the same for us in Normandy?

All the tanks were firing into the village at that point; enemy machine-gun fire responded. We had positioned ourselves right behind the second *Panther*. It was getting too uncomfortable on the road for my taste. We moved to the right and worked our way forward along the ditch.

I tripped over a dead Canadian. A small armored carrier was smoking along the road embankment. As we moved on, I heard somebody groaning. A wounded man was on the road over to our left; machine-gun bursts were ripping down the street. Additional *Panthers* with *Panzergrenadiere* mounted on them were pushing into the village.

We advanced along the ditch in a series of bounds. I reached the wounded man. He was lying on his back, groaning and in pain. My God, it was von Büttner! The reconnaissance company commander had been shot in the stomach. I felt for and found the severe wound. Büttner recognized me and squeezed my hand. As an old frontline soldier he knew this was his last battle. "Tell my wife I loved her very much!" came from his lips, slowly but clearly. I knelt by his side as the doctor dressed his wound. Helmut Belke was covering us, kneeling a few meters away.

I heard a sound from the other side of the road; a shadow ran across it. Friend or foe? Belke fired as he dove and hit a Canadian in the head; Belke also fell at the side of the road. My companion for so many years did not stand up; he had also fought his last battle, killed by a round to the stomach. I tried to give him hope, but he would hear none of it: "No, no I know what this means. This is the end. Please tell my parents!"

Panzergrenadiere were storming past us. *SS-Oberscharführer* Sander, son of the mayor of Dessau, clasped Helmut's hand. Sander was also killed in action a few hours later. Tears were running down my face; the old comrades were thinning out. I jumped on the motorcycle in order to regain contact with our forces. A few seconds later I was in flames; the fuel tank had been penetrated and was burning like a torch. The *Panzergrenadiere* dragged me off and smothered the flames with dirt.

There was firing from all directions in the village. We had reached the center, but the lead tank had been hit. The command post of the "Regina Rifles" had been overrun. The surprise attack was successful, but where were the *Panzergrenadiere* of *SS-Panzer-Grenadier-Regiment 26*? We could not hold out there on our own, and we were too weak to capture all of Bretteville. With heavy heart I decided to withdraw the units at dawn to the high ground east of Rots. The result of the operations up to that point was that Montgomery had not achieved his objective. According to the operation plan, Caen should had been taken by the Allies on 6 June.

At about midday the *I./SS-Panzer-Grenadier-Regiment 26* assumed the Rots sector. *SS-Obersturmbannführer* Wünsche was wounded during the night attack and *SS-Obersturmführer* Fuß, platoon leader in the reconnaissance company, was missing. Von Büttner's successor was killed in the early afternoon. The *15. (Aufklärung)/SS-Panzer-Grenadier-Regiment 25* had lost two company commanders in 24 hours.

I encountered the divisional intelligence officer at the regimental headquarters. We were both aware that the German High Command would have to act rather quickly if the Allied bridgehead were to be destroyed. Until then all the *Panzer-Divisionen* had been committed as soon as they had reached the combat zone. None of them had been able to carry out a deliberate attack. Practically all the *Panzer* units were already forced onto the defensive; the urgently needed infantry divisions were positioned east of the Seine and remained inactive. Day by day we were becoming weaker and the Allies stronger.

During the afternoon the Commander of *Panzergruppe West*, *General* Geyr von Schweppenburg, arrived at the command post. The general knew our division quite well; he had frequently inspected it during training. Our shortcomings and strong points did not go unnoticed by him. He expressed his appreciation to us for all we had done.

We climbed up to the observation point in the Ardenne monastery. The general asked for my opinion of the situation. I explained my regiment's situation in a few words and expressed my fears that the war would be decided in the next few days. Von Schweppenburg looked at me quickly and said: "My dear *Herr* Meyer, the war can now only be won by the politicians."

Von Schweppenburg told me he had decided to mount an attack with the *21. Panzer-Division*, the *12. SS-Panzer-Division "Hitlerjugend"* and the *Panzer-Lehr-Division*. He wanted to break through to the coast with those divisions. It was intended for the *Panzer-Lehr-Division* to assume the St. Manvieu — Putot — Bronay sector; *SS-Panzer-Grenadier-Regiment 26*, which would be freed up by that action, was to be inserted in the sector east of the Muc. The planned date of the attack was the night of 10/11 June. I thought a night attack with those forces promised success. In any event, the attack had to start very early, so that we would be inside the enemy's positions by daybreak, thus negating the heavy naval gunfire. A daylight attack seemed hopeless to me because of the Allied artillery and air superiority. The preparations for the deliberate attack began immediately, but the difficulties were enormous. Ammunition and fuel supplies could only be brought up at night. For example, the ammunition had to be brought up from the woods north of Paris.

The sector in front of *SS-Panzer-Grenadier-Regiment 25* was quiet; the enemy had probably received too much of a shock on 7 and 8 June to undertake additional attacks on Caen. There was a constant series of defensive and offensive operations in the sector to our left. The *Panzer-Lehr-Division* had taken Melon and was holding that sector.

*

I had already been told at the outset of the invasion that the Allies did not take the Geneva Convention very seriously and the divisions that had already landed had taken few prisoners. On the morning of 9 June I found a group of German soldiers on the railway line south of Rots. They had obviously not been killed in action as they were all laying next to the road and all had been shot through the head. They were soldiers from the *21. Panzer-Division* and the staff of the *12. SS-Panzer-Division "Hitlerjugend"*. The incident was immediately reported to the division which, in turn, passed it back to corps.

The regimental commander of *Panzer-Artillerie-Regiment 130 (Panzer-Lehr-Division)* became a British prisoner of war together with a battalion commander, *Major* Zeißler, *Hauptmann Graf* Clary-Aldringen, and about 6 noncommissioned officers and enlisted men on the morning of 8 June. They had been captured by armor from the British "Inns of Court" Regiment, which had penetrated behind the German lines.

After the German officers had refused to voluntarily act as human shields, the badly injured *Oberst* Luxemburger was fettered by two British officers, beaten unconscious and, in that blood-covered condition, tied to a British tank as a shield. After receiving orders from their superiors, the British tanks fired on *Major* Zeißler, *Hauptmann Graf* Clary-Aldringen and the other soldiers as they moved off.

Hauptmann Graf Clary was found by members of *SS-Bataillon Siebken* and brought to the battalion command post. The British tank on which *Oberst* Luxemburger had been tied as a shield was hit by a German antitank gun; he died two days later in a military hospital. *Hauptmann Graf* Clary received first aid from *SS-Sanitäter* Klöden.

On 7 June, a notebook was found on a Canadian captain with notes about the orders given before the invasion. Besides the tactical instructions, there were also instructions on how to fight. It read: "No prisoners are to be taken." The notebook was given to the Commander-in-Chief of the *7. Armee*, *Generaloberst* Dollmann, by the Operations Officer of the *12. SS-Panzer-Division* on 8 June. The officers and men of the Canadian 3rd Division, who had been interrogated by the *12. SS-Panzer-Division*, confirmed they had received orders from their superiors not to take any prisoners. A soldier stated that they were not supposed to take prisoners if they were hindering operations.

During Bernhard Siebken's war crimes trial in November 1948, *Oberst i.G.* Meyer-Detering, the Intelligence Officer to the Commander-in-Chief West, *Feldmarschall* von Rundstedt, stated: "Right at the onset of the invasion, I twice received documents captured in the Canadian Army sector on the eastern flank of the invasion front which showed that no prisoners were to be taken."

At the same trial *Oberstleutnant i.G.* von Zastrow, who was the intelligence officer to the Commanding General of *Panzergruppe West*, *General* Geyr von Schweppenburg, stated that he obtained information at that time of crimes committed by Allied troops in violation of the Hague Declaration on Land Warfare and of the Geneva Convention. He then told about one incident when a number of German soldiers had been shot by Canadian soldiers after they had already been taken prisoner.

Furthermore, *Oberstleutnant* von Zastrow gave evidence concerning a Canadian captain. This captain had been taken prisoner during the course of the fighting in the Somme area. Because he belonged to the same unit in which the orders were found at the beginning of the invasion and where crimes had been committed, von Zastrow accused him of those violations of international law. When asked directly whether he had heard about the shooting of German prisoners, the captain answered: "I wasn't involved in the invasion; I've only been serving with the unit as a replacement for a short time. I had heard about atrocities, however. But then very strict orders were issued that any such incidents would be punished."

I outline these incidents here, because they illustrate how the Allies conducted their fighting and because they decisively influenced my own personal destiny. I want to briefly cover the subject of "war criminals" here before it is again covered later.

*

The preparations for the deliberate attack were in full swing. There were only a couple of hours available to accomplish the only hammer blow available against the bridgehead. If this blow were not successful there wouldn't be another chance to drive the Allies into the sea. After the attack ended the three *Panzer-Divisionen* would be burnt out and not able to repeat such an operation.

After the orders of *Panzergruppe West* had been issued in the presence of the Commander-in-Chief of *Heeresgruppe B*, *Feldmarschall* Rommel, the *Panzer-Lehr-Division* reported an enemy breakthrough from the west. Shortly after that report the operations staff of *Panzergruppe West* was virtually annihilated by a carpet-bomb attack. *Panzergruppe West* lost its Chief of Staff, *General* Ritter und Edler von Dawans, the operations officer, and a number of other officers. During that attack, we also lost our liaison officer, *SS-Hauptsturmführer* Wilhelm Beck, a tank company commander who had participated in every engagement since 1939. He had been awarded the Knight's Cross for his part in the recapture of Kharkov. The signals battalion was put out of action. The commanding general, who was lightly injured, escaped with a few men. He was out of action until 26 June and, as a result, there was no coordinated counterattack during the next few days. The armor units had to revert to the defensive as a result of the increasing daily pressure of the English armored divisions.

On 11 June, the front came to life on the left flank of *SS-Panzer-Grenadier-Regiment 25*. I immediately went to the *I./SS-Panzer-Grenadier-Regiment 26* and found Bernhard Krause at his battalion's observation post. Together we watched an attack on the positions of *SS-Pionier-Bataillon 12* on the high ground north of Cheux. The ground was prepared by all calibers of weapons and was converted into a smoking, smoldering mass in a couple of minutes. We were looking through a scissors telescope and could not see any movement, no matter how hard we tried. The fire then started to fall on the rear slope of the hill and slowly shifted towards the village of Cheux. Canadian infantry jumped out of the ditches and from behind vegetation and attacked the engineer's position with tank support.

Bernhard Krause reacted immediately. In a short period of time he had directed his heavy infantry weapons at identified targets and the fire fell into the midst of the attackers. This flanking fire was not without impact. The enemy infantry bogged down on the forward slope of the high ground and suffered considerable losses. Some of the attacking tanks had entered the battalion's minefields and were disabled. The Canadian "Queen's Own" Infantry Regiment and the 1st Hussars Tank Regiment had to pull back to their original positions. They did not repeat the attack.

On the way back to the command post of *SS-Panzer-Grenadier-Regiment 25*, I passed through Le Bourg/Rots in order to get a quick look at Bretteville and Norrey and meet up with *SS-Hauptsturmführer* Pfeiffer. Pfeiffer had been Hitler's adjutant for a long time and was then commanding a *Panther* company. His successor at the *Führer's* headquarters was then *SS-Hauptsturmführer* Günsche, who had been the commander of the *III./SS-Panzer-Grenadier-Regiment 25*.

Between St. Mauvieu and Rots, I was fired on from the direction of Norrey. Machine- and antitank guns were trying to hit me, but my good old *BMW* was a poor target. I moved quickly on to Rots, but what was I to encounter there? Was I dreaming? Weren't those enemy tanks moving only 50 meters in front of me. Indeed they were! The two Shermans slowly moved out of the town and halted on the left-hand side of the road. The barrels were directed at my motorcycle! It would have been no use to turn round. I would not have been able to do that. So there was nothing left to do but to move right on through them! I accelerated to get past the two tanks; perhaps I could slink away into the wood on the right.

Suddenly there was a bang off to my right behind me. I saw the first tank rocking on its tracks and a crewman leaping out of the turret as if he had been bitten by a tarantula. Following my instincts, I rolled to the right into a ditch and took cover. Before I find out what was going on, the second tank exploded. Immediately my companion and I dashed towards the rear and took cover again. Only then did we realize what had actually happened. We had raced past the last defensive outposts just at the very moment they were trying to fire at the two tanks. Our troops could not warn us. Good old Pfeiffer was in his Panther about 100 meters away from the first enemy tank. He had blasted both of them into the next world. *SS-Hauptsturmführer* Pfeiffer died a soldier's death 24 hours later, when he was fatally wounded by shrapnel.

At the dressing station of *SS-Panzer-Grenadier-Regiment 25* I again found a lot of wounded. The constant arrival of wounded, without any fighting taking place, gave us all pause to think. The conduct of operations was such that the *Panzer-Divisionen* were being decimated by naval gunfire and low-flying aircraft without being able to conduct offensive operations. It couldn't — it mustn't — go on like that! The *Panzer-Divisionen* had to regain freedom of movement.

On 11 June, the *II./SS-Panzer-Grenadier-Regiment 26* was attacked by the Canadian "Queen's Own Rifles" Regiment, reinforced by the 1st Hussars Tank Regiment and supported by strong artillery. The attack was repulsed. The battalion held its ground. The enemy lost 12 tanks while we forfeited 3 during those operations. Units of the Canadian 8th Brigade, supported by the British 46th Royal Marine Commando and tanks, attacked Rots. Several enemy tanks were destroyed. The village was evacuated under superior enemy pressure.

A threatening situation had developed near Tilly-sur-Seulles on the right flank of the *Panzer-Lehr-Division*. On orders of the *I. SS-Panzer-Korps*, the last reserve of the *12. SS-Panzer-Division*, the divisional *Begleitkompanie*, was attached to the *Panzer-Lehr-Division* and deployed on its right flank, where the situation was stabilized.

The situation on the Caen front became ever more threatening on a daily basis. The *Schwerpunkt* of the fighting had moved to our neighbor on the left, the *Panzer-Lehr-Division*. The enemy managed to break through to Caumont at Balleroy. The 7th Armoured Division was pushing forward to Livry; the encirclement of the *I. SS-Panzer-Korps* started to loom large.

I was ordered to the division. There I met the commanding general of the *I. SS-Panzer-Korps*, *Generaloberst der Waffen-SS* Sepp Dietrich. Dietrich provided an overview of the entire situation on the invasion front and frankly told us that there were no more reserves and he no longer believed a deliberate attack was possible. He railed: "They want to defend everything...but with what? Those who defend everything, defend nothing."

This axiom was originally formulated by Frederick the Great as: "Small minds want to defend everything; clever people keep their eyes on the main objective."

I visited the regiment's battalions with the onset of nightfall. Waldmüller had nestled down by an antitank ditch and was pleased in a wicked way that all of the British salvoes were being received by the village behind his command post. It had been cleared of troops. The soldiers grinned as I groped my way through the positions. Forward outposts had been established in knocked-out enemy tanks — the poor guys could only be relieved after dark. There was continuous reconnaissance patrol activity by both sides. The left flank of the British 3rd Division was in front of the *I./SS-Panzer-Grenadier-Regiment 25*; next to it was the Canadian 3rd Division.

The advance guard of the 7th Armoured Division reached Hill 213, two kilometers east of Villers-Bocage, on the morning of 13 June.

The commander of the *2./SS-schwere Panzer-Abteilung 101*, *SS-Obersturmführer* Michael Wittmann, who had already destroyed 119 tanks in Russia and been decorated with the Oakleaves to the Knights Cross, had moved in front of his company to conduct terrain reconnaissance. He encountered a column of enemy tanks near Hill 213. The enemy was completely unprepared. He hesitated for a few moments and was undecided whether he should withdraw or attack the superior enemy forces. His unit had taken heavy losses through bombing attacks during the past few days. He knew every single tank was important and that he could not act carelessly. He had to attack or the advance guard of the 7th Armoured Division would drive into the rear of the *Panzer-Lehr-Division*. If a breakthrough of the German front succeeded, the defense of Caen would be lifted off its hinges.

He stood by himself as the 22nd Tank Brigade under its commander, Brigadier Hinde, moved into Villers-Bocage meeting no resistance. Because of this unexpected development, the lead armor element carelessly rolled further along the road to Caen towards its objective, Hill 213. The motorized infantry company that followed took a break on the road. At that point, the thunder of a main gun cut through the morning silence. The lead vehicle stood there burning and, at a distance of 100 meters, a *Tiger* came roaring out of the woods. It turned onto the road, rolled along the line of the half-tracks and fired them all up, one after another, in quick succession. That was followed by a dozen armored vehicles of the regimental headquarters and the artillery observers and a reconnaissance platoon, that happened to be behind the row of vehicles. A 75 mm round from a Cromwell tank, fired at point-blank range, bounced off the *Tiger* without effect. Within a few minutes the road was like an inferno. Twenty-five tanks were in flames, all victims of this single *Tiger*.

In the meantime, the enemy tanks providing cover on Hill 213 were attacked and destroyed by Wittmann's remaining *Tigers*. Wittmann's tank was hit on the tracks and disabled. The crew dismounted and fought its way back to the company. Villers-Bocage was evacuated by the enemy, who pulled back to Livry. Wittmann was decorated with the Swords to his Oakleaves of the Knight's Cross of the Iron Cross. At that point, his kills totaled 138 enemy tanks and 132 antitank guns.

Formations of the British 49th Division attacked the positions on the left flank of the *12. SS-Panzer-Division*. The attack was repulsed, the positions held. The division asked corps for permission to withdraw from the salient in the sector south of Putot and Bronay to a line from St. Manvieu through Fontenay to Tilly. Electronic intelligence established the enemy's intention to eliminate this salient by attacking it. The corps refused to grant permission.

The Allied bridgehead became stronger hour by the hour. By 18 June, 500,000 soldiers with 77,000 vehicles had been landed in France. We were still waiting for reinforcements and for the *Luftwaffe* support which we had been promised. But we waited in vain. Our *Panzer-Divisionen* were bleeding to death in their positions. No major decisions were made; the fighting was limited to tactical "repair work".

On 16 June, I was urgently requested to come to the division's command post. The divisional operations officer, Hubert Meyer, was on the telephone and indicated the matter was urgent. I could not get any details out of him. I rushed immediately to headquarters. Tall trees that had been shot to pieces still obscured the road. The ripped-open roofs on both sides of the road demonstrated the intense enemy naval gunfire, which had ceased about half an hour before. I sensed nothing good. I found the staff in a state of uproar.

The faces of the soldiers told me everything. The chief-of-staff came forward and reported: "The divisional commander was killed in action half an hour ago. On order of the corps, you are the acting commander of the division." We stood there speechless, but in our handshake there was an obligation and mutual promise: Our actions would honor our comrade Fritz Witt and his work all the more!

I sought the patch of grass on which Fritz Witt was laid to his final rest. I stood in front of what used to be my comrade. I could not look into his face; it didn't exist anymore.

Fritz Witt had been killed because he made sure the soldiers were the first ones into the trenches. When he leapt in, a round hit the ground immediately in front of the trench and killed him on the spot. The loss had affected us to the marrow. Fritz Witt was revered by all ranks of the division. He had close friendly relations with most of the officers in both peace and war. The death of the commander probably bonds the division even more closely together in its cohesiveness and increases its will to confront the enemy all the more.

SS-Obersturmbannführer Milius assumed command of my regiment; he had been the commander of the *III./SS-Panzer-Grenadier-Regiment 25*. I gave his battalion to *SS-Hauptsturmführer* Fritz Steger. The divisional command post was moved to Verson.

Late in the evening, the chief-of-staff gave me a detailed report on the situation of the division. The division's casualties were considerable and, in some cases, irreplaceable. First of all, the large number of officer and non-commissioned officer casualties had reached the danger point. Most of the company commanders and platoon leaders had already been killed in action or were wounded. The *Panzergrenadier-Bataillone* had, on average, the combat strength of two companies. There was no immediate prospect of receiving replacements for these losses.

In such a situation, one could calculate the day the division would finally be annihilated. Continuous enemy artillery fire, together with the fighter-bomber and bomber sorties, devoured our division. The British 3rd and 49th and the Canadian 3rd Infantry Divisions and several tank brigades were

240

fighting in the sector of the *12. SS-Panzer-Division*.

There was no doubt among the staff of the division that the enemy would attempt to pinch off the frontal salient on the left flank of the division within the next day or so in order to secure a favorable jump-off position for the expected offensive.

At dawn I visited the positions of the *II.* and *III./SS-Panzer-Grenadier-Regiment 26* and the *SS-Aufklärungs-Abteilung 12*. I established their new positions. The mood among the grenadiers was good. Bernhard Siebken, the commander of the *II./SS-Panzer-Grenadier-Regiment 26*, who had lost a finger, accompanied me through his sector. He was unwilling to leave the battalion and didn't take the loss of his finger too seriously.

Burnt-out enemy tanks were at Putot and Bronay. Bremer's reconnaissance battalion had captured some enemy tanks and incorporated them in the defense. The battalion had suffered considerable casualties and needed to be pulled out of the lines as quickly as possible.

The corps granted permission to withdraw the front to positions south of the Muc. The *SS-Aufklärungs-Abteilung 12* was withdrawn from the lines and placed into the divisional reserve.

In the sector of the *Panzer-Lehr-Division*, intensive offensive operations were being conducted by the British. The British 50th Infantry Division and several tank brigades relentlessly attacked the positions north of Tilly.

In the early morning hours of 18 June, the earth trembled on both sides of Cristot as the British 49th Infantry Division threw itself at the deserted positions on the left flank of *SS-Panzer-Grenadier-Regiment 26*. I stood next to the commander of the *III./SS-Panzer-Grenadier-Regiment 26,* and we watched the rounds plough up the terrain which had only been evacuated the day before. We congratulated ourselves on our early withdrawal.

Cristot was occupied fairly quickly by the forces of the 49th Infantry Division. After a short time, the attack on Fontenay continued. The earth shook. The *10./SS-Panzer-Grenadier-Regiment 26* fighting in front of us, was covered by masses of steel. In a very short time more than 3,000 artillery rounds exploded on its positions. The company was fighting desperately on the northern edge of the Parc de Boislonde. The tall trees were uprooted and flung into the defenders lines. A wave of tanks followed the barrage. The tanks fired round after round into the woods. Enemy infantry closely advanced behind the tanks and the company's smashed positions were overrun. Some of the company's strongpoints continued fighting, but they were soon silenced. Our artillery was not in a position to halt the attack. Units of the 49th Infantry Division pushed through to the southern edge of the woods, but they were stopped there by the *9./SS-Panzer-Grenadier-Regiment 26.*

Under the battalion commander's leadership, the *9./SS-Panzer-*

Grenadier-Regiment 26 attacked the British, supported by a brief barrage from our artillery. It advanced to the northern edge of the woods, breaking into the old positions. Fighting hand-to-hand, it rolled up the position and pushed the English back towards the north. The patch of woods was firmly in the hands of the *III./SS-Panzer-Grenadier-Regiment 26* again. But the victory did not last long. Concentrated artillery fire and well-aimed fire from the tanks inflicted heavy losses on the company. The survivors of the company were pushed back to the northern edge of Fontenay.

The division's front was stable but the losses had reached critical proportions. There was no divisional reserve to speak of. From electronic intelligence we concluded there would be a new attack by strong enemy forces against the left flank of the division. In the sector of the brave *Panzer-Lehr-Division*, Tilly had been lost. It had been fighting for Tilly for days.

In the early hours of 21 June, I left the divisional command post and moved to the sector of *SS-Panzer-Grenadier-Regiment 26*. I met its commander, *SS-Standartenführer* Möhnke, in the command post of the *III./SS-Panzer-Grenadier-Regiment 26* in a farm north of Fontenay. The ruins of the farm were surrounded by high hedges; wounded lay behind an earthen wall and were waiting for evacuation. Dead comrades were buried in the orchard. Thick ground mist cloaked the destruction. Tanks and antitank guns had taken up firing positions behind the hedges, and the crews were busy camouflaging their weapons. Enemy harassing fire fell on the entire sector of the regiment, but especially on the Rauray — Fontenay road.

After a short conference I carefully moved close to Fontenay, accompanied by *Dr.* Stift. The outskirts were blocked by rubble and acrid smoke mixed with the thick mist. Suddenly, two *Panzergrenadiere* of the *II./SS-Panzer-Grenadier-Regiment 26* appeared in front of us. They had mess tins and rations in their hands and wanted to find the rest of their section. We crossed the village's main street and crept northwards. In front of us was the patch of woods that had been so hotly contested.

Climbing over dead English bodies, I reached the positions of the *15./SS-Panzer-Grenadier-Regiment 26*. *Panzergrenadiere* with emaciated faces were crouching in foxholes and shell craters. There were no officers left: They were all dead or wounded. The dead company commander was brought back during the night by a patrol. What a wonderful relationship must have existed between the company commander and his soldiers. Even his remains were not allowed to fall into enemy hands.

The attitude of the young soldiers was incomprehensible to us. Briefly they told me about the recent fighting and their operations. Their heroic fight was self-evident to them. They talked about the enemy without hatred; they repeatedly emphasized the enemy's outstanding morale. There was also a feeling of bitterness in their comments when they spoke about overwhelming enemy equipment. Again and again I heard: "Damn it, where would we be

242

now if we had the enemy's equipment?" We still hadn't seen a single German plane in the sky.

The young *Panzergrenadiere* asked me to leave as it was growing light and the enemy artillery fire was growing heavier every minute. We were caught by a heavy barrage in the middle of the village. I leapt behind some stone stairs and waited for this unpleasant early morning greeting to pass. One of my companions lay on the street with smashed limbs; a direct hit had torn him to pieces. We dashed through the village pursued by the heavy artillery fire falling on Fontenay and were glad to leave the ruins. The artillery fire was also falling on the sector of the *III./SS-Panzer-Grenadier-Regiment 26*. Was it the overture to the expected offensive?

The battalion command post was only about a hundred meters away, but the distance seemed endless. We finally reached it by dashing from one bit of cover to another.

In the meantime, daylight had arrived and the first reports had come in. Fontenay was under attack from the direction of St. Pierre and Cristot. Enemy tanks were moving over the low earth bank and starting to push toward Fontenay. A tank engagement began. Our camouflaged *Panthers* with their superior guns had the advantage. Burning enemy tanks were scattered over the battlefield. If only we had more ammunition! Our artillery had to fire sparingly; supply had become almost impossible.

Telephone contact with the division's staff remained intact so the operations officer was able to tell me that all was quiet on the division's right flank. I stayed with the battalion until late afternoon and experienced first hand the unit's indescribably good morale. All the attacks on Fontenay were repulsed in hard fighting. The British 49th Division could not shake the foundation of our defense on the left flank.

I moved back to the divisional headquarters with the onset of darkness. The burning front presented a ghostly picture. The remnants of some trucks were smoldering on the Caen — Villers-Bocage road. They were supposed to be bringing ammunition to the front. Fighter-bombers relieved them of the responsibility. The ammunition exploded some kilometers behind the front.

I saw worried faces during the situation update. Without talking about it openly we knew we were approaching a catastrophe. The static type of fighting in the murderous bridgehead north of the Orne would inevitably lead to the destruction of the *Panzer-Division* deployed there. Faced with the enemy's enormous naval and air superiority, we could predict when the front would collapse. The tactical "fixing" along the battlefield had cost us the irreplaceable blood of our best soldiers and was destroying precious equipment. We were already feeding on ourselves. Up to that point we had not received a single replacement for our wounded or killed soldiers or a single tank or artillery piece.

After a few hours of sleep, we were dragged back to reality by the noise of the front. Calls of alarm arrived at divisional headquarters. The British 49th Division, supported by heavy artillery fire, was attacking the right flank of the *Panzer-Lehr-Division*. The woods west of Tesel-Bretteville had already been lost and the momentum of the British 49th Division had not yet been broken.

With a sense of foreboding, I moved to the *III./SS-Panzer-Grenadier-Regiment 26* in Fontenay. The division's left flank was in danger. Heavy fire was falling on the whole battalion sector. You could hardly recognize Fontenay. Screaming rounds were tearing the last buildings to pieces. Each attack on the village had been repulsed until that point. Communications with the companies was broken. The smoke of exploding projectiles obscured the view. It was impossible to determine the main defensive line. The artillery fired barrage after barrage.

The village was like a simmering cauldron. Heavy rounds drilled deeply into the earth and left smoking craters behind. Based on an old soldier's saying — "A round does not hit the same crater twice" — I jumped into a crater and watched the enemy tanks attack Fontenay. Firing continuously, feeling secure, the steel colossi were moving slowly towards the rubble of Fontenay. Our antitank guns were destroyed by the insane artillery fire.

The *Panzergrenadiere* held their *Panzerfäuste* tightly. Man against tank! What a contrast! And what a heroic spirit this contrast revealed! The first tank was smoking by that point. I could see how the soldiers were leaping onto the vehicles. The enemy artillery was firing over our heads. The commander of the *I./SS-Panzer-Regiment 12* jumped into my crater and reported an immediate counterattack by a tank company. The tanks were about 100 meters behind us and were moving out. Tank rounds were spitting over our heads. The lead enemy tanks were fighting the *Panzergrenadiere* in the rubble, and the tanks further back had not yet noticed our tanks. The counterattacking company had to cross the Fontenay-Cheux road to get better firing positions. The tank-versus-tank engagement started. There were casualties on both sides. Thick, black, oily smoke rolled over the battlefield. I wanted to see the soldiers in Fontenay, so I dashed along behind the company commander's tank.

Battle-weary soldiers waved to me. They yelled out humorous comments, their eyes shining. It mystified me where these young soldiers were getting the strength to live through such a storm of steel. They assured me again and again they would defend the rubble to the last round and hold their positions against all comers. The company commander's tank received a hit. It turned a few meters to the left; the hatch flew open, smoke gushed out of the turret, and the company commander forced himself through the hatch. He staggered towards us, stumbled and collapsed. The *Panzergrenadiere* pulled him behind the remains of a wall. We then realized that *SS-Obersturmführer* Ruckdeschel had lost an arm. The bleeding stump was bound up and a medic summoned.

That attack was also repulsed. The *III./SS-Panzer-Grenadier-Regiment 26* had held firm in its positions and was preparing itself for the next British assault. The fighting continued to rage in the sector to our left. The enemy spearhead was already in Juvigny by the afternoon and, therefore, deep in the flank of the division. I had just arrived in Rauray with the commander of *SS-Panzer-Grenadier-Regiment 26*, *SS-Standartenführer* Möhnke, when the chief of staff informed me about the situation of the *Panzer-Lehr-Division* and pointed out the danger of a British breakthrough.

The corps ordered the employment of one of the tank battalions of the division at about 1400 hours to eliminate the penetration in the right flank of the *Panzer-Lehr-Division*. There were only the exhausted remnants of *SS-Aufklärungs-Abteilung 12* available for infantry support during this counterattack. The *Panther-Abteilung* and that part of the reconnaissance battalion still available for operations immediately launched the counterattack.

The attack moved off via Tessel and Bretteville after a short barrage by friendly artillery towards a patch of woods one and a half kilometers to the west. We threw the enemy out of the woods by the onset of darkness but did not reach our old main line of defense. The enemy lost several tanks. The brave *III./SS-Panzer-Grenadier-Regiment 26* had suffered heavy casualties during the course of the fighting and was withdrawn to positions north of Rauray.

That evening the corps ordered the deployment of our last tank battalion to restore the situation in that sector the next morning. The *Panzer-Lehr-Division* was to be assisted at all costs.

I vainly asked for that order to be rescinded. The chief of staff's graphic situation report that friendly reconnaissance had identified the staging of strong enemy forces — especially armor — in the sector of *SS-Panzer-Grenadier-Regiment 26* did not influence the corps to change its order. My remark that an enemy tank attack was expected at any moment and the *II./SS-Panzer-Regiment 12* was in very favorable defensive positions was also dismissed. And so it was that on 26 June there was not a single tank in the divisional sector.

The men of *SS-Panzer-Grenadier-Regiment 26* waited in their foxholes throughout the night, exhausted by the intense defensive fighting of the recent past. They were waiting for the next attack. Thick, humid fog lay over the hedges and fields.

Morning dawned; everything was still quiet. Max Wünsche and I were at Rauray, watching the last tank roll into its jump-off position. It became lighter and lighter. It wouldn't be much longer at that point before the dance of death continued. The German batteries then fired their barrages. Low-flying British planes roared overhead and fired their rockets into Rauray. The hell of attrition warfare had begun.

The first tanks rattled forward, tracks clanking. The attack initially gained ground but was stopped by an English counterattack. A bitterly contested tank-versus-tank engagement started. The impenetrable hedgerows didn't allow our tanks to take advantage of their longer range. The lack of infantry proved to be especially disadvantageous. Intense artillery fire made coordination enormously difficult and effective command and control virtually impossible.

We could hear nothing east of Rauray. The entire fight had shifted to the west where the tanks pounded each other stubbornly. The tell-tale columns of oily smoke hung in the sky again. Each column indicated a tank's grave. I was uneasy because of the situation in the sector of *SS-Panzer-Grenadier-Regiment 26*. There wasn't a single artillery impact to the right of Rauray. It had started to rain, thank God. That meant we were protected from the fighter-bombers.

But what happened then? The earth seemed to open and gobble us all up. All hell had been let loose. All that remained of Rauray were fragments of smashed trees and buildings. I laid in a roadside ditch listening to the noise of battle. There was no let up to the artillery barrage. The fog mixed with the smoke of the bursting shells. I couldn't make out anything. All telephone lines had been destroyed and communications with divisional headquarters and units at the front no longer existed. A runner from the *II./SS-Panzer-Grenadier-Regiment 26* ran up to me and shouted: "Tank after tank on the battalion's right flank!" His message was swallowed by bursting shells. My ears tried unsuccessfully to analyze the sounds of battle and all I heard was the permanent spitting, cracking and booming of the bursting shells, mixed with the noise of tank tracks.

This was the offensive I had expected! The cornerstone of the German front in Normandy was at stake. Caen, the target, was to be smothered by an enveloping attack. Caen was to be Montgomery's prize, bringing about the collapse of the German front. We all stared spellbound at the murderous spectacle. Fiery steel hurtled over us and drilled into the ground.

I shouted for Wünsche. Runners crossed the road and vanished into the green hedgerow. Wünsche was at my side a short time later. I didn't need to give that experienced soldier a lengthy explanation. We had lain side by side too often; he knew me and knew what I wanted.

I outlined my estimate of the situation in a few words to Wünsche:

The enemy is trying to break through in the sector of *SS-Panzer-Grenadier-Regiment 26* with massed tank forces in order to capture Caen. 1. The attack on Juvigny is to be called off immediately. 2. Rauray will be held at all costs and is the cornerstone of the defense. 3. You are responsible for Rauray.

I moved towards Fontenay once again and, after a few hundred meters, encountered elements of the *III./SS-Panzer-Grenadier-Regiment 26*. Enemy

246

tank fire covered the road. It was impossible to move any further north but there was no need to feel my way forward. The battlefield was spread out in front of me. I could take in everything and found my judgment confirmed. This was the expected offensive. Tanks and half-tracks were advancing into *SS-Panzer-Grenadier-Regiment 26*. The barrage rolled over the earth like an enormous steel roller, crushing everything that lived. Only rarely did I see movement from the brave *Panzergrenadiere*. They held their ground stubbornly, fighting with the courage born of despair. Flashes of dazzling destruction from Rauray hit the oncoming tanks. British tanks were burning north of Rauray.

In front of us were two Tommies on the ground who had been carried slightly too far by the momentum of their attack. They had been disarmed and were told to climb quickly on my vehicle. A wounded Tommy was passed onto the dressing station at Rauray.

I moved like mad towards Verson. My place was at the divisional command post. The concentrated enemy artillery fire was shifted south with increasing intensity. It was already impacting in the Colleville area. Our own artillery hammered the attackers relentlessly.

I reached the divisional headquarters in a few minutes. The chief of staff was still holding the telephone handset in his hand and reported: "That was our last conversation with the engineer battalion commander." The commander had reported: "Enemy artillery has destroyed my antitank defenses. The battalion is being overrun by British tanks. Individual positions are still holding out in and around Cheux. Enemy tanks are trying to crush my dugout. Where are our tanks? I need a counterattack from the direction of Rau...." At that point, the line was cut. Radio communications had also been destroyed.

There was also an urgent report from the *I./SS-Panzer-Grenadier-Regiment 26*. The battalion was being attacked by strong forces. All attacks on St. Mauvieu had been repulsed up to that point. More reports arrived in rapid sequence. The whole front line was erupting.

Nothing else could be done for the time other than concentrate divisional artillery fire on those enemy units that had broken through. The only available resources were the divisional *Begleitkompanie* and the decimated reconnaissance company of *SS-Panzer-Grenadier-Regiment 25*.

I concentrated everything on defending Verson. Monitored radio traffic and prisoner statements told us that an armored division and two infantry divisions, each reinforced by an armored brigade, had mounted an attack across a 5-kilometer front. Those formations hadn't been employed yet; that meant they were fully rested. These fresh divisions with about 600 tanks were attacking three friendly battalions with diminished combat power. The enemy's main strengths were his enormous artillery reserves and masses of armor.

The danger of a breakthrough was obvious. My chief of staff desperately pointed out the hopeless situation. The corps gave only one answer: "The positions must be defended to the last round! We have to fight for time. The *II. SS-Panzer-Korps* is on its way to the front."

As so often in the past, command and control was being exercised from a tactical perspective and not strategic considerations. Important decisions were not made. Mobile defense had been abandoned. We had no other choice than to sell our lives as dearly as possible.

The sky seemed to adapt itself to what was happening on the ground; torrential rain accompanied our every step.

Northeast of Verson, I could see countless British tanks from the position of the divisional *Begleitkompanie*. They were advancing towards Grainville. The front had been penetrated and only odd pockets of resistance hampered the enemy flood.

My God! The division had to stop the attack. It had to prevent this deep breakthrough and fight for time for the German High Command!

Once again I raced over to the divisional command post and tried to get through on the line to Wünsche. It worked! Our brave signals soldiers had just repaired the line. How often had these men raced through Hell? What nameless heroism hid behind the phrase "wire dog"!

Max Wünsche reported strong tank units on both sides of Rauray. Every attack on Rauray up to that point had been fended off with heavy enemy losses. The divisional linchpin stood unshaken. I returned to the *Begleitkompanie* a few minutes later. All command and control had become impossible. At that point, I could only be a soldier among soldiers. The eyes of the grenadiers lit up I when they noticed me moving from section to section. These soldiers were unshakeable. They would not waver or give way.

Soon there was no piece of ground where a round had not exploded. Enemy tank rounds exploded in our lines. Our defensive area was reinforced by two tanks and an antitank gun. We clasped the few remaining *Panzerfäuste* tightly to our bodies.

A *Panzer IV* exploded and two Shermans were burning in front of us. The masses of enemy armor gave me the willies. Didn't it border on madness to try to stop this army of steel with a handful of soldiers and a few guns? It was too late to speculate; there was only one thing left to do — Fight!

Two Shermans pushed closer down a defile. Some grenadiers lay feverishly in wait with their *Panzerfäuste* behind the blackberry bushes. They seemed to be one with mother earth.

I held my breath, and the exploding rounds had suddenly lost their terror. Spellbound, we watched the soldiers as they got ready to jump. The lead tank advanced further and further down the sunken road with the covering tank rolling slowly behind him. It rolled past at that point; the second tank was as far as our soldiers. The gun barrels were pointed at Verson, but they would never fire again. A soldier rushed at the second tank like an arrow from a well-strung bow. His *Panzerfaust* smashed into the Sherman's side while he was still jumping. The tank rolled on a few meters, then stopped, smoking. The lead tank had also been halted; it lost its tracks on mines. Two survivors surrendered.

We breathed freely for a moment in relief. It was an uplifting feeling to see these steel monsters destroyed by individual courage. The incident of the two tanks was forgotten a few seconds later, however.

The reconnaissance company fought for its life off to my right; I could no longer identify its position. Wild artillery fire whirled the muddy earth high into the air. The antitank gun was still in position; it fired round after round into the British 11th Armoured Division's column of tanks. A new barrage converted the gun into a heap of scrap metal. There were no more serviceable antitank weapons. Tank rounds had shredded the company. The first foxholes were overrun. Here and there soldiers were trying unsuccessfully to destroy the tanks with *Panzerfäuste* but in vain. The accompanying infantry warded off every attack on its tanks.

I vainly tried to obtain artillery support. The specter of "lack of ammunition" had been plaguing us for a long time. A couple of German artillery rounds were not enough to check the onslaught. The British tank attack continued.

I felt a burning emptiness in my heart for the first time and cursed this endless slaughter. What was happening at that point had nothing to do with war — it was outright murder. I knew every single one of these young soldiers. The oldest was barely 18. The boys had not learned how to live but, by God, they knew how to die!

Grating tank tracks ended their young lives. Tears ran down my face; I started hating the war.

The rain poured relentlessly down. Heavy clouds moved way above the tortured earth. The British tanks were rolling towards our position at that point. Escape was impossible. We had to stay. Our hands gripped the shafts of our *Panzerfäuste*; we did not want to die without defending ourselves.

A new sound suddenly mixed into the hellish concert. A lone *Tiger* was giving us room to breathe. Its 88 mm rounds gave the Shermans an unmistakable command to halt. The British turned away; they called off their attack in the direction of Mouen.

We found two British tanks on our return to the divisional command post; runners had destroyed them with their *Panzerfäuste*. The wrecks were less than 200 meters from the command post. The staff had dug in for an all-around defense.

The heavy fighting had caused high and irreplaceable losses. A break-through could not be prevented unless we had new units. Our own corps held out the prospect of reinforcements from the *II. SS-Panzer-Korps* for the next day. The corps was very keen that the command post be moved further to the rear. I refused this request. Hubert Meyer supported me. In such a critical situation the place for the commander was up front.

I was informed by the *I./SS-Panzer-Grenadier-Regiment 26* that it had been under continuous attack since morning. There was little of the battalion's combat strength left. In the course of the night, remnants of the battalion fought their way back to Carpiquet airfield. Young *SS-Unterscharführer* Emil Dürr destroyed two enemy tanks with hollow charges during the night. The tanks had been in the Chateau garden. When the hollow charge fell off the second tank, he went back to the tank and held the primed charge in place. The tank was destroyed, and he was fatally wounded. He was posthumously awarded the Knight's Cross. His deed opened the escape route for the rest of the battalion.

The British had still not stopped their attack. I could hear our tanks firing from the direction of Grainville. One *Panzer IV* company , under the command of *SS-Hauptsturmführer* Siegel, covered the change of position of the *II./SS-Panzer-Artillerie-Regiment 12*. The British had already broken into a battery position. The battalion was pulled back through the Salbey sector. The artillery battalion commander — Müller —was killed in hand-to-hand fighting.

Time had lost meaning for us. We worked on a situation map by the glimmer of candlelight and prepared new defensive positions. I waited desperately for reinforcements.

At midnight I had a pleasant surprise. Michel, my loyal Cossack, was suddenly standing in front of me, grinning from ear to ear. I had sent him on leave for a few days. He brought me a letter from my wife telling me of the conception of our fifth child. Michel had been stopped at a movement control center. He was suspected of being a Russian, but who could ever stop my magnificent Michel? In response to my question: "How did you get here?" He replied "Skedaddled!"

The enemy renewed its infantry and armor attack in the direction of Grainville as it was getting light. *Kompanie Siegel* had fended off four attacks by 0900 hours. Several burning enemy tanks littered the countryside. Unfortunately, Siegel's tank was knocked out and he sustained severe burns to his face and hands. Our own tank advance towards Cheux failed because of the strong enemy antitank screen. Although the attack was unsuccessful, a group of 20 men from *SS-Pionier-Bataillon 12*, under the command of *SS-Sturmbannführer* Müller, was rescued by the lead elements and saved from certain capture. It was all that was left of the battalion.

Müller was standing in front of me a few minutes later. His deep, sunken eyes told all. He no longer had an undamaged bit of uniform on him. His knees were bloody and lacerated; his face was hardly recognizable under the dust. One arm was in an improvised sling.

He relayed the drama of his battalion in a few short words. After an enormous artillery barrage by 600 guns against the left wing of the division, the battalion was overrun by tanks of the British 2nd Armoured Division. The battalion fought until it was annihilated. Only a handful of men survived the murderous fight.

Müller himself defended his command post against all the enemy infantry attacks, but he was powerless against the masses of tanks. By midday he and a few men were encircled in his command post. Some tanks were firing into the earth bunker while others were trying to crush it, without success. The engineers had built an ingenious bulwark. It defied all attempts to destroy it.

A captured combat engineer was finally sent into the bunker to ask his comrades to surrender. He preferred to stay there and share his comrades' fate. The attack continued past the command post after demolition attempts had badly shaken the bunker and it looked like a mass grave. The survivors finally fought their way to our lines at about midnight. They were found completely exhausted at Le Haut du Bosq, after having decided to take a short break.

During the course of the day Rauray was lost. The *II./SS–Panzer-Artillerie-Regiment 12* had exhausted its ammunition. The enemy managed to create a bridgehead over the Ordon at Buron in the afternoon. Our radio intelligence intercepted the inquiry: "Do you still insist on a quick operation against Verson?" The enemy was obviously well informed about the location of the division's command post. We did not hear the reply. All available men from the divisional staff were then employed at Fontaine.

A stranger reported to me as I entered the divisional command post. He introduced himself as a civilian official from the *Reich* Foreign Minister's staff and asked me to give him precise information on the situation. The minister could no longer understand the constant withdrawals!

Before I could digest this, tank rounds crashed into our ruin. Enemy tanks were in front of our command post once again. Our command post was empty in no time. Everyone was crouching in a trench with his *Panzerfaust*, waiting for further surprises. I never saw the alleged messenger from the *Reich* Foreign Minister again! What could he have reported to his superiors?

The situation became more and more critical by the hour. The British had managed to establish another bridgehead at Garrus. The enemy was feeling his way slowly but surely south. Up to that point we had only been able to employ outposts of *SS-Panzer-Regiment 12* and patrols of *SS-Panzer-Aufklärungs-Abteilung 12*. Two tanks were destroyed at close quarters by a

Nebelwerfer battery when they tried to overrun the rocket regiment that had been attached to us.

It was obvious that Montgomery intended to cross the Orne at St. Andre and then perhaps advance to the Falaise — Caen road. The hotly contested town of Caen would fall into his lap like a ripe plum through such a maneuver. We hoped to spoil his plan. We had to hold out for a few more days. The *II. SS-Panzer-Korps* was moving up. It had been pulled out of the Eastern Front.

SS-Panzer-Regiment 12 received the order to occupy Hill 112 and prevent a breakthrough to the Orne bridges. A battered tank company was all that was available for that task.

Some tanks and the remnants of the *15. (Aufklärung)/SS-Panzer-Grenadier-Regiment 25* secured the area near Fontaine. The divisional command post was transferred from Verson to Caen. The sector encompassing Verson, Hill 112 and Evrey was assumed by the *II. SS-Panzer-Korps* on 28 June.

Four *SS-Panzer-Divisionen* were at our disposal at that point. They were divisions in name only, however. None of them had the combat power of a division any more. The *9.* and *10. SS-Panzer-Divisionen* were involved in offensive operations in Poland, when they received the order for deployment in Normandy. They had rolled into the battle of attrition for the west from the mud of Poland. The *1. SS-Panzer-Division "Leibstandarte"* reached the front on 28 June. This division was also only a shadow of its former self. It had been pulled out of Russia two weeks before and was supposed to be refitting in Belgium. It was not up to strength, either in equipment or personnel. The commanding general of the *II. SS-Panzer-Korps*, *General der Waffen-SS* Paul Hausser, was supposed to carry out the counterattack on 29 June using those forces. During the night the divisions moved into the assembly areas.

There was calm in the sector of the 12. SS-Panzer-Division. The positions around Carpiquet were improved by the survivors of the *I./SS-Panzer-Grenadier-Regiment 26*.

We were awakened by heavy fire from the battleships on the morning of 29 June. Caen was under fire once again. Fighter-bombers hung like hornets in the clear sky and hurled themselves down at every vehicle. Around 0700 hours I was lying on the road in Verson, a fighter-bomber has set an artillery vehicle on fire and exploding ammunition was flying in all directions. The street was too narrow to pass around it, and we had to wait for the vehicle to burn out. If only we could get out of that one-horse-town! We sat like a rat in a trap among the old walls of the village. An ambulance was on fire. We could not rescue the wounded inside. We watched them burn up before our eyes.

Enemy artillery fire explored the ground around Verson and Hill 112.

Shortly afterwards a massive barrage hit Hill 112. Would the British antici-
pate our plans and attack before we did? With an uncomfortable feeling, I
watched tanks of the British 11th Armoured Division climb the slope south
of the Odon and take Hill 112 in a pincer movement. The summit could no
longer be identified. The impact of heavy gunfire was tossing the Norman soil
around, meter by meter. There was no longer any doubt. The British had
launched a pre-emptive attack. Our divisions were engaged in their assembly
areas by rolling artillery fire and enemy bombs.

The *II. SS-Panzer-Korps* lost Hill 112. The British 11th Armoured
Division's tanks had seized the key to further operations against the Orne
bridges. The whole area could be observed from Hill 112. No movement
escaped British observation. It was not long before we noticed a fire direction
center on top of the hill. The heavy artillery battalion of *SS-Panzer-Artillerie-
Regiment 12* fired on the British forward position on the hill. Our field of
view was magnificent. We were north of the hill and could observe the entire
slope from the Odon to the summit.

The sector in front of the *12. SS-Panzer-Division* was remarkably quiet.
There was only the usual naval gunfire on the road intersections. The tank
accompanying me received a direct hit. A second 380 mm round killed the
attending doctor.

Late afternoon brought with it the certainty that the deliberate attack of
the *II. SS-Panzer-Korps* could not succeed. It was impossible to gain ground
against this superior artillery fire, not to mention the absolute air supremacy.
The attack was bound to fail because of the impossibility of the demands and
the insufficient means at our disposal.

During the night, we made preparations for the all-round defense of
Caen. We could count on the British breaking through west of the city. In the
middle of these preparations orders arrived from the *II. SS-Panzer-Korps* to
resume the attack on Hill 112 and the British 7th Armoured Division.

Nothing surprised us anymore. Max Wünsche quickly joined me and
received orders for the attack. His tanks and the remainder of *SS-Panzer-
Aufklärungs-Abteilung 12* had to execute the attack. We started to treat our
tanks with kid gloves. We had received no replacement tanks up to that point
and our strength melted away daily. The constant use of piecemeal tactics
enraged me. Where had happened to the days of the big armor offensives?

Concentrated fire hammered Hill 112 at dawn. Our tanks pushed close
to the hill in the light mist and took cover before the final assault. It would
start in a few minutes. Wünsche and I smoked a last cigarette. A handshake
— and the dance began!

According to standard practice, the tanks advanced forward to the formerly tree-covered hill, firing their high-explosive rounds into the chaos. The
enemy artillery tried to smash the assault with intense fire, but it failed and

the hill fell into our hands once more.

We soon reached the summit and cut off the withdrawal of a British company in Bren Gun Carriers. They were captured. Burning tanks were from both sides of the hill. There was hardly a square meter of earth on the hill that had not been ploughed up by shells and bombs.

The conquest of Hill 112 gave the *II. SS-Panzer-Korps* some respite. Directed artillery fire from the hill had been eliminated.

The Final Fighting Around Caen

During the recent fighting the British VIII Corps, consisting of the Scottish 15th, the British 43rd and 49th Infantry and 11th Armoured Divisions, had lived off the marrow of the *12. SS-Panzer-Division*. *SS-Panzer-Grenadier-Regiment 26* had been reduced to the combat strength of a weak battalion. *SS-Panzer-Regiment 12* had also suffered severe losses. *SS-Panzer-Aufklärungs-Abteilung 12* had only a mixed company at its disposal, and *SS-Panzer-Pionier-Bataillon 12* was as good as annihilated. The division had lost one battalion of its artillery regiment through naval bombardment and constant fighter-bomber attacks.

The *12. SS-Panzer-Division* could no longer be considered fully operational. The remnants had, at best, the combat power of a *Kampfgruppe*. Despite the enormous stress and heavy losses to which the young soldiers had been exposed during the previous fighting, the formation was considered a fully operational *Panzer-Division* and charged with the defense of Caen.

The British VII Corps was poised on the deep flank of the *12. SS-Panzer-Division*. Montgomery's pincer movement was clearly identifiable. Even a layman could see that the next target for the Allied attack would be Caen. To avoid the risk of being separated from the division and stranded outside the encirclement, the divisional command post was moved to the middle of the town. I wanted to share the fate of my soldiers.

The chief of staff and I did not have any illusions. We knew that executing the *Führer's* order —"Caen is to be defended to the last round." — meant the end of the division. We wanted to fight. We were prepared to give our lives, but the fighting had to have a purpose. I bristled at the thought of allowing my young soldiers to bleed to death in the city's rubble. The division has to be preserved for a more flexible form of combat.

There was only minor activity in the divisional sector, but Canadian tactical battlefield reconnaissance was very noticeable. Reconnaissance patrols were constantly checking Carpiquet and the western edge of the airfield. There was an impression within the division that the Canadians were planning an attack on Carpiquet to break open the front north of Caen.

Carpiquet was an old Normandy farming village with houses built out of dressed stone. The village follows the shape of the terrain and forms a long funnel bordered by the Caen to Bayeux railway line on one side and the airfield on the other. The entire length of the village was visible from the observation post at Ardenne.

I went to visit Carpiquet's defenders. The streets and farmsteads were empty of people. The roads were still negotiable, with rubble only blocking the way here and there. The empty village seemed eerie. I found the "defenders" at the western outskirts. About 50 *Panzergrenadiere* had taken refuge in some abandoned trenches and bomb dugouts left by the former defenders of the airfield. These 50 were the survivors of the *I./SS-Panzer-Grenadier-Regiment 26*. The rest of the battalion had occupied the far side of the airfield. The entire strength of the defense was between 150 and 200 soldiers.

The defenders of Carpiquet no longer had any tank destroying weapons at their disposal. This battalion's antitank guns had been destroyed a few days previously. There were minefields, however, to its front. The *Panzergrenadiere* knew their mission. The platoon leader and his soldiers were to withdraw, fighting a delaying action, to the eastern outskirts of Carpiquet and tempt the attacking Canadians to enter the village. There were 88 mm guns positioned in ambush positions just to the east of Carpiquet. Furthermore, the outskirts of the village were within the fields of fire of positioned tanks.

As a result of the previous fighting, it was no longer possible to reinforce the infantry forces in this sector. The only option for defense was the concentration of all heavy weapons. Our artillery and mortars had already zeroed in on the village.

Lively Canadian radio activity was reported following my return to the divisional command post. Evaluation of it led to the conclusion that enemy forces were assembled at Norrey and St. Manvieu. The radio traffic increased significantly on 3 July.

To take advantage of the possibility of wrecking the enemy attacking units' preparations and, at the very least, of inflicting severe damage upon an enemy presumably assembled in a very confined space, the concentrated fire of the artillery was directed into this area at 0600 hours. We hit their assembly areas with good effect.

While the rockets howled over the airfield, leaving their long, fiery trails behind them, I scrambled over the rubble of the airfield buildings to find Bernhard Krause. Bernhard had chosen a bomb dugout as his command post. From there he could observe the airfield and Carpiquet.

I encountered the forward observer of *Werferbrigade 7* in the bunker. A few grenadiers were about 75 to 100 meters ahead of us. Five tanks were positioned in the ruined airfield buildings. They had to be completely under cover; the fighter-bombers were very active in this area. Bernhard Krause had hardly started to report on the fighting for St. Mauvieu when there were crashes and shrieks all round us. We crowded together in the bunker entrance. The bunker shook as the 38 and 40 cm rounds from the battleships exploded nearby.

The Canadian 3rd Infantry Division had moved out to attack. Carpiquet and the airfield were the targets. What a huge expenditure of resources to destroy a handful of soldiers!

The concentrated artillery fire smashed any intention to defend. There were probably several artillery regiments involved. The naval rounds spun entire hangars into the air. The village could not be identified at the moment. Thick clouds of smoke lay to the west. Above us the Typhoons were looking for their victims. Their rockets could hardly be heard above the explosions of the heavy rounds. The forward observers seemed not to be affected and remained at their scissors telescopes; they were requesting final protective fires.

The Canadian 8th Brigade, reinforced by the Winnipeg Rifles and supported by the Fort Gary Horse Tank Regiment, threw itself at the remnants of Krause's battalion. The artillery fire shifted. Pale faces gazed at me. Nobody said a word. We could only hear the voices of the forward observers. Enemy tanks were rolling forward out of Marcelet. The haze was very thick. Our artillery rounds landed among the advancing tanks, but the fire scarcely seemed to disturb them. They rolled slowly towards us. The grenadiers were waiting in their bomb dugouts for the order to occupy their firing positions. Everyone had taken cover during the barrage. The first of the enemy infantry was starting to appear out of the woods. Our artillery fired at the edge of the woods, inflicting heavy casualties on the Winnipeg Rifles.

We were still crouched under cover and only our tanks had started to retaliate. Our tanks could not be identified; they were seemingly covered by rubble of the airfield buildings. We could hardly endure the tension any longer as we listened to the noise of battle and awaited the first burst of fire from our forward machine-gun posts. I spotted flamethrowers being employed offensively at the western outskirts of the village. Mounted on small tanks, they were advancing under cover of the Shermans. One of them was caught in the minefield and was ablaze.

The 50 *Panzergrenadiere* at Carpiquet were under attack by three infantry battalions supported by tanks. The fighting in the village was bitter. Rubble obstructed the progress of the enemy tanks. A tank assault across the airfield failed as the area was covered by our well-camouflaged tanks and a battery of 88's. Only a few minutes had passed. The Winnipeg Rifles advanced hesitantly, not seeming to trust the empty battlefield. They moved slowly towards the first airfield building. At that point they were still about 150 meters from the hall. They had left the protective woods and were exposed on the airfield at that point. Then we heard the "voice" we had been waiting for:

257

"Rat-tat-tat-tat-tat-tat". Our *MG 42* mowed the enemy down.

I jumped into a corner. The soldiers dashed out of the bunker. Not a word was spoken — they all leapt up and ran to take up their old positions. Infantry fighting was the order of the day. With their sleeves rolled up and eyes directed to the front, they loaded and fired their weapons automatically.

The attackers must have sustained heavy losses. The momentum of their attack had been broken and their tanks had started to take cover. *Bataillon Krause* had also taken casualties. The wounded were dragged into the bunkers and taken care of. Things were not going well on the far side of the airfield. The Canadians were gaining ground and fighting was already taking place in the middle of the village. Our artillery had been concentrating on the western part of the town. I telephoned the chief of staff and prepared him for the loss of Carpiquet, but I had no worries about the southern part of the airfield. The remnants of the *I./SS-Panzer-Grenadier-Regiment* 26 would be able to hold their positions. The battalion commander was once again the backbone of the defense as so often in the past. Bernhard Krause was the premier grenadier of his battalion. He spoke to his soldiers like a father in his deep, quiet voice. There were no unpleasant surprises to expect from this section of the battle line.

I took my leave and worked my way over to the demolished hangars on the eastern edge of the airfield. Erich Holsten was waiting for me there. In a few minutes we were back at the divisional command post and we breathed easier again. It really wasn't much fun to move through enemy artillery fire in a *Volkswagen*.

Our signals intelligence section was working superbly. Those guys had earned some praise. As a result of their monitoring we were well informed about enemy movements. This was especially true in the fighting for Carpiquet. The commander of the de la Chaudiere Regiment reported the capture of the town by radio to his brigade from the center of town. He was ordered to return but our artillery and mortar fire held him fast. Every time that he announced his departure another bombardment followed.

Only about 20 of the *Panzergrenadiere* who had defended Carpiquet so obstinately were still fit for action. Not a single noncommissioned officer had survived. The survivors had taken over as security for the 88 mm battery that was positioned just to the east of Carpiquet. *SS-Bataillon Weidenhaupt* attempted a counterattack on Carpiquet during the night of 4/5 July but failed. The enemy in Carpiquet suffered considerable losses through several high-explosive and napalm barrages. Even during the offensive launched against Caen on 8 July, the enemy at Carpiquet remained on the defensive. We held the airfield until 8 of July.

After the enemy had failed to enlarge the Odon bridgehead and break through to the Orne, he had also failed in his attack from the west to take the airfield. We therefore assumed that he would then try to collapse the cornerstone of the German defense by a frontal attack and drive deep into the Normandy countryside. We prepared ourselves for the final fighting for Caen.

After having visited all the units in their positions during the past few days and spoken extensively with enlisted personnel, noncommissioned officers and officers about the continued defense of Caen, I was certain that the town would become our brave division's coffin. Defense of the town was no longer possible. The force ratios were too unequal. The weakened German forces were incapable of a defense in depth and there were no readily available reserves.

The division alerted the corps in no uncertain terms that the worn-out divisional forces were insufficient to hold their ground against such superior enemy forces. The corps could not place any additional forces at our disposal, however.

We made all necessary preparations to meet the expected attack as effectively as possible but had no answer to the question: What would happen if airborne troops were dropped to the rear of the division and entered the undefended part of the city?

The division was convinced that the offensive against Caen would be initiated by an airborne operation south of the town with a simultaneous advance from the Odon bridgehead via the Orne in the direction of the Caen — Falaise road. A complete breakthrough of the German frontline could not be prevented and the road to Paris would then be open.

On the evening of 7 July we realized that the next 24 hours would decide the fate of Caen. About 500 Lancaster and Halifax bombers joined the attack during the late evening and dropped 2,500 tons of bombs on the town's northern outskirts. The close-flying aircraft formations suffered negligible losses from our *Flak*, but we also had no casualties to report. The units themselves not been affected by the attack. However, the streets of Caen were blocked and the civilian population again had to make a dreadful sacrifice. The hospitals were overflowing.

It must be said here that there was friendship and a willingness to help each other in the relationship between the German forces and the French population. Up to that point there had been no rancor or animosity on the part of the French. They looked at the rubble of their homes in confusion and with a shaking of their heads, unable to comprehend the destruction of their town. They knew that Caen had not had any German units within its walls on that day or on 6 June. During all the fighting in the area the units did not have to detail a single soldier for security duties. The French took care of discipline and order themselves.

The air raid seemed to be the prelude to the main assault. Every last soldier was at the ready. The artillery stood by for orders to lay down a curtain of fire in front of our own lines. We waited. The phones were silent. We stared tensely out into the night and waited for the enemy ground force to attack. Minutes passed without the silence being broken. It was inconceivable, but true. The Allies had made no attempt to exploit the tremendous bombing operation.

I visited the commanders once again to determine the effect of the bombing raid for myself and found, with astonishment, that I had overestimated its impact on morale. The troops hated the fighter-bomber attacks far more than the mass bombardment of those cumbersome juggernauts. Indeed, the front line was so sparsely manned that a bomber attack couldn't cause much damage. Two thousand five hundred tons of bombs had merely succeeded in overturning a few *SPW*.

The troops were expecting a big attack and readied themselves for the inevitable. We had no false illusions as to the outcome of the fighting. The corps was again informed of the division's hopeless situation. I waited for the already dawning day with anxious misgivings. Hubert Meyer had fallen asleep at the map table. What an excellent chief of staff I had found in that comrade!

*

Artillery fire of unimaginable intensity from both ground and naval forces fell on the front line of the *12. SS-Panzer Division*. Our cellar shook at all its corners. Plaster and dust settled on the candlelit map. Our artillery and mortar formations laid down final protective fires. We had been "procuring" ammunition for days and were trying to give tangible aid to the heavily engaged infantry. Fighter-bombers were throwing themselves at our artillery positions and attacking every vehicle on the roads. The Orne bridges were continuously under attack.

The first reports came in. All battalions were heavily engaged in defensive fighting. The enemy was attacking with strong tank support across the entire front. Our neighbor to the right, the *16. Luftwaffenfelddivision*, was not equal to the task. It was shattered by a renewed bombing attack and this improvised division's will to resist collapsed under the weight of the enemy's destructive power. The British 3rd Infantry Division pushed into the *Luftwaffe* division's lines and soon threatened our division's deep flank.

Four dazed battalions defended our division's sector while the enemy attacked with the British 59th and Canadian 3rd Infantry Divisions, which were reinforced by brigades of tanks.

The *Schwerpunkt* of the attack seemed to lie with the 59th Infantry Division in the sector of the *I./SS-Panzer-Grenadier-Regiment 25*. Furthermore, this battalion was under attack by elements of the British 3rd Infantry Division. The battalion lost almost all of its company-grade officers in the first hour of the attack. *SS-Sturmbannführer* Waldmüller, the battalion commander, positioned himself in the midst of his unit and was the heart of

260

its resistance. The *1./SS-Panzer-Grenadier-Regiment 25* was screening the right flank, and its fire was disrupting the relatively easy advance of the Canadian 3rd Infantry Division into the sector of the decimated *Luftwaffe* division.

The brave *I./SS-Panzer-Grenadier-Regiment 25* stood like a breakwater on the battlefield. Unshaken, despite the enemy's enormous superiority in men and equipment, it warded off every attack. The enemy failed to overrun the battalion in this first assault.

The *II./SS-Panzer-Grenadier-Regiment 25* also put up a heroic resistance; its antitank guns having long been destroyed by artillery fire. It only had *Panzerfäuste* at its disposal. All of its company commanders had also been killed. *SS-Hauptsturmführer Dr. Tiray* had destroyed three Sherman tanks by himself and was killed while trying to dispatch a fourth.

The *III./SS-Panzer-Grenadier-Regiment 25* was under attack by the Canadian 3rd Infantry Division, fighting in the ruins of Buron and Authie, where the battle was especially hard and bitter. The Canadians had not forgotten that their advance was halted at Buron and Authie on 7 July, and they had had to pay a heavy and bloody price. The *Panzergrenadiere* of the battalion clung to the ruins, fighting fiercely for every inch of ground.

I did not understand why the Canadians did not continue their attack from the direction of Carpiquet. We had only an 88 mm battery and the remnants of the *I./SS-Panzer-Grenadier-Regiment 26* opposite Carpiquet. That battalion had been bled white. An energetic thrust from there towards the Orne Bridge in Caen would have sealed the fate of the *12. SS-Panzer-Division* within a few hours. The division's only immediate reserve was a newly arrived tank company with 15 *Panthers*.

Urgent reports from the front started to stack up. The *16. Luftwaffenfelddivision* seemed to had been swept off the face of the earth. Our division immediately sent part of the *II./SS-Panzer-Regiment 12* and the divisional *Begleitkompanie* to secure the area at Cabaret to the northeast of Caen. The *I./SS-Panzer-Regiment 12* was fighting at the northern outskirts of the city.

Fighter-bombers continuously attacked the Orne bridges and the approach roads south of Caen. Any movement into Caen had become impossible. We could not evacuate our wounded or receive supplies. The roads had become death runs. The bombers roared through the sky once more from the north, aiming for the town. We could hardly believe that this tormented town was to suffer yet again.

The enemy expenditure of men and equipment in capturing the town was scarcely imaginable. The gods only know why this unoccupied town was being razed to the ground. With the exception of the staff of the *12. SS-Panzer-Division*, which had only been there for the past couple of days, there had been no formations in the town. The first wave of bombers made for the bridges, causing fires south of the Orne. The town center was once again blanketed with bombs. Caen was enveloped in flames, smoke and ashes.

We suddenly saw the last wave of bombers heading for the garrison church and releasing its bombs. I jumped through the cellar entrance of our command post and threw myself into the farthest corner of the cellar. A tremendous noise shook the vault and the candles went out. I couldn't breathe. I could barely see my hand in front of my face through the dense dust cloud. Hubert Meyer called out to me and more voices could be heard. Suddenly a soldier screamed out: "We've been buried alive, we've been buried alive!" The young soldier could only be calmed down with some effort. The outside concussion had hurled him through the open door into the cellar.

The garrison church, only 50 meters from the command post, had been completely destroyed. A big crater was all that we could see at that point. Stone blocks had whirled through the air, falling on the camouflage netting under which were our radio vehicles, thus destroying all our radio communications. We soon overcame this disruption to our communications network. The civilian population had suffered heavy losses once again.

A few minutes after the bombing, the commander-in-chief of the *5. Panzer-Armee, General der Panzertruppen* Eberbach, appeared at the divisional command post. *General* Eberbach was *General* Geyr von Schweppenburg's successor.

He had managed to get through in the lull between bombing raids on the Orne bridges. The commander-in-chief expressed his appreciation for the division's performance. He was still unaware of the catastrophe regarding the *16. Luftwaffenfelddivision. General* Eberbach immediately recognized the seriousness of the situation and ordered the commitment of the *21. Panzer Division* in the *Luftwaffe* division's sector. A reinforced battalion of the *21. Panzer-Division* was all that crossed the Orne that day.

Alarming reports arrived while the commander-in-chief was still with us. The enemy had broken through between the *II.* and *III./SS-Panzer-Grenadier-Regiment 25*, that is, between Galmanche and Buron. He had taken Contest and his weapons were controlling the approach to the Ardenne monastery.

The *II./SS-Panzer-Regiment 12*, with the exception of those units employed east of the railway in the sector of the *16. Luftwaffenfelddivision*, was immediately committed into a counterattack. It threw the enemy back but could not retake Contest. The enemy's tank superiority halted the counterattack.

General der Panzertruppen Eberbach took his leave. I was convinced he would do everything to prevent further deaths in the rubble of Caen. The fighting continued with the same intensity. It was a mystery to me why the

Canadians and British were advancing so hesitantly. Their enormous tank superiority was hardly fulfilling its potential. Instead of pushing their tank formations quickly and deeply into our defenses and creating a bridgehead over the Orne, they were only using tanks to support their infantry operations.

With the exception of the extremely agile and well-led artillery, the attacker lacked momentum and initiative on the battlefield and was conducting the assault on Caen along tactical principles employed in World War I. You could only afford to conduct such warfare against an army that had already been bled white.

The commander of the *16. Luftwaffenfelddivision, Generalleutnant* Sievers, appeared at the command post of the *12. SS-Panzer-Division* and asked for a briefing on the situation. He had been out of contact with his units for some hours. The report hit him hard. He immediately went to the northeastern outskirts of Caen to make his own, on the spot, assessment. He tried to reassemble his scattered division and form a stable front. The fighting morale of his units was too far gone for that, however. The steamroller trundled on. Slowly but steadily, the battlefield was being turned into a cratered landscape.

During the afternoon the enemy took Gruchy. After a long and bloody fight, the *16. (Pionier)/SS-Panzer-Grenadier-Regiment 25*, which had defended it, was entirely wiped out. The brave engineer company, under the command of *SS-Obersturmführer* Werner, was annihilated. The only one from the company I ever saw again was a runner. The engineers died in their positions.

After seesaw fighting, Authie and Franqueville were lost. During an immediate counterattack, the commander of the *III./SS-Panzer-Grenadier-Regiment 1, SS-Obersturmführer* Weidenhaupt, was wounded. The rest of the battalion brought the enemy' attack to a halt north of Ardenne.

The division's situation was extremely serious. The three battalions of *SS-Panzer-Grenadier-Regiment 25* were on their own, almost encircled and fighting bitterly in Malon, Galmanche and Buron. Wire communications had been destroyed. The only means of communication was the radio. The division's front was stretched to the breaking point; reserves were no longer available. Only the 15 *Panthers* of von Ribbentrop's company were on the reverse slope just north of the town.

I couldn't stand it any more! I wanted to see the situation for myself and make the necessary decisions in the middle of the fighting. The *Führer's* order not to give up Caen could no longer be executed. We could hold out for perhaps a couple of hours longer, but there wouldn't be any survivors from the division. I struggled against allowing the division to be destroyed. Hubert Meyer supported my intention to leave the destroyed town to the Allies without further fighting and to withdraw the division to the eastern bank of the Orne.

Erich Holsten had saddled his swift "horse". Our good *Volkswagen* was ready to go. Michel, my loyal Cossack, was already sitting in it when I got in. We all knew we would have a wild ride ahead of us. In a few minutes we reached von Ribbentrop's *Panther* company. The foremost tanks were already in action against Shermans in Contest.

In front of us was the Ardenne monastery. The entire complex was under artillery fire and the tall towers no longer existed. Their stumps stretched accusingly towards the sky. On the reverse slope I suddenly felt worried and took the wheel. In this situation, there was no stopping or turning. Impacting rounds had torn up the trail to the monastery; bomb craters covered the battlefield.

We had scarcely left the last high ground behind when rounds flew around us. The tanks in Contest had taken us under fire. Cold sweat came out of every pore as the vehicle virtually flew over the ground. If only the dammed chirping of enemy machine guns had not been there! Only a few meters lay between us and the ruins. We did it. Direct fire could no longer reach us.

The monastery's orchard looked like an inferno. Round after round exploded in front of the regimental command post. We hesitated for a few seconds before we set out for our final dash. Taking advantage of a pause in the shelling, we rushed toward the building. Dead soldiers were scattered around the vicinity of the headquarters. While leaping out of the car, I recognized the body of the commander of the headquarters company. Shrapnel had killed him.

We stumbled gasping into the old building to find the commander of *SS-Panzer-Grenadier-Regiment 25* in the cellar. He was wounded and speaking to *SS-Hauptsturmführer* Steger, the commander of *III./SS-Panzer-Grenadier-Regiment 25*. The radio was the only means of communication with the battalions. The ceiling seemed to be moving above us, even though it was deeply embedded in the earth and was supported by enormous arches. We heard a continuous booming. I spoke to *SS-Hauptsturmführer* Steger in Buron. He reported most of his battalion had been killed in action and enemy tanks were outside the village. He requested urgent help. All available tanks were sent to Buron in order to break a hole in the encirclement. The attack failed. From the church tower I observed the seesaw tank engagement. Both sides sustained heavy losses.

Enemy tanks pushed towards Ardenne from Authie. Von Ribbentrop's company destroyed three tanks in the defense of the regimental command post. The burning tanks were 100 meters west of Ardenne.

More and more wounded dragged themselves into the monastery's big cellar. Medics performed superhuman feats in saving their wounded comrades. My long-time fighting comrade, *Dr.* Erich Gatternig, worked tirelessly to overcome the misery and the pain. One could hardly stand the wailing in the old vaults. The stream of wounded did not stop.

We couldn't give up the fight! We had to wait for night to evacuate our wounded comrades under the cover of darkness and give our forward elements a chance to break out.

I sat in a *Panther* and was moving towards Cussy. Cussy was being defended by the battery commander of the *1./SS-Flak-Abteilung 12*, *SS-Hauptsturmführer* Ritzel. This small town was only a heap of rubble. Three burning Shermans were in front of the battery's position. The battery's losses were high. One gun had been put out of action by artillery fire. *SS-Hauptsturmführer* Ritzel served as a gunner on a piece. He promised me he would do everything to hold the position until nightfall and thus enable the evacuation of the wounded from Ardenne.

Soon after that, I was back in the monastery. Enemy infantry and tanks had broken into Buron at that point. I could not identify Steger's command post due to the smoke and explosions. Flamethrower tanks were raging around the positions of the *III./SS-Panzer-Grenadier-Regiment 25* . Burning *Panzergrenadiere* jumped into the air and then collapsed. Flamethrower tanks were the most feared weapons. Because the small Bren Gun Carriers only worked under the covering fire of their big brothers, they were very difficult to engage.

Enemy tanks overran Steger's command post and the battalion staff ceased to exist. The fighting only continued in the western part of the village. *SS-Standartenführer* Milius, commander of *Panzer-Grenadier-Regiment* 25, was given orders to evacuate the monastery after moving out the wounded and occupy positions on the outskirts of the city. It was my intention to withdraw the remnants of the division to the eastern bank of the Orne during the night.

The enemy target practice at our *Volkswagen* started again. Erich Holsten drove and I clung on to the side. With a lot of luck, we reached the rubble of the city. After my return from Ardenne, I reported the critical situation to the corps and urgently asked for permission to withdraw the remnants of the division to the eastern bank of the Orne. I left no doubt that Caen could not be held without the remnants of the division bleeding to death. The corps turned down the request. The *Führer* had ordered that the town had to be held at all costs! All protests and my reference to the senselessness of further sacrifices achieved nothing. We were to die in Caen!

I flew into an enormous fury when I thought of the brave soldiers who had been fighting day and night for four weeks and who were to be needlessly sacrificed. I refused to carry out the untenable order and started the evacuation of the city. The heavy weapons immediately occupied new positions on the eastern bank of the Orne. After the onset of darkness, the battalions were withdrawn to the edge of Caen. The way had to be cleared for them by the employment of a small group of tanks. The *III./SS-Panzer-Grenadier-Regiment 25* had some 100 enlisted soldiers and noncommissioned officers left.

All of the officers had been killed, wounded or missing in action.

The battlefield had become quiet. We were happy the Canadians were inactive. Had they continued their attack at night, the division would have been completely annihilated. The Canadians entered the Ardenne monastery and prevented the continued evacuation of the wounded. Soon after midnight, *SS-Standartenführer* Milius requested artillery fire on Ardenne in order to gain some breathing space. As this measure was the only option for forcing a way clear to our wounded comrades, I give permission to the rocket battery to fire two salvos at the monastery. Our observer in Ardenne directed the fire. The enemy withdrew. After all the wounded had been evacuated, the monastery was abandoned.

SS-Standartenführer Milius reported the evacuation of Ardenne at midnight. The survivors had occupied new positions at the edge of the city. The crews of the 88 mm battery at Cussy died at their guns. *SS- Hauptsturmführer* Ritzel died in hand-to-hand fighting at his battery position. Their heroic fight enabled the evacuation of their wounded comrades.

Shortly after midnight, I assembled all the commanders and told them of my decision to evacuate the town during the course of the night and occupy new positions east of the Orne. The commanders were relieved. They unanimously supported the intention of the division to evacuate destroyed Caen without fighting.

At 0200 hours I was searching for the *I./SS-Panzer-Grenadier-Regiment 25* on the northern outskirts of Caen. The rest of the battalion had to fight its way through enemy forces, leaving a bloody path behind them. The battalion's losses were shocking. *SS-Obersturmführer* Schünemann's platoon was defending itself in a group of farmhouses. It was impossible for the *Panzergrenadiere* to break through to the rear. According to radio intercepts, this lost band was still fighting 48 hours later. It was then annihilated in a fighter-bomber attack.

I found the survivors of the *I./SS-Panzer-Grenadier-Regiment 25* in a bomb dugout at the edge of the city. The soldiers, totally exhausted by the fighting, had fallen into a deep sleep. The officers had taken over guard duty. Stragglers stumbled into the bunker and collapsed into whatever small space there was left. What luck that the English and Canadians were not in pursuit! The soldiers of the *12. SS-Panzer-Division* were at the end of their physical endurance. They had fought for weeks in the front line without replacements and had suffered the hammer blows of modern attrition warfare.

They had gone to war weeks before with fresh, blooming faces. At this point, camouflaged, muddy steel helmets cast shade on emaciated faces whose eyes had, all too often, looked into another world. The men presented a picture of deep human misery. But it was immaterial; they couldn't be allowed to rest any longer. They had to defend the eastern bank of the Orne. Waldmüller received his new orders and tore his men out of their leaden sleep. Every

Panzergrenadier had to be woken individually. They staggered drowsily out of the bunker and hung their ammunition around their necks once again; the heavy machine-gun belts dragged the half-awake soldiers forward. Swearing, they hitched themselves to two heavy infantry guns and turned back towards the burning town. Two German tanks guarded the northern approaches.

During my absence, the chief of staff had vainly tried again and again to obtain permission from corps to evacuate Caen. The corps finally ordered the evacuation of the city around 0300 hours.

Because the withdrawal had already started and the heavy weapons had already occupied their new firing positions east of the Orne, the evacuation could be carried out quietly and undisturbed by the enemy.

New positions were occupied in the sector stretching from the Caen railway station to the bend in the Orne at Fleur sur Orne. The men were exhausted and unable to begin improving the new positions. After crossing the Orne and reaching their new positions, the soldiers fell into a deep sleep, relying on their comrades guarding the northern outskirts of the city.

In the morning, the *2./SS-Flak-Abteilung 12* left its position on the western outskirts of Caen, overlooking Carpiquet. Even there, only a few hundred meters east of Carpiquet, there was still no enemy contact. At 0440 hours the division staff left Caen and established its command post at Carcelles. The relief of the division by the *272. Infanterie-Division* was expected. The new command post lay hidden between some ancient beech, oak and elm trees. A neat Norman mansion, dreamily sheltered in the park, offered us peace and quiet. Unfortunately, we did not have any opportunity for rest. At least I was able to enjoy a couple of buckets of water and scrubbed myself from head to toe.

By 0800 hours I was again visiting units in the southern part of Caen. The soldiers and officers lay like corpses in the gardens on the bank of the Orne. They had sunk into a death-like sleep. The units had reached the end of their strength.

It was only in the afternoon that enemy reconnaissance patrols felt their way towards the city. At midday the last outposts of the *12. SS-Panzer-Division* and the 21. Panzer-Division had crossed the Orne. After the commander of the III./*SS-Panzer-Grenadier-Regiment 26* — Olboetter — had crossed the Orne, the last bridge was demolished. Towards evening the first rounds were exchanged across the Orne. Three Allied divisions had taken the northern part of Caen.

On 11 July the division was relieved by the *272. Infanterie-Division*. This was carried out without enemy interference. The enemy had only conducted reconnaissance patrols in the division's sector. It was only for that reason that the completely exhausted formations were able to establish defensive positions.

From the start of the invasion to the evacuation of Caen on 9 July, the division had suffered heavy losses of men and equipment. More than 20% of the soldiers had been killed in action and more than 40% reported as wounded or missing. The military esprit of the young soldiers cannot be better summarized than by the words of a former opponent: "The *12. SS-Panzer-Division*, which defended this sector, fought with a toughness and intensity which was not encountered anywhere else during the entire campaign."

From The Evacuation Of Caen To The Falaise Pocket

After the bloody fighting around Caen the *12. SS-Panzer-Division* was transferred to the area around Potigny, north of Falaise, for refitting. *SS-Panzer-Artillery-Regiment 12* and *SS-Flak-Abteilung 12* were attached to support the *272. Infanterie-Division*.

As a lengthy refitting in an area close to the front was out of the question, the staffs of the *SS-Panzer-Grenadier-Regimenter* were moved to the area around Vimouthiers. They had the mission of creating provisional companies out of the replacements who were arriving and those wounded who were again fit for duty.

The remnants of the *Panzer-Grenadier-Bataillone* were consolidated into two *Kampfgruppen*. Some *Panzerkompanien* were transferred to the Le Neubourg area for refitting. There was little in the way of recuperation. We worked feverishly on the units' combat readiness and planned and implemented resupply.

I was ordered to report to the *I. SS-Panzer-Korps*. Erich Holsten had left me a few days before in order to have an operation. The young soldiers wanted to give me a long-time comrade as Erich's successor. They formed the idea of having Max Bornhöft transferred from *SS-Panzer-Aufklärungs-Abteilung 1* to our division. He had assisted me from 1940 to 1943. The surprise inaugurated by my magnificent soldiers was complete. Accompanied by a hail of greetings from the runners, Max and I shook hands. We were sitting side by side again after exactly one year's separation.

Fighter-bombers hunted us on the journey to the *I. SS-Panzer-Korps*. The dead-straight road from Falaise to Caen was permanently patrolled by fighter-bombers and was only used by a few dispatch riders. Logistics traffic was non-existent. The formations could only be supplied by night.

The *I. SS-Panzer-Korps* had moved its command post into a densely vegetated and wooded area south of Bretteville sur Laize. I reported to the commanding general of the corps. I was more than an hour late. Suddenly, in complete surprise, I was standing face-to-face with the commander-in-chief in the west, *Feldmarschall* von Rundstedt. The commander-in-chief and Sepp Dietrich were sitting in the shade of a tree and had harsh words regarding the continuous interference from the *Oberkommando der Wehrmacht*.

The elderly *Feldmarschall* expressed his gratitude to the *12. SS–Panzer-Division*. He noted with regret the division's irreplaceable losses and again expressed his admiration for the young soldier's professional bearing. In a few words he compared the youth of Langemarck with the youth of Caen. He said: "Your soldiers had the passion of the young regiments of Langemarck but they are far superior to them in training, especially in that they are led by veteran officers and noncommissioned officers. It is terrible that these trusting youths are being sacrificed in a hopeless situation."

During lunch, I listened with astonishment to the *Feldmarschall* and Sepp Dietrich openly condemning the conduct of the war in Normandy. During the course of the conversation it became apparent that there was agreement between the commander-in-chief, the commanding general and myself on the impossibility of the present situation.

On 17 July the division was alerted; the enemy had broken through the positions of the *272. Infanterie-Division* between Maltot and Vendes. The enemy was repulsed with a counterattack and denied a breakthrough to the Orne. About 50 prisoners were taken during the operation.

During the early afternoon, I was surprised to be ordered to the *I. SS-Panzer-Korps* to report to *Feldmarschall* Rommel. The *Feldmarschall* expressed his recognition of the division and regretted he could not visit us due to lack of time. At the end of the visit, he asked me for an evaluation of the situation. I replied:

A British offensive south of Caen can be expected in the near future. The objective of the attack will be to smash the right wing — the critical point of the front in Normandy — to enable them to advance into the heart of France. The units will fight and the soldiers will continue to die in their positions, but they will not prevent the British tanks from rolling over their bodies and marching on Paris. The enemy's overwhelming enemy air supremacy makes tactical maneuver virtually impossible. The fighter-bombers even attack individual dispatch riders. Redeployment of the smallest units, let alone the formation of a *Schwerpunkt*, cannot be executed without serious losses because of continuous air coverage. The road network is under their control day and night. A few fighter-bombers are enough to delay or even stop movements. *Herr Feldmarschall*, give us an air umbrella, give us some fighter units! We are not afraid of the enemy ground forces; we are powerless against the massed employment of the air force, however.

It would have been better not to have made that last request. I saw that I had touched on a sensitive area. The *Feldmarschall* said excitedly:

Who are you telling this to? Do you think I move around the countryside with my eyes closed?...I have written report after report. I was already pointing out the destructive effectiveness of the fighter-bombers in Africa...But the higher-ups know better, of course...They don't believe my reports any more!...Something has to happen!...The war in the west has to end!...But what will happen in the east?"

The *Feldmarschall* and I walked back and forth for a few minutes before he bid me a fond farewell. Sepp Dietrich asked the *Feldmarschall* to be careful and avoid the main road. He suggested his big car be exchanged for a *Kübelwagen*. The *Feldmarschall* waved away the suggestion with a smile and drove away. He was attacked and wounded shortly afterwards at Foy de Montgomery

South of the Orne were Faubourg de Vaucelles and Colombelles, suburbs of Caen. They were modern industrial complexes surrounded by housing areas for the workers. Immediately south of those housing areas were the rich, fertile fields of Normandy. They stretched as far as the old town of Falaise, the birthplace of William the Conqueror.

The terrain between the two towns climbed slowly but steadily and reached a height of 200 meters on both sides of Potigny. The heights were covered with woods and allowed a view to the north. Immediately south of the range of hills the Laison River cut through the countryside. Caen and Falaise were connected by Route National 158, a straight road that bent slightly at Potigny. Scattered woods lined both sides of the road.

It was Montgomery's intention after Caen's capture to break out of the bridgehead and reach the heights between Falaise and Caen. In order to realize this plan, the British VIII Corps with three armored divisions and the Canadian II Corps with two infantry divisions and a tank brigade were staged. The attack was to be supported by the US 8th Air Force and the 2nd Tactical Air Force.

Opposing these superior forces were the 272. Infanterie-Division (without a single tank or heavy antitank gun), the badly battered *21. Panzer-Division* with remnants of the *16. Luftwaffenfelddivision* and elements of the *1. SS-Panzer-Division*. The two *Kampfgruppen* of the *12. SS-Panzer-Division* were in reserve at Potigny.

The German leadership expected the enemy's big offensive south of Caen in the near future. The attack at Maltot was only viewed as a diversionary maneuver. In order to oppose an enemy breakthrough to the east, one *Kampfgruppe* of the *12. SS-Panzer-Division* was moved to the vicinity of Lisieux.

I visited the commander-in-chief of the *5. Panzerarmee* on the evening of 17 July. *General der Panzertruppen* Eberbach was convinced that the expected attack would take place during the next few hours. All units in the Caen sector were put on alert.

In the early hours of 18 July the earth south of Caen started to tremble. The Allied air force had launched the offensive with the dropping of 7,700 tons of bombs. Fighter-bombers attacked the artillery positions and the roads immediately behind the front. The first bombs tripped the alarm in the *Kampfgruppe*. The *Panzergrenadiere* leapt onto their vehicles and rubbed the last sleep from their eyes. They didn't ask questions. There was hardly any conversation. They prepared themselves silently for the next fight. We had no illusions. Officers and men knew the futility of fighting. They awaited their operations orders in silence but with a will to fulfill their duty to the bitter end.

The *Kampfgruppe* was employed on both sides of the Cagny — Vimont road in the sector of the tenaciously fighting *21. Panzer-Division*. The enemy tanks were halted at Frenouville and all further attacks were fended off with heavy enemy losses. The rest of the *12. SS-Panzer-Division* had to take over the sector of the *21. Panzer-Division* on both sides of the Cagny — Vimont road during the following night.

In the neighboring sector, the *1. SS-Panzer-Division* destroyed more than 100 tanks of the British 11th Armoured Division. Jochen Peiper had saved the day again with his *Panthers*. Montgomery's large-scale offensive had not achieved its aim. The high ground, which had been his objective, was still under the defenders' control.

The fighting being conducted resembled exactly that of the previous fighting in Normandy. Magnificent planning and enormous amounts of equipment followed by a hesitant tank attack with no momentum or drive. Up to that point, British tank units had only occupied broken terrain. Where was the spirit of the Light Brigade at Balaclava in the Crimean War? The enemy tanks crawled across the terrain like turtles; their massed power was not exploited.

The division's positions were improved as quickly as possible. The enemy did not continue his attacks in our sector. On 20 July I visited the division's positions with the commander of *SS-Panzer-Aufklärungs-Abteilung 12*. I reconnoitered a fall-back position on a line from Vimont to St. Sylvain. The new position was immediately dotted with strongpoints. We could no longer allow ourselves the luxury of a continuous defensive system. The division's combat strength was, at best, that of a reinforced regiment.

I returned to the divisional command post at about 1900 hours and was informed about the attempted assassination at the *Führer's* headquarters. (It is not correct that we were informed by a military office in Paris, as was later maintained. We were informed by neither side. The information came exclusively from radio news reports).

The attempt had no influence on the relationship between the Army and *Waffen-SS* units. There was no difference of opinion among the combat units. The terrorist act was rejected equally by all units. The soldiers had no sympathy for the 20 July conspirators. They were longing for an end to the war and were themselves searching for ways and means to end the futile struggle. However, at no time were they ready to break their soldier's oath.

Early on the morning of 21 July, the commander of *Kampfgruppe Waldmüller* reported that *Feldmarschall* von Kluge had almost driven beyond the frontlines in the sector of the *Kampfgruppe* and was inspecting frontline positions. *Feldmarschall* von Kluge was trying to form his own impression of the state of his forces and had chosen the *12. SS-Panzer-Division* for that purpose.

The *Feldmarschall* familiarized himself with the situation and agreed with my assessment. He expressed his gratitude for the admirable soldierly bearing of our young soldiers and announced we would be relieved by an infantry division soon.

Von Kluge revealed himself to be very open minded and was completely candid with me. He considered the situation in Normandy to be very critical. He sharply criticized the static defense of the shattered Normandy countryside. The *Feldmarschall* stayed at command post for a few hours and spoke to the commander-in-chief of the *5. Panzer-Armee, General der Panzertruppen* Eberbach, as well as to the commanding general of the *I. SS-Panzer-Korps*, Sepp Dietrich, and to the commander of the *21. Panzer-Division*, Feuchtinger, who had all arrived in the meantime. After inspecting the front, von Kluge sent a comprehensive report on the true situation to Hitler.

During the previous week the enemy had conducted raids in the division's sector. Radio intercepts gave us reason to expect an attack along the road to Vimont. *Generalmajor* Peltz, the commander of all combat aircraft on the Western front, unexpectedly visited the division in order to coordinate *Luftwaffe* and ground troop operations for operations against the enemy's front line. The fighter units had to take off from airfields in Holland and Belgium. There were no forward air controllers available to direct the aircraft. Communications could only be achieved via signal flares. The formations had to reach the front flying at low altitude. We had major concerns whether such measures could be accomplished.

A few days after this reconnaissance, the first operational sortie of 20 to 30 machines followed. The units could hardly believe that German aircraft had finally appeared almost two months after the invasion. The machines flew in over the front at about 50 meters. Unfortunately, the second wave dropped its bombs in the middle of the positions of the *I./SS-Panzer-Grenadier-Regiment 25*. *Generalmajor* Peltz and I had the pleasure of laying beneath our own rain of bombs. Luckily, there were no casualties. The operation was never repeated.

The division was relieved by the *272. Infanterie-Division* during the night of 4/5 August for refitting in the area east of Falaise. However, due to the latest developments, the order was withdrawn and the division was kept on standby north of Falaise. We awaited major reinforcements in vain and only received a *Panzerjäger-Kompanie* that was only partially motorized. The *Panzer-Grenadier-Regimenter* did not receive a single man.

On a visit to the *I. SS-Panzer-Korps*, I noticed with trepidation that all of the *Panzer-Divisionen* that had been employed east of the Orne were now west of it. previously in action east of the Orne were now west of it. The *2., 116.* and *21. Panzer-Divisionen*, as well as the *1.* and *9. SS-Panzer-Divisionen*, were all assembled west of the Orne.

The remnants of my division — with about 50 combat vehicles — were the only armor forces east of the Orne. That meant that the two *Kampfgruppen* of the *12. SS-Panzer-Division* were the only operational reserves east of the river. That exposed the German front south of Caen and caused great concern. In the event of a renewed Allied attack, the eastern flank of the German front would inevitably cave in and open the way to the interior of France. With only 50 combat vehicles left, it was impossible to hope that we could stop the three armored divisions and three infantry divisions of the English and Canadians. We foresaw the collapse of the German eastern flank and prepared ourselves for our last fight.

The British 59th Division successfully forced a bridgehead over the Orne at Thury Harcourt on the evening of 6 August. *Kampfgruppe Krause*, in conjunction with elements of the *89. Infanterie-Division*, was immediately ordered to eliminate the bridgehead .

Moving out from St. Laurent, it managed to clear the Foret de Grimbosq of the enemy but was pinned by concentrated artillery fire as it left the wooded terrain for the open countryside that sloped gently to the Orne. The enemy had excellent observation positions from the heights of the west bank.

I went to *Kampfgruppe Krause* early on 7 August and found its command post in a forester's house in the Foret de Grimbosq. Wounded soldiers of the *89. Infanterie-Division* and the *Kampfgruppe* lay in the shadow of high trees waiting for their evacuation. Enemy artillery fire was falling on the road and the edge of the woods south of Grimbosq.

Despite the enemy's enormous artillery superiority, we were successful in entering Grimbosq and reducing the enemy bridgehead. Again, the losses were frightfully high. On returning to Krause's command post I rarely saw a single unwounded soldier. The artillery fire was devastating in the woods. Before the bridgehead could be entirely eliminated, events took place which rendered the operation secondary and caused the immediate withdrawal of *Kampfgruppe Krause*.

The Allies were aware of the switch of the *Panzer-Divisionen* to the western sector and, apart from the 50 combat vehicles of the burnt-out *12. SS-Panzer-Division*, there were only two infantry divisions south of Caen. What was more obvious than to smash the weak German eastern flank and push on towards the south via Falaise? By doing so, they would encircle and destroy the German armies in Normandy in a big pincer movement in conjunction with the American forces.

On 4 August Montgomery ordered the Canadian 1st Army to launch an attack in the direction of Falaise to speed the collapse of the German Army. The commanding general of the Canadian II Corps, Lieutenant General Simonds, was charged with the execution of this task. Simonds was the youngest commanding general in the Canadian Army and, without doubt, a very able and chivalrous opponent. He had commanded an armored division in Italy for a short period and was an excellent planner and tactician. He was probably the Canadians' most distinguished staff officer, but I dare not judge whether he was an equally able combat commander.

The fighting south of Caen clearly demonstrated that the Canadians did not have a dashing armored commander at their disposal. Furthermore, that battle was conducted with enormous superiority in troops and equipment. At no time, however, did the unit commanders dare to make instant decisions or take advantage of favorable situations. The combat commanders lacked the initiative to seize a chance when offered and lead their tanks into the depth of the enemy's rear. The Canadians slugged their way south — hesitant, with trepidation and waiting for orders from "above".

General Simonds had the following forces at his disposal for Operation "Totalize":

British 51st Infantry Division
Polish 1st Armored Division
Canadian 4th Armoured Division
Canadian 2nd Infantry Division.
Canadian 33rd Armoured Brigade
Canadian 2nd Armoured Brigade.
Canadian 3rd Infantry Brigade (reserve).

General Simonds intended to smash the German defense with these forces and reach the town of Falaise. According to General Crerar, commander –in-chief of the Canadian 1st Army, 8 August 1944 was to become an even blacker day for the German Army than 8 August 1918 had been east of Amiens.

General Simonds' plan was to attack in darkness without artillery preparation and break through the defensive strongpoints using long, dense tank columns. The accompanying infantry was to follow in special armored vehicles and attack the assumed second line of defense. The night attack included the use of a large British night-bomber formation. The second phase of the attack was to be launched in the early afternoon with an operation by the American 8th Air Force to open the way for the tank armada. The third phase was to end in the late afternoon with the encirclement of Falaise.

The concentrated power of the Canadian II Corps assembled according to plan late in the evening of 7 August. The tanks were closely packed together and were, in and of themselves, a deadly spear in the hands of the Canadian commanders. In all probability, such a concentrated tank force could simply not be stopped. It would utterly crush our defense into the ground.

The Canadian deployment would seem to have guaranteed victory over the German eastern flank in Normandy. Crerar's words were justified. The God of War, however, had decided differently. Despite the enormous accumulation of equipment, it was the human who remained victorious. The advancing tank squadrons were stopped by a group of men who were not afraid to look death in the face. The Canadian II Corps' objective was reached eight days later than planned. The rubble of Falaise only fell into Canadian hands on 16 August.

How did it look on the German side that 7 August? Opposing the seven major Allied formations of several hundred tanks and hundreds of heavy bombers and fighter-bombers was the *89. Infanterie-Division*. That division had neither tanks nor heavy antitank guns or any mobile reserves. The artillery was horse drawn and could be hopelessly out-maneuvered. East of the Orne only the two *Kampfgruppen* of the *12. SS-Panzer-Division* were available as a reserve. *Kampfgruppe Krause* was, however, involved in the attack on the bridgehead at Thury Harcourt on 7 August and, as a result, was about 20 kilometers away from the area of the Canadian offensive. The *12. SS-Panzer-Division,* together with the *schwere SS-Panzer-Abteilung 101*, the corps' *Tiger* battalion, had about 50 tanks at its disposal, nothing more. Moreover, the other infantry formation in the area, the *85. Infanterie-Division*, was on the move. Its lead elements had only reached the area around Trun. It could not be expected to be employed until 10 August at the earliest.

The division gave an extensive situation report to the corps following my return from the Thury Harcourt bridgehead. It urgently warned against the withdrawal of the last tanks south of Caen. I had also found out that my two *Kampfgruppen* were also to be turned towards the west.

*

A continuous booming and rumbling north of Bretteville announced the anticipated Allied offensive shortly before midnight. Air attacks hammered the positions of the *89. Infanterie-Division* and created a fiery glow in the sky. The front was on fire!

The first bombs automatically tripped the alarm for the units. Reconnaissance units moved north and tried to contact the engaged regiments of the *89. Infanterie-Division*. Hour after hour passed in gloomy expectation of the coming day. The giant hammer blows of the enemy bombers told us more than any mortal could. The drumming of the bombs and shells drew our attention. There was no point in wanting to escape this hellfire; its throat had already opened to receive us.

I raced towards Bretteville with some dispatch riders before daylight to obtain an overview of the previous night's events. For a split second I delighted in the lovely green of the woods and thought about the weeks we spent on that quiet wooded road in 1942 during the division's refitting. The sound of the front pulled me back to reality. Death and destruction allowed no memories of happy times; the rumbling battle sounded like a dull roaring of the drums of destruction. There were no blaring victory fanfares to be heard.

I talked to Mohnke in Urville and received the first reports on the night's events. The positions of the *89. Infanterie-Division* had been overrun; the division was as good as destroyed. Only a few individual strongpoints were still intact; they were like islands in the stream of battle, giving the attacking Canadians a hot reception time and again.

There was no communications whatsoever with the units up front and the surviving pockets of resistance fought on independently. There was no cohesion to the defense; they had to rely on their own resources. Our brave soldiers stood like rocks against the wild flood of the Canadian tank armada and forced it to halt again and again.

A lucky coincidence was that I knew the terrain in great detail. I had been there with my old reconnaissance battalion in the fall of 1942, and we had conducted plenty of exercises. I knew, therefore, that the high ground at Potigny dominated the terrain and the Laison sector was a natural tank obstacle. The Canadian attack had to be halted north of Potigny or the fate of the *7.* and *5. Armeen* would have been sealed. I moved towards Bretteville sur Laize with the clear intention of holding the Laison sector.

Bretteville was impassable. The bombs had blocked the streets with rubble. We moved across open fields to try to reach Cintheaux that way. Cintheaux was a large estate and was located right on the Caen — Falaise road. There were hardly any movements to be seen on the main road. And who would have been moving around anyway? The *89. Infanterie-Division* was north of Cintheaux. There was a huge gap from there south to Falaise. The Allied objective, which they desired so passionately, was spread out in front of them undefended and unoccupied.

I found a platoon of *Panzerjäger* from *Kampfgruppe Waldmüller* at Cintheaux. With foresight, Waldmüller had already moved the platoon there during the night. The place was under artillery fire.

I couldn't believe my eyes. Groups of German soldiers were running south in panic down both sides of the Caen — Falaise road. I was seeing German soldiers running away for the first time during those long, gruesome years of genocide. They were unresponsive. They had been through hellfire and stumbled past us with fear-filled eyes. I looked at the leaderless groups in fascination. My uniform stuck to my body; the heavy burden of responsibility made me break out in a sweat. I suddenly realized that the fate of Falaise and the safety of both armies depended on my decision.

I stood up in the *Volkswagen* and moved in the direction of Caen. More and more confused soldiers approached me fleeing southwards. I vainly tried to stabilize the collapsing front. The appalling bombardment had unnerved the units of the *89. Infanterie-Division*. Rounds landed on the road, sweeping it empty. The retreat could only continue off to the sides of the road. I jumped out of the car and was alone in the middle of the road.

I slowly approached the front and addressed the fleeing soldiers. They were startled and stopped. They looked at me incredulously, wondering how I could stand on the road armed with just a carbine. The young soldiers probably thought I had cracked. But then they recognized me, turned round, and waved to their comrades to come and organize the defense around Cintheaux. The place had to be held at all costs to gain time for the *Kampfgruppen;* speed was imperative.

I reached Mohnke's headquarters after a bombing attack. He was sitting on top of a radio vehicle in the rubble and looked the worse for wear. His head was in his hands; he could not hear. The dispatch riders had suffered casualties.

While with Mohnke, I saw the commander-in-chief of the *5. Panzer-Armee, General der Panzertruppen* Eberbach. The *General* had come to see for himself the effects of the earlier Allied attacks and make decisions based on personal observation. The commander-in-chief gave me full freedom of action and agreed with my estimate of the situation. In the meantime, Hubert Meyer had directed *Kampfgruppe Waldmüller* to Bretteville le Rabet. From there it could be employed based on the situation.

I issued the following orders:

1. *Kampfgruppe Waldmüller*, reinforced by the *I./SS-Panzer-Regiment 12* and the remnants of *schwere SS-Panzer-Abteilung 101*, counterattacks to seize the high ground south of St. Aignan.

2. Divisional *Begleitkompanie*, reinforced by the *1./SS-Panzerjäger-Abteilung 12* (with self-propelled antitank guns) advances through Estrees and takes the high ground west of St. Sylvain.

3. *Kampfgruppe Krause*, reinforced by the *II./SS-Panzer-Regiment 12*, disengages from the enemy, occupies the high ground west of Potigny and defends the area between Laison and Laize. (At that point involved in the attack against the enemy bridgehead at Grimbosq.)

4. Divisional command post at Potigny; I will be with *Kampfgruppe Waldmüller*.

I met Waldmüller north of Bretteville le Rabat and we moved to Cintheaux together to orient ourselves. Wittmann's *Tigers* were already east of Cintheaux, hidden behind the hedgerows. They had not engaged in the firefight up to that point.

Cintheaux was under artillery fire. The open terrain did not seem to be receiving any fire, however. From the northern outskirts of the village we saw the dense columns of tanks north of the road to Bretteville sur Laize. The tanks were clumped together. It was the same south of Garcelles and at the edge of the woods southeast of it. The massed tanks almost took our breath away. We could not understand the Canadians' behavior. Why didn't that overwhelming tank force pursue the attack? Why did the Canadian command give us the time and opportunity to take countermeasures? The much-feared fighter-bombers were missing. The systematic use of the fighter-bombers alone would have destroyed the remnants of my division on Route Nationale 158 and forced a breakthrough for the Canadian II Corps. Nothing could have then prevented the Canadians from taking Falaise that evening. Only the gods know why that did not happen.

Waldmüller and I knew we couldn't let the enemy tank squadrons run up against us. The enemy tanks could not be allowed to conduct another attack. Enemy armored divisions stood ready to attack down all of the roads. The attack could not be launched. We had to try to gain the initiative.

I decided to defend Cintheaux with those forces already employed and to launch an attack east of the road with lightning speed and all available units. By doing that, I hoped to disrupt the enemy's intent. I designated the woods southeast of Garcelles as the objective. Because a large quarry made a tank attack south of Cintheaux unlikely, I had no fears there. We had to risk the attack to gain time for the Laison sector. The attack was planned to start at 1230 hours.

During my last conference with Waldmüller and Wittmann, we observed a lone bomber flying over us a couple of times dropping flares. It seemed to us that it was some sort of flying command post and I ordered an immediate attack to get the units out of the bombing zone. I shook Michel Wittmann's hand once again and indicated to him the extremely critical situation. Good Michael laughed his youthful laugh and climbed into his *Tiger*. One-hundred-thirty-eight enemy tanks had fallen victim to him. Would he increase that count or would he become a victim himself?

The tanks rolled out rapidly to the north. They crossed the open terrain at speed and used folds in the terrain to engage the enemy. The tank attack helped sweep along the *Panzergrenadiere*. They approached the attack objective widely dispersed. I was at the northern outskirts of Cintheaux; the enemy artillery was laying down a barrage on the attacking tanks. Michael Wittmann's *Tiger* raced into the enemy fire. I knew how he operated in such situations: Keep going! Don't stop. Move through the muck and create some breathing room for yourself. All of the tanks advanced through the steely inferno. They had to prevent the enemy from attacking. They had to throw off his timetable. Waldmüller followed with his *Panzergrenadiere*. The magnificent soldiers followed their officers.

A machine-gunner cried out to me in the all-destructive artillery fire. He pointed to the northwest. Speechless when confronted with the overwhelming power of the Allies, I observed an endless chain of large 4-engined bombers approaching us. The ironic remarks of a few soldiers allowed us to forget the great danger for a fraction of a second. A young soldier from Berlin cried out: "What an honor for Churchill to send us a bomber for each one of us!" Actually, he was quite right. More bombers were approaching than we had soldiers lying on the ground.

There was only one way to save ourselves at that point: Get out of the estate and move out into the open terrain! As fast as lightning, the defenders of Cintheaux left the estate and waited for the discharge of the bombs in the green fields north of it. We had been right: Village after village was being flattened. It did not take very long before large fires sent flames skyward. We noted with pleasure that the American bomber fleet had also covered the Canadians. Based on an error of the pathfinder, the bombs were also landing on the attack groups. General Keller, the commander of the Canadian 3rd Infantry Division, was put out of action. Severely wounded, he was forced to leave his division.

The final bomber waves flew over the vigorously attacking *Kampfgruppe Waldmüller* without dropping a single bomb on an armored vehicle. The aircrew had engaged the targets they had been assigned without worrying about the situation that had changed in the meantime. Apparently the Canadian armor divisions were fighting without air-force liaison officers and, therefore, were not able to influence the attacking bombers.

At that point it became clear to me why the leading elements of the Canadian ground troops had not continued their attack and we had received the necessary time to take countermeasures. Not realizing the true nature of the situation, the attacking divisions had stuck to the timetable of the Canadian II Corps. As a result, they were cheated out of a victory. A tank battle cannot be led from a map table. The responsible commander must be with the forward-most elements of his attacking troops in order to make decisions appropriate to the situation and deliver decisive blows. A tank attack broken into phases is similar to a cavalry charge in which a feed break for the horses has been planned.

The employment of the American 8th Air Force was not able to effect the counterattack. *Kampfgruppe Waldmüller* had approached the patch of woods and was engaged with Polish infantry. The grim duel of tank against tank was being conducted between the vehicles of the Canadian 4th Armoured Division and the *Tigers* of Michael Wittmann. Occasionally, the *Tigers* could hardly be recognized. Very flexible artillery fire was being directed against the *Tigers* and the *Panthers*.

In the meantime, we had reoccupied our old positions at Cintheaux. The estate was being attacked from due north and lay under the direct fire of Canadian tanks. Flanking fire from a few of Wittmann's *Tigers* helped to keep the Shermans away from Cintheaux. We observed strong enemy movements one kilometer in front of us. They were headed in the direction of Brettville-sur-Laize. Attack after attack collapsed in front of us. We had incomparable luck. Our opponents did not launch a single concentrated attack against us. The divisional *Begleitkompanie* reported its location as west of St. Sylvain. It was fighting the lead elements of the Polish 1st Armored Division and had destroyed several armored vehicles. The Poles no longer attempted to move out of the woods at Cramesnil. Later on we discovered that this was the first operation of the Polish 1st Armored Division.

The fighting had lasted several hours. Wounded were collected south of Cintheaux. They were evacuated under fire from the enemy. Late in the afternoon I discovered that neither the army nor the corps were in a position to send reinforcements. A few *Tigers* were a possibility. I was hoping *Kampfgruppe Krause* had reached the Potigny sector in time and had set up a blocking position. At that point I had not received a single report from Krause.

Combat reconnaissance reported the loss of Brettville-sur-Laize in the late afternoon. The Canadian 2nd Infantry Division, under the command of Major General L. Foulkes, had overwhelmed the scattered elements of the *89. Infanterie-Division* there. A deliberate defense of the village had been out of the question. The defenders had neither antitank weapons nor artillery at their disposal.

The fighting north and east of Cintheaux lasted until it got dark. It was practically a miracle the overwhelming superiority hadn't overrun us a long time ago. Our armored vehicles cut through the heavy earth like battleships. Their main guns must have taught a little respect to the enemy attack units.

The Canadian 4th Armoured Division, under the command of General Kitching, had not been able to overrun a lost band of German grenadiers. Cintheaux was still in the hands of a dozen nameless soldiers. After Brettville-sur-Laize was lost, the enemy was deep on the flanks of *Kampfgruppe Waldmüller* and the heroic defenders of Cintheaux.

As a result, the *12. SS-Panzer-Division* decided to bring the *Kampfgruppe* back to the Laison sector under the cover of darkness and hold that position until the *85. Infanterie-Division* arrived. The defenders of Cintheaux and the armored vehicles of *Kampfgruppe Waldmüller* were able to disengage themselves from the enemy without difficulty. The tanks covered the withdrawal and were staged in the woods at Chateau Quesnay by the division.

I linked up with the commander of the *89. Infanterie-Division* just south of Cintheaux. The general had probably experienced the most demanding day of his career. It was difficult for him to understand how his division consisted of only a few scattered elements at that point. We left the village of Bretteville-le-Rabat together shortly after midnight. Together with the tanks providing cover, we were the last German soldiers to leave the engagement area of 8 August.

When I got to the command post, Hubert Meyer reported to me that *Kampfgruppe Krause* had only been able to disengage from the enemy late in the afternoon of 8 August and was only then reaching the position it had been ordered to occupy.

The enlisted soldiers and the officers presented a pitiful picture. The soldiers had been continuously engaged in hard fighting since 6 June and were at the end of their tether. Emaciated bodies sought a few hours of sleep on the hard Norman soil.

We had talked about the inability to win the war during the previous weeks and had cursed the conflict with all of its terribleness for humanity. Why didn't we call it quits? Why did we continue the senseless struggle? Full of despair, we sought an answer to those questions.

The soldiers and the officers could see how things were going to turn out. Despite that realization, however, no one thought about laying down his arms or trying to get out of harm's way. The political goals of the Allies were seen as much more terrible than the most gruesome death. Death had long lost its power to terrorize. We saw in death a portion of God's creation and, as a result, release from all worries. We continued to fight in good conscience. Even in that hopeless situation, we believed we had to fulfill our duty to our homeland.

By the light of a candle I wrote a birthday greeting to my daughter. She would turn one-year old in a few days.

The division issued the following orders for the defense of the Laison sector:

1. *Kampfgruppe Krause* defends the high ground north of Maizières and Rouves, to include Hill 132.

2. *Kampfgruppe Waldmüller* defends the sector from Hill 140 to Hill 183 on the Falaise — Caen road.

3. The *III./SS-Panzer-Grenadier-Regiment 26* (Olboetter) defends Hill 195 (2 kilometers northwest of Potigny) and collects all scattered soldiers of the *89. Infanterie-Division.*

4. All tanks of the *12. SS-Panzer-Division* and *schwere SS-Panzer-Abteilung 101* are to be staged in the woods at Quesnay under the commander of *SS-Panzer-Regiment 12 (Wünsche).*

5. Divisional artillery occupies positions south of the Laison to effectively support the entire division.

6. Divisional *Begleitkompanie* remains under divisional control at Potigny.

7. Divisional command post 1 kilometer east of Potigny below Tambeau de Marie Joly.

Gruppe Olboetter had established itself in a defensive position in the afternoon of 8 August on Hill 195. It had reinforced itself considerably with stragglers from the *89. Infanterie-Division.* The artillery was also in position by

2200 hours the same day. The commander of *SS-Panzer-Regiment 12* reported the assembly of the tanks at Quesnay at 0300 hours. No report had yet been received from *Kampfgruppe Waldmüller* or the divisional headquarters company.

I climbed the high ground at Tambeau de Marie Joly and listened as dawn broke. The Laison sector was still quiet. The beautiful countryside was still peaceful. I observed the opposite high ground with binoculars. Green grain fields lay sleepily on the reverse slope. Slim spruces waved in the first beams of sunlight on the hilltops. Even the glittering dew on the grass shone in such a wonderful way as to make me forget the war for a few moments. The sun broke through and the first morning greeting came from the throats of a thousand tiny birds.

However, the silence was misleading. Although I could see no movement, I knew that *Tigers* and *Panthers* were poised in position to destroy young human lives in the Quesnay woods. The exhausted bodies of my soldiers were lying somewhere in the grain fields at that moment awaiting the enemy attack. On my right were the slim barrels of an 88 mm battery on the reverse slope; they also awaited their victims.

Maybe the barrels of guns and mortars were already aiming at the division's exhausted soldiers from the other side. Perhaps the enemy tank squadrons' engines were purring into life at that second and the first firing orders were being given to extinguish our lives! Yes, the silence was misleading; the dance of death would soon start.

A small armored car moved out of the valley and slowly moved towards Hill 140. It soon reached the highest point of the crest. It was a captured English armored car that was being used to pass reports. A whiplash crack tore the morning silence as the armored car was fired at by an enemy tank from its position in the trees. I watched the encounter breathlessly. The armored car accelerated southwards and raced across the field at breakneck speed. The ground fell away and soon it was out of the tank's field of fire. I witnessed this confrontation in complete surprise. I was faced with a puzzle. How could that enemy tank have been on the hill? Thinking something was wrong, I dashed to the telephone and called Wünsche.

Wünsche had already alerted his tanks and was waiting for the return of *SS-Obersturmführer* Meitzel who was supposed to establish contact with *Kampfgruppe Waldmüller*. Meitzel reported: "There are no German forces on the hill. There are enemy tanks on the high ground."

An icy shock went through my bones. If the report were true, then all of *Kampfgruppe Waldmüller* and the divisional *Begleitkompanie* were lost. That couldn't be true. One thing was certain, however, neither of them had reported up to then.

Meitzel moved back in his armored car to gain a more accurate picture of the enemy. As soon as he crossed over the ridge, his car received a hit. He was thrown out of the open turret. He was quickly surrounded by enemy infantry and captured.

Reconnaissance soon clarified the situation. An enemy combat team had occupied the high ground and dominated the Laison valley with its weapons. That menace had to be eliminated at once if we were to hold that sector for the rapidly approaching *85. Infanterie-Division*. The Laison sector offered the only option for defense north of Falaise. The situation called for rapid action. The high ground had to belong to us once again!

With the exception of *Kampfgruppe Krause*, which was not even at company strength and was in position east of Hill 140 and Hill 183, not a single German soldier was on the range of high ground. Furthermore, the main Caen — Falaise road was only covered by a couple of tanks. Once again, Falaise was inadequately defended.

Wünsche shouted a few words to his veteran tank crews and pointed to Hill 190. It was our intent to attack with some *Tiger* tanks from the west and with 15 *Panthers* from the east. While the *Tigers* slowly left the woods and approached the ridge, the *Panthers* rattled down the valley road towards Krause's sector so that they could wheel inward there. During the movement of the two tank groups, the hill came under artillery and mortar fire. Our only 88 mm battery waited for targets in vain. The enemy tanks wouldn't venture beyond the ridge. Two *Tigers* took up firing positions even with one another. They had snuck through the undergrowth unnoticed by the enemy and were on his flank. The first 88 mm rounds slammed out of the barrels. Two Shermans exploded noisily. The enemy hammered at the *Tigers* they had spotted. Five *Tigers* took part in the engagement and pinned the enemy. The *Tigers* had chosen to pin the enemy with fire; they exploited their greater firepower. More and more enemy tanks were burning, sending telltale smoke into the sky.

I was with the *Tiger* section and suddenly saw Jürgensen's first *Panthers*. The enemy tanks were cornered at that point. Death and destruction hit them from the east. Pinning them through superior firepower would guarantee us success! Each thicket and perilous spot was peppered with gunfire. The entire ridgeline was systematically covered. Smoke cloud after smoke cloud merged together. We could hardly believe that each cloud represented a tank's grave. The lack of foot soldiers prevented us from penetrating into the tree-encrusted northern slope of the ridge. Two bicycle companies of the *85. Infanterie-Division* were expected at any moment.

At that point we could see fighter-bombers in the heavens. Did they want to attack us or did they have other targets? I was concerned for the tanks. They were in the open. They looked like targets on a range. In a flash, they were above us. They aircraft described a curve and then dove on the Canadian combat team. They were above us like lightning, made a banking turn and attacked the Canadian battle group. Not a single aircraft attacked a *Tiger* or a *Panther*. The hill was covered in the smoke of exploding tanks in a few moments. *Tigers* and *Panthers* took advantage of the chaos and took possession of the ridge. The ridgeline looked like a tank cemetery.

I saw two half-tracks break out of the woods and race towards the north at around 1100 hours. One *Tiger* in my vicinity opened fire, but the vehicles were able to get away. Fire could not be opened against the vehicles until they were far away due to the thick vegetation. According to prisoner statements, one of the half-tracks carried the wounded Lieutenant Colonel A. J. Hay of the Algonquin Regiment. The Canadian combat team commander, Lieutenant Colonel D. G. Worthington, had been killed in action that afternoon.

The tanks pushed onto the trails in the woods with the bicycle companies of the *85. Infanterie-Division*, which had just arrived, and increasingly pressured the Canadian positions. At that critical juncture and taking advantage of the air attack, *SS-Obersturmführer* Meitzel suggested to his captors that they surrender. Meitzel had broken his arm when thrown out of his armored car. He was with the Canadians in the center of the inferno on the northern edge of the ridgeline. The Canadians had bandaged Meitzel and treated him chivalrously. His first suggestion was politely rejected. However, when the air attacks and the artillery fire caused more and more casualties among the Canadian infantry, the offer was accepted.

Meitzel led 21 Canadian soldiers and 2 officers into the positions of *Kampfgruppe Krause*. He reported at division headquarters around 1500 hours with the 23 Canadians. Among the prisoners was a Captain J. A. Renwick of the 28th Tank Regiment (British Columbia Regiment). I talked with Renwick for about half an hour about the madness of the war. He made a good impression. He said nothing about the fighting that was occurring then. Based on prisoner interrogations and the questions posed to Meitzel while he was a prisoner, the following picture of the situation was formed:

Our own counterattack in the afternoon of 8 August had brought the enemy's attack to a standstill. He then set up defensive positions with the Polish 1st Armored Division at St. Sylvain and the Canadian 4th Armoured Division at Cintheaux. To regain the initiative, the Canadian 4th Armoured Division had launched a night attack on Hill 195 northwest of Potigny using the 28th Armoured regiment (British Columbia regiment) and two infantry regiments of the Algonquin Regiment. The narrow area between the Laison and Laize was to be opened as a result and a swift breakthrough to Falaise enabled. As a result of faulty navigation during the night, the combat team took unoccupied Hill 140 instead of Hill 195.

Meitzel was asked about the "big asphalt road" they had been looking for in vain. The enemy tank group had passed *Gruppe Waldmüller*, which had dislodged itself from Cintheaux to occupy Hill 140. Waldmüller had been pushed eastwards and waited for darkness to reach our lines. The divisional *Begleitkompanie*, which had been passed by the Poles, was in a similar situation.

During the night the survivors of the Canadian combat team fought their way through to the Polish 1st Armored Division. The 28th Tank Regiment left 47 knocked-out tanks on the battlefield, all of which had been knocked out by the guns of the *Tigers* and *Panthers*. We did not lose a single tank.

Our defensive accomplishments during the previous 48 hours had cost us heavy losses even though they were far smaller than those of the enemy. We discovered on 9 August that our brave comrade Michael Wittmann had made his last tank attack. Moving ahead of his tanks, he and his loyal crew destroyed some Shermans east of Cintheaux. He then led his tank section forward. His impetuous tank attack had probably dampened the momentum of the Canadian 4th Armoured Division's attack and had bought some time and space for the defenders along the Laison. Michael Wittmann died the way he had lived — brave, inspirational and, as always, an example to his soldiers. He displayed a true Prussian attitude to duty until his death. The flames of his *Tiger* marked his last fight and the end of a good comrade and soldier. However, the spirit of this brave officer lived on in his young tankers who fought and died with the same bravery as their old commander until the end of the struggle.

Kampfgruppe Krause had suffered critical losses in the fighting for the Thury — Harcourt bridgehead. It only had the combat power of a weak infantry company.

In the course of the night, *Kampfgruppe Waldmüller* and the divisional *Begleitkompanie* reached our lines and assumed their sector. *Kampfgruppe Waldmüller* also only had the strength of a weak company. If only the *85. Infanterie-Division* could take over the sector. We survivors of the *12. SS-Panzer-Division* could hardly keep going any longer. A further attack on the part of the Canadians would lead to a catastrophe; we were no longer able to fight. The past ten weeks had sucked the marrow from our bones. Completely worn out, the soldiers sank to the ground to find some sleep. But that night also brought us no rest.

A firestorm raced over Hill 195 and swept through Olboetter's positions. Tracer rounds from the supporting tanks hit the attacking enemy infantry. The dull cracks of hand grenades mixing with the defenders' angry cries shook us out of our leaden sleep. The Argyll and Sutherland Highlanders Regiment was attacking Hill 195. When I reached the hill Olboetter was in the middle of his soldiers, leading them in a counterattack. The enemy had broken into the widely dispersed positions and was just about to capture the entire hill. The *Panzergrenadiere* attacked the enemy spearheads in shock-troop fashion and threw them back into the darkness.

The high ground could be held with the assistance of the tanks that were providing cover. The enemy was made to suffer heavy casualties. By dawn he was exposed to flanking fire from the tanks positioned in the Quesnay woods. His attack on this key terrain failed. A few could hold out against many there.

The attack that had failed on Hill 195 was continued a few hours later by the Polish 1st Armored Division at Maizieres. The Polish armored division attempted to cross the Laison at Condé by bypassing *Kampfgruppe Krause.*

The tank spearhead had been halted by a single antitank gun the previous day. Nine Polish tanks remained in front of the German antitank gun; their burning wrecks glowing until the morning. One unfortunate direct hit also killed the crew of our antitank gun.

After the destruction of that single antitank gun, the way was open for the Polish division. There were simply no more troops available to prevent the crossing of the Laison. But the Poles also lacked the momentum needed; they withdrew to the north.

The tanks on Hill 195 had to be switched to the division's right flank in a great hurry to attack the flank of the Polish advance. Half a dozen tanks raced east along the concealed road. Would they make it in time?

We were lucky. A freshly arrived self-propelled *Panzerjäger-Kompanie* under the command of *SS-Obersturmführer* Hurdelbrink made contact with the Polish spearhead. It was the company's first contact with the enemy using the newly issued *Jagdpanzer IV.* Forty Polish tanks were destroyed in short order. *SS-Obersturmführer* Hurdelbrink himself knocked out 11 tanks. The breakthrough had been prevented.

The division's right wing was relieved by elements of the *85. Infanterie-Division* during the course of 11 August. *Kampfgruppe Krause* could finally be moved out of its positions. Before *Kampfgruppe Waldmüller* was able to hand over its sector, the Canadian 8th Infantry Brigade attacked the tank group in the Quesnay woods. That attack was also repulsed with heavy losses for the Canadians. On 12 August the *12. SS-Panzer-Division* was able to hand the Potigny sector over to the *85. Infanterie-Division.*

Some one hundred young soldiers, completely exhausted and shattered by the previous fighting, had resisted an overwhelming superiority in men and materiel. Two fresh armored divisions and one infantry brigade were unable to break the 17- and 18-year-old soldiers' will to resist or overrun them.

In postwar literature, the failure of the Canadian II Corps was attributed to the presence of defensive positions in depth and sizable *Flak* formations from the *III. Flak-Korps* under *Generalleutnant* Pickert. This argument is not true. Yes, a "position" consisting of foxholes had indeed been prepared on a line running from St. Sylvain to Bretteville-sur-Laize. It was to be used as a fall-back position by the *89. Infanterie-Division* in the event of a planned withdrawal. However, the course of the fighting on 8 August must show, even to the layman, that the use of this prepared "position" was not possible. Who might have occupied that "position"? Perhaps a few hundred men from *Kampfgruppe Waldmüller*? This so-called "position" did not influence the course of the fighting in any way. There were simply not enough troops to occupy it.

Furthermore, it must be noted that the units of the *III. Flak-Korps* were scattered along the entire Normandy front and its guns were mainly used against the enemy bomber formations. Not a single gun from the *III. Flak-Korps* was employed against enemy tanks within the *12. SS-Panzer-Division* sector from the beginning of the invasion to the Falaise pocket. I saw the last battery of the corps on the morning of 8 August south of Bretteville-sur-Laize. The battery then went into position west of Falaise. The 88 mm guns could, without doubt, have rendered good service in the antitank role, but they were under *Luftwaffe* command and not that of the combat divisions.

Complete success was denied to the Canadian II Corps because the leadership of the two divisions conducting the assault was inexperienced and used its tanks piecemeal and indecisively. An experienced tank commander would have led the Canadian 4th Armoured Division to victory on the first day of Operation "Totalize". The piecemeal attacks of 9 and 10 August were as incomprehensible as the hesitant advance on 8 August.

Our division occupied a blocking position between Perriéres and Falaise. Some *Tigers* of *schwere SS-Panzer-Abteilung 502* were placed under the operational control of the *85. Infanterie-Division* and employed on both sides of Potigny.

The situation on the western flank of the Normandy front was not entirely known to me, but it seemed inevitable that the front would have to be pulled back to the Seine in short order. The commander-in-chief in the west had neither the troops nor the materiel to conduct a delaying action. It was impossible to operate against modern armored divisions with burnt-out horse-drawn infantry divisions.

Wherever it had not yet occurred, the headquarters and cadre of the burnt-out elements of the division, as well as the logistical elements, were moved to the areas around Evreux and Bernay. Preparations were made to move the combat-support elements to the eastern bank of the Seine.

The survivors of *Kampfgruppe Waldmüller* were incorporated into *Kampfgruppe Krause*. The division's combat strength on 13 August consisted of the following:

20	armored fighting vehicles (including *Panzerjäger*)
1	platoon of *Panzergrenadiere* in *SPW*
1	armored reconnaissance section
300	dismounted *Panzergrenadiere*
1	88 mm *Flak* battery with 4 guns
1	37 mm self-propelled *Flak* battery with 9 guns

1	20 mm self-propelled *Flak* company (*14./SS-Panzer-Grenadier-Regiment 26*)
3	batteries of heavy field howitzers
1	battery of 10 cm guns

The artillery's change of position had to take place in leapfrog fashion due to the lack of prime movers. No ammunition had reached the division since the day before, and its firepower cold not be used to maximum advantage.

The division's total strength was 500 junior enlisted personnel, noncommissioned officers and officers.

We all knew that the fighting would only end with death or capture, but nobody was ready to stop fighting. The thought of the call for Germany's unconditional surrender formulated by the Allies at Casablanca kept us motivated to continue fighting. Germany's war was surely lost but the front had to be held. The Allies had to be convinced that the absurd decision to demand an "unconditional surrender" would not pay off and a different basis for negotiating a peace had to be found.

My comrades were not fanatics; they wanted to live and, if possible, return home in good health. No, no, it was not that fanaticism so often claimed by the enemy that compelled us to fight on! We did not throw away our weapons because we still believed we had to fight for our homeland.

The division received an approximate picture of the current situation during 13 August. The position of the German armies had become untenable. There was a large pocket of decimated German divisions between Argentan, Falaise and the high ground from Trun to Chambois. The threat of complete annihilation was clearly visible; the jaws of death had already been set loose. The only usable withdrawal route went via Trun and wound up the hill in sharp bends. However, the road was no longer in any condition to take all the troops and guarantee their escape. The hastily thrown-together infantry divisions with their horse-drawn equipment were the greatest obstacle to still-mobile armor formations. The catastrophe continued to develop.

We finally got some sleep during the night of 13/14 August. That was the last quiet night with their companions for a lot of my comrades. The muffled noise of fighting raging in the west kept us awake for a long time, but nature eventually took its course.

On the morning of 14 August Wünsche, Krause, Olboetter and I moved to the sector northwest of Falaise to draw up the new positions. Hill 159 north of Falaise controlled the sector and we immediately occupied it, setting up a series of strongpoints. Other prominent terrain features east of Hill 159 as far as the Dives River characterized our "front".

We did not trust the Canadians' "peaceful" behavior and started to reinforce our strongpoints at once. Based on an estimate of the entire situation and the lay of the land, the Allied attack had to be between Jort and Falaise.

The Canadians were the northern claw of the encircling pincer; the southern claw was formed by the Americans at Argentan. The death struggle of the two German armies would begin as soon as the two claws met. It was with this certainty in mind that I prepared the soldiers and officers of our once so proud division for the final battle. I was not surprised when my brave comrades accepted my judgment as self-evident. They had just lived through this crisis and knew exactly what the result would be.

We experienced the same old story at about 1400 hours. Hundreds of Halifax and Lancaster bombers turned the position of the *85. Infanterie-Division* into a cemetery. More and more bombers and fighter-bombers attacked and dove on the *85. Infanterie-Division*, breaking the backbone of the defense. Artillery and antitank defenses were destroyed by the bombs or blinded by smoke. The ground attack against the *85. Infanterie-Division* was executed by the Polish 1st Armored Division, the Canadian 4th Armoured Division and the Canadian 3rd Infantry Division.

The Canadian II Corps' tanks lined up in parade-ground formation — tank next to tank. They waited for their commanders' signals; they intended to break a way through the defense zone for the Canadian Corps with steamroller tactics. It was a mystery why the Canadians had chosen such an inflexible battle formation. Instead of leading their tanks close to the enemy in a dispersed formation, affording the opportunity to use the effect of their guns and maneuverability to smash the positions and make a swift and deep advance into the enemy rear, those steel monsters rolled clumsily and sluggishly over the terrain. Precious time was lost as the tanks crossed the Laison area, since they could not negotiate the marshy terrain in their clumsy battle formation.

The Canadian divisions — well equipped and outfitted with modern equipment — were still north of their objectives on the evening of the first day of the attack. Even during that phase of the operation, the Canadian armor was used as infantry support. Neither the enormous firepower nor the speed of the formations was effectively used.

The Canadian leadership failed to use imaginative planning. Not one of the Canadian attacks showed the genius of a great commander. Their planning always got stuck in the tactics of attrition warfare. The successful elimination of the defending German divisions was never exploited with an effective breakthrough. As soon as the attacking spearheads encountered an enemy outside the main engagement area, the lead elements lost their momentum and started to dissipate their energies in small, piecemeal operations. The course of the fighting confirmed my observations.

The first wave of Canadian tanks ran up against the sparsely-occupied line of resistance north of Falaise late in the afternoon. The attack of the Canadian 4th Armoured Division and the Canadian 3rd Infantry Division got bogged down in front of the remains of our once powerful division. The attack of two divisions failed because of the fighting spirit of 500 soldiers. Hill 159 remained in the possession of a handful of German soldiers.

I moved to all the strongpoints along our front during the night and explained the situation of our two armies to the young soldiers. At that point, they knew they were holding the northern flank of a big funnel and their holding of the position made the withdrawal of exhausted units possible.

At dawn, 20 to 30 men from the *85. Infanterie-Division* reached the positions on Hill 159 and voluntarily joined in the defense. That group had marched through the enemy outposts during the night. We approached some stragglers who had a wounded man with them in a *Kübelwagen*. They shook our hands.

The Canadians continued their attack and, within a short time, Hill 159 was on fire. Round after round impacted into the earth and eviscerated it. Our tanks had been dispersed in ambush positions. They were waiting for the dark shadows which would soon come out of the dark wall of smoke and dust. The first enemy tanks were burning. Enemy infantry was nailed to the ground by well-aimed bursts of machine-gun fire. Did we still have any nerves; could we still be recognized as human beings? Our eyes wandered again and again into the wall of shellfire. We did not hear the bursting, exploding and repulsive howling of the rounds anymore. Each movement in the wall took our breath away, however. Would there be a mass of tanks suddenly appearing out of the wall of fire? Would yesterday's spectacle to be repeated? Would we be lying under creaking tank tracks in the next few moments? Nothing of the kind happened. The enemy tanks kept their distance and didn't overrun us. They stopped in front of Hill 159.

The enemy attacked repeatedly at Jort and Perriéres, trying to force a crossing over the Dives. The few tanks that were still operational were thrown at the most threatened points and brought the attack to a standstill.

The *III./SS-Panzer-Artillerie-Regiment 12* contributed in no small part to our success. It had found a small ammo dump at Falaise by chance and no longer had to be thrifty in its employment of ammunition. The positions north of Falaise remained in the hands of German soldiers.

It was still long before sunrise, but we expected new enemy attacks east of Falaise. We didn't understand the enemy. Why did he waste such an enormous amount of bombs and rounds on the poor remnant of the *12. SS-Panzer-Division*? His vastly superior numbers of tanks only had to run over us at full speed to finish us off. But nothing happened. Each attack was repulsed until the afternoon. During the operations around Hill 159, the commander of the *II./SS-Panzer-Regiment 12*, SS-Sturmbannführer Prinz, was killed. Once again I was witness to the last battle of a warrior friend. Prinz had been with me on all fronts since 1940. He was a victim of the artillery bombardment.

Fighter-bombers dove on the little patch of woods at Bois du Roi, unleashing their rockets into the long-destroyed woods. Some tanks east of Hill 159 fell victim to the Typhoon attacks. I met Max Wünsche between Versainville and Hill 159. He informed me of the hopeless situation on the hill. Enemy tanks were racing towards us. Their rounds exploded on the road; Max Wünsche disappeared. I felt a burning hot pain; blood ran down my face. I dove head first into a little hedge; shrapnel had opened up my skull. I looked dizzily at the road. Our *Kübelwagen* had disappeared and Max Bornhöft was no longer to be seen. I was alone but at no time did I feel deserted; I knew my comrades would not abandon me.

The tanks moved closer and closer. I crawled along the ditch to get out of the enemy's axis of advance. The Shermans were in action against some tanks occupying a good position on the reverse slope. This had to be Max Wünsche's work. Tank rounds screamed overhead.

I did not believe my eyes; Max Bornhöft had returned. Under the covering fire of the tanks, he was racing down the road to get me. I desperately waved at him. The road could be observed along its entire length and ran across the enemy's front. Rounds exploded all round Max but that didn't dissuade him. The steering wheel remained firmly in his hands. I was waiting in the ditch, ready to leap into the car. We were back on the reverse slope like lightning. Wünsche welcomed me. He had directed the tanks' fire. I continued the fight with a half-shaved head and a couple of stitches.

Hill 159, so fiercely defended, fell in the afternoon; the survivors moved back to the Aute sector. *SS-Obersturmführer* Hauck, leader of an armored reconnaissance patrol, reported an attack by the Polish 1st Armored Division at Jort. The Poles were trying to force a crossing of the Dives again but, up to that point, all their attacks had been repulsed. Units of the Canadian 2nd Infantry Division entered Falaise late in the afternoon. The 6th Brigade, commanded by Brigadier H. Young, had finally managed to overrun the city of William the Conqueror. The fighting in the ruins of the totally destroyed city continued.

The division moved out of the blocking position after nightfall and retired to the Aute sector; the new line ran from Morteaux — Damlainville to Falaise.

17 August began with further attacks on Jort by the Polish 1st Armored Division. The *3./SS-Flak-Abteilung 12* was practically destroyed. The commander, *SS-Untersturmführer* Hartwig, was mortally wounded; the rest of the battery was driven back eastwards.

The enemy crossed the Dives and advanced southeast. From that point on the Polish 1st Armored Division did not have a cohesive combat formation in front of it. The road to Trun and Chambois was open to the Poles. The Falaise pocket could be closed.

SS-Obersturmführer Hauck's recon section was eliminated late in the afternoon. He was wounded and captured, but he managed to escape and report to the division about the threat of the enemy's strong tank forces pushing deep into our right flank toward Trun.

About 60 soldiers of the division were still locked in a hard grim fight in Falaise. These men, exhausted after having been in continuous combat since 6 June, were fighting the Canadian 6th Infantry Brigade. Two *Tigers* were the backbone of the defense, but the soldiers already had the mark of death on them. Late that evening, two grenadiers who had been selected by lot — none of the comrades wanted to leave the group — brought the last report and messages from the brave band. They died shortly after midnight in the rubble of the Ecole Supérieure.

The rest of the division conducted a desperate defense along the line Dives — Nécy (8 kilometers southeast of the Falaise — Argentan road). In Nécy two damaged *Tigers* prevented the British 53rd Infantry Division's armored spearhead from advancing. On 19 August, at about 0200 hours, the two Tigers were destroyed. *SS-Obersturmführer* Meitzel and the other survivors were captured; all the *Tiger* crewmen had been wounded.

During the night of 18/19 August we got rid of the last radio transmitter and all non-essential vehicles. We only retained some *Kübelwagen*, *SPW* and prime movers. The divisional command post at Necy was overrun by enemy tanks and infantry shortly before sunrise. My messenger collapsed, taking a round in the stomach. We took the young soldier with us. Making use of the dim light of dawn, we fought our way south with the rest of *Kampfgruppe Krause* and occupied a new line southeast of the railway tracks.

During the night *Kampfgruppe Wünsche's* staff drove into advanced enemy elements. Most of it was wounded and taken prisoner. Max Wünsche and two other officers were eventually captured six days later. We stumbled on into misery, numb to the inhuman tragedy in the pocket.

Towards midday, the commanding general of the *LXXXIV. Armee-Korps*, *General* Elfeld, and his chief of staff, *Oberstleutnant* von Kriegern, appeared at our command post. Our division no longer had any communications with higher headquarters. The staff of the *85. Infanterie-Division* (*Generalmajor* Fiebig) had been detached from the corps. That left *General* Elfeld with only the remnants of the *12. SS-Panzer-Division* to command.

The misery around us screamed to high heaven. Refugees and soldiers from the defeated German armies looked helplessly at the bombers flying continuously overhead. It was useless to take cover from the bursting shells and bombs. Concentrated in such a confined space, we offered once-in-a-lifetime targets to the enemy air power. The wooded areas were full of wounded soldiers and the sundered bodies of horses. Death shadowed us at every step. We stood out like targets on a range. The guns of the Canadian 4th Armoured and Polish 1st Armored Division could take us under open sights.

It was impossible to miss.

By chance we found the command post of the *7. Armee* in an orchard one kilometer southwest of Trun. The commanding general and I went to the army headquarters. The roads were impassable. They had been blocked by motorized units and the horse-drawn trains of the infantry division. Burning vehicles and exploding ammunition — newly impacting rounds landing among them — marked the course of the road.

We ran, stumbled and jumped by stages towards the headquarters. The area was under constant artillery fire. Swarms of fighter-bombers had masses of targets. We found the staff of the *7. Armee* in a ditch behind a farm. Our respected *Generaloberst der Waffen-SS* Hausser was sitting on the edge of a trench studying a map. With the commander-in-chief were his chief of staff, *Oberst* von Gersdorf, *Oberstleutnant i.G.* von Kluge and *Major i.G.* Guderian.

An exploding ammunition vehicle tossed its "greetings" on over to us; *Oberst* von Gersdorf was wounded. The issuing of orders continued. The *Oberst* remained at the side of the general. *General* Hausser gave the order to break out in the coming night.

The *Panzergruppe* of the *1. SS-Panzer-Division* was to force the breakout at daybreak at Chambois, and the *3. Fallschirmjägerdivision* was to break out at St. Lambert after midnight. Initially, it was not to use its weapons. The remnants of the *12. SS-Panzer-Division* were to hold the northwest edge of the pocket until midnight and then join the *3. Fallschirmjägerdivision*. We said goodbye to the commander-in-chief with a final handshake. He looked at us gravely with his one good eye; he had lost the other one in the battle for Moscow.

We ran through the hail of rounds again and took cover in a quarry. Countless soldiers were lying crowded together in the shadows of the steep walls. They were waiting for the protection of night to jump out of hell.

A direct hit struck a group of infantry next to us; several soldiers were mortally wounded. A *Feldwebel lost* his right leg above the knee. We quickly pulled him closer to the wall; cries for a medic were lost in roaring shell fire.

We encountered the commanding general of the II *Fallschirmjägerkorps*, Meindl, and the commander of the *3. Fallschirmjägerdivision*, *Generalleutnant* Schimpf, in a cottage. The paratroopers discussed the coming night's break-out with us. Two *Tigers* were to support the *3. Fallschirmjägerdivision*.

We reached our command post in complete exhaustion. Command and control of the division during the breakout was not possible. The roads were completely jammed and there were no means of communications any more. The division formed into two groups. Those motorized units that still existed were to break out behind the *1. SS-Panzer-Division* via Chambois. That group was to be commanded by the divisional artillery commander, Drechsler. The divisional staff, joined by *General* Elfeld and the rest of *Kampfgruppe*

Krause, were to follow the *3. Fallschirmjägerdivision*. I subdivided our group into several sections so they could act independently if need be. Guns for which there were no more prime movers were blown up at midnight.

At midnight I had assembled all that were still in the pocket around a group of farm buildings. A liaison party was with the *3. Fallschirmjägerdivision*. Because the reconnaissance unit did not return and no noise of fighting was heard from St. Lambert, we assumed the paratroopers' breakout was a success. We started to move.

General Elfeld, *Oberstleutnant* von Kriegern and Hubert Meyer followed with the lead element. I went in the direction of Chambois. We had to march cross-country if we were to make progress. The few roads and trails were impassable, hopelessly blocked. Enemy artillery brought down harassing fire. There were fires flaring and the bright colors of exploding ammunition at all points of the compass. Exhausted soldiers wandered to and fro; the confusion in the pocket made orientation impossible.

We were west of Chambois by dawn and linked up with the *Panzergruppe* of the *1. SS-Panzer-Division* which was just about to launch an attack. We joined in as infantry. I jumped on the back deck of a tank and, in order to get on board, I grabbed the belt of a comrade lying behind the turret. I let go in horror; he was dead. He had been killed by shrapnel. Antitank, tank and artillery fire impacted among the attackers. I had no means of communication. The tanks hesitated and withdrew under enemy fire.

We reassembled behind some willow bushes next to the eroded bed of the Dives. The riverbed was about 2 meters deep and 3 to 4 meters wide. We were witness to a horrible tragedy. Galloping horses tumbled with carriage and riders into the ditch. The horses and men struggled in the mud of the almost waterless brook. Agitated men climbed over the wrecks, and were then torn to pieces by Canadian artillery rounds.

Several hundred prisoners lay helplessly under the fire of their own guns. They could not leave the pocket.

After crossing the Dives, I assembled the infantry element between Chambois and Trun. The entire area was covered with dead and dying German soldiers. The enemy was on the slopes and fired continuously into the pocket. Most of the victims belonged to the support units of the infantry divisions who had remained in the pocket with their horse-drawn transport. Leaderless, they ran for their lives.

General Elfeld and *Oberstleutnant* von Kriegern were missing. They were not at the rallying point. To restore order in the now swollen group, I got the men to assemble under cover of a farm. While nearby columns of unarmed soldiers made their way towards the enemy to get out of the cauldron, a lot of officers and soldiers without weapons joined my group. Only those who managed to find weapons were allowed to come with us. Most of them complied.

I knew every tree and bush between Trun and Chambois. Units of my regiment had been stationed in both villages before the invasion. I took the lead; Bernhard Krause led the other half of our group. We were probably 200 strong all together.

Hubert Meyer, *SS-Obersturmführer* Köln and my loyal Michel were by my side. We had to bound across the Trun — Chambois road. Enemy tanks were racing up and down the road. The tanks we had had in the pocket could not cross the Dives and could not assist. Its bed was too deep.

Countless dead soldiers were lying behind the bushes and walls of the orchards. They had all run into the waiting weapons of the Canadian 4th Armoured Division. Would we be able to break through that iron ring of encirclement? The enemy had occupied the high ground and was firing down into the roads and fields.

Michel took the white bandage from my head. The brave Cossack from Dnjepropetrowsk said: "Bandage not good; I make new later on!" In his opinion, the white bandage would give me away.

I jumped from cover to cover with pistol in hand. The ditches were filled with dead. They must have been overrun by tanks. Looted vehicles stood in the fields and behind the hedges. We worked ourselves closer and closer to the slope. Machine-gun fire whistled above us. We had landed between two tanks! They were about 150 meters apart and were firing into the pocket. A Bren Gun Carrier was moving back and forth off to our right. Suddenly, it disappeared and we dashed eastwards between the two tanks like a shot from a gun.

Machine-gun fire raked our lines, but we could not be halted. We overran the Canadian infantry in a few seconds. It all happened in almost complete silence; only the whistling projectiles and exploding rounds overhead could be heard. We would have to break through the enemy blocking position soon. I could not go on. The sweat burned in my inflamed eyes; the head wound had reopened, but it was impossible to halt. We had to get through the enemy.

Suddenly there was a Sherman tank 30 meters off to our right. Hubert Meyer yelled at me. I had practically run into the tank's weapons. We ran, leaping and jumping over the ground like weasels, the hedgerows protecting us from being seen. I could not go on anymore; the last few days had taken too much out of me. Hubert Meyer took over command and motivated everyone to continue. Soon all had passed me. *SS-Obersturmführer* Köln and Michel remained with me. Machine-gun fire flew around our ears. Tears were running down Michel's face; he couldn't get me out of there fast enough. He encouraged me like a mother does a child. Repeatedly I heard: "Commander, come! Commander, come! Only a few hundred meters to go. Please, commander, come!"

We ran alone across a field; I had given up trying to keep under cover or running doubled over. I stumbled slowly eastwards and fell into a ditch. My comrades were lying there waiting for me. We scrambled across the road and struggled up the ridgeline that ran from Chambois to the northeast. We gazed speechlessly into the pocket behind us and cursed the men who had so precipitately sacrificed two German armies.

We had discussed shortening the front at the Seine ever since the loss of Caen. We thought that a pre-emptive evacuation of western France to a position on the Seine would have been possible. Behind the Seine the infantry divisions that had been so carelessly sacrificed might have been able to prove their worth and the *Panzer-Divisionen* might have had time to get refitted.

We marched along the ridgeline; we were subjected to intermittent shell fire there. In complete ignorance of the situation, we didn't expect to link up with our units until we were across the Seine. In the area south of Vimoutiers we encountered outposts of *SS-Panzer-Aufklärungs-Abteilung 2* of the *2. SS-Panzer-Division "Das Reich"*. At the command post of *SS-Panzer-Grenadier-Regiment "Deutschland"* we were told that the regiment and other formations were taking part in an attack against Chambois to open the pocket. The initial attack with small forces failed but, on 21 August, it was successful. The way was opened for numerous motorized and non-motorized formations to escape. I was able to give a useful situation report to the regiment before the attack.

Part of the division's motorized group managed to escape the pocket on the afternoon of 20 August; additional elements followed the next day. The artillery had lost several of its heavy guns; the 37 mm battery managed to fight its way out almost intact. The commander of *SS-Nachrichten-Abteilung 12*, *SS-Sturmbannführer* Pandel, was killed trying to save a valuable signals vehicle.

The battle for Normandy was over.

German soldiers had once again accomplished the seemingly impossible. They did not deserve the terrible defeat of the Falaise Pocket. Officers, non-commissioned officers and junior enlisted personnel had carried out their duty to the very end. The defeat could not be blamed on the front-line soldiers. The bitter cup had been served to them by gamblers at the map table. The performance of the German soldier in Normandy would be immortalized forever in the history books.

Our former enemy's judgment speaks for itself: "The only guys who really earned medals in this war," a rifleman said, "were those *SS* guys. Everyone of them deserved a VC. They were a bad bunch of bastards, but were they ever soldiers! They made us fellows look like amateurs."

"The fighting record of the *12. SS-Panzer-Division* in Normandy was probably excelled by few divisions, either Allied or German."

I have already frequently mentioned that the Canadian commanders were very hesitant and would only attack with vastly superior forces. That applied especially to the Canadian Operations "Totalize" and "Tractable". During those operations the Canadians not only lost the initiative on the battlefield but also lost a great chance to eliminate the smashed German armies completely.

After 4 August the Canadian II Corps only fought against the *12. SS-Panzer-Division*. Our division had hardly the strength of a battalion. Precious time had been lost. A single Canadian division would had been sufficient to hold the northern flank of the pocket and exert pressure on the concentrated German troops. The three remaining divisions — including among them two extremely well-equipped armor divisions — could have closed the pocket at Trun and Chambois on 16 August at the latest. Our units would not then have been in a position to break through such a ring.

Had the Canadian command done this the battle of the Ardennes would, in all probability, not have taken place. The *Panzer-Divisionen* which were the decisive formations in that offensive could not had been refitted in such an astonishingly short time had the core of veteran personnel not succeeded in breaking through the blocking positions between Trun and Falaise. In my opinion, that escape can be traced back to hesitant and indecisive Canadian leadership.

I believe the Allies drew the same conclusion concerning the leadership of the Canadian 4th Armoured Division when they carried out a change of command. But the fault must not lie with Brigadier Kitching alone. The Canadians could not have been unaware of the nature of their opponent. Their intelligence was good and air reconnaissance was constantly overhead.

The commander of the advance-guard battalion, Major D. V. Currie, was awarded the Victoria Cross, which he fully deserved. Major Currie had advanced as far as St. Lambert sur Dives on 19 August, thus effectively blocking the only escape route from the pocket. That battalion's performance was outstanding. Its dead and living soldiers deserve our respect.

*

I reported to the *I. SS-Panzer-Korps* on the afternoon of 20 August with that part of the division that had broken through. We were greeted with joy and thanks; They had thought we were dead. I could not help but cry when I rendered my report. Thousands of my comrades were resting in the soil of Normandy.

The situation briefing indicated there was no stable front west of the Seine and no defense lines existed to the east of it. The prospects were catastrophic, and we could only put our trust in the *Westwall*.

I enjoyed hearing that those elements of the division that had been pulled out earlier for refitting had fended off the enemy along the line Laigle — Verneuil — Dreux and thereby prevented the formation of a new pocket west of the Seine. Those formations had acted entirely on their own. *SS-Sturmbannführer* Gerd Bremer was decorated with the Oakleaves to the Knight's cross for that.

The division's headquarters staff reached Louviers via Le Neubourg. From there it would assume command of the elements of the division that were still operational. The rest of the division was involved in mobile rearguard actions with American tank formations advancing northwards from Dreux and Verneuil. Meanwhile, the troops who had escaped from the pocket crossed the Seine below Rouen. The staff had crossed the Seine at Elbeuf without suffering any losses.

I reported to the Commander-in-Chief West, *Feldmarschall* Model, in Rouen. The *Feldmarschall* harbored no illusions about the situation and talked about needing 35 to 40 divisions if the front were to be stabilized. Because we all knew that 40 divisions were not available, we turned again and again to the *Westwall*.

One improvised divisional *Kampfgruppe* held Elbeuf until 26 August. After the evacuation, the *Kampfgruppe* defended the loop of the Seine south of Rouen at the Forêt de la Londe, thus allowing our forces to disengage from the enemy.

Our soldiers fought Canadian units for the last time in the Forêt de la Londe. They held up the Canadian 2nd Infantry Division until 29 August. The *Kampfgruppe*, under the command of W. Mohnke, finally withdrew that afternoon.

After a two-day stay in the Beauvais area, the division was moved to the area around Hirson. Refitting so close to the front was impossible. We marched over World War I's blood-soaked fields under the cover of darkness, taking the same roads we raced down in 1940 when we were heading west. Our march column looked miserable; the convoys rolled through the night with each operational vehicle towing several others.

In Hirson the division was attached to the *General der Panzertruppen West*, *General* Stumpf, who was personally briefed about the division's personnel and equipment situation. *General* Stumpf told me the news of the award of the Swords to the Oak Leaves of my Knight's Cross.

The division started to reconstitute itself at once and re-equip the decimated units. Equipment was supposed to be issued from Verdun and Metz. The losses in men and equipment were frightening. The combat elements had lost more than 80% of the manpower with which they had started the campaign. The combat-support elements had also suffered unusually high casualties through enemy air action.

The division had lost more than 80% of its tanks in combat and during the withdrawal. About 70% of its armored cars and *SPW*, 60% of its guns and 50% of its vehicles were lost. Those enormous losses could not be made good in a few days, but we had no other choice — the division had to be combat ready as soon as possible.

We did not like the area around Hirson nor the overall situation. The combat-support elements and those units not ready for combat were immediately transferred east of the Meuse. By 31 August the Americans had reached Soissons and Laon and were advancing northeast. A *Kampfgruppe* from the division delayed them on the Thaon until the night of 1/2 September. In the meantime, *Gruppe Mohnke* had arrived at the division.

Because the division was threatened in its rear, it pulled back to the northeast and occupied a blocking position at Anor. We had to fight for time to enable the infantry to cross the Meuse. During the move to that position, the commander of the *III./SS-Panzer-Grenadier-Regiment 26*, Knight's Cross holder Erich Olboetter, was wounded while driving over a mine laid in the road by partisans. Both his legs were ripped off; he died during the night in the military hospital at Charleville. In Erich Olboetter I had once again lost an old warrior friend who had constantly fought at my side since 1939. He was an aggressive soldier and an ideal commander.

During the night of 1/2 September we held the blocking position at Benumont together with the remnants of the *116. Panzer-Division*. The division pulled back to Florennes via Philippeville under enemy pressure. Shortly before reaching Florennes, the commander of the *II./SS-Panzer-Grenadier-Regiment 26*, *SS-Hauptsturmführer* Heinz Schrott, was killed by perfidious partisans.

The "glorious" fight of the so-called partisans was nothing more than mean, common murder. The originators of partisan war were the real war criminals of that war. They acted against all humanity and appealed to the basest of instincts. I had never previously experienced partisan warfare; nor had I felt the frequently claimed French or Belgian hatred. On the contrary, I was always able to observe good relationships between the units and the population of the occupied territories. This observation was especially valid for the population of Normandy which had faced so much suffering.

The so-called partisans only raised their heads when they did not have to fear for their own bodies and lives. They did not fight; instead, they treacherously murdered individual members of the German Army. Seen from a military point of view, the actions of the partisans did not have any influence on the German conduct of the war. The population that was not involved suffered the most harm as a result of the retaliations of German troops. It was not the advocates of partisan warfare, a violation of international law, who suffered. The hatred between the nations was stirred according to a plan. It was deepened for a long time through the criminal activities of the partisans. Nor can one deny that the Allies actively promoted communism in Western Europe with their partisan policy. Without the perfidious actions of the "brave" partisans there would have been no cause for "war crimes trials".

We crossed the Meuse at Yvoir on 4 September to occupy a defensive position beyond that sector. The division assumed the sector Godinne — Houx. The *2. SS-Panzer-Division* assumed the sector on both sides of Dinant.

The combat strength of the division was roughly 600 infantry; it was divided into two *Kampfgruppen*. Tanks were no longer available; the remaining tanks were in Lüttich being repaired. There was no ammunition available for the heavy field-howitzer battery. One 88 mm *Flak* battery was deployed at the crossroads northwest of Spontin in a ground-support role.

The Americans immediately tried to cross the Meuse at Godinne and Yvoir. They were repulsed with heavy losses. However, they succeeded in creating a bridgehead at Houx. They advanced into the woods and established positions there. During the counterattack the bridgehead was reduced and was supposed to be eliminated before the onset of darkness on 6 September.

I covered the entire front and discussed the further defense of the Meuse with Milius and Siebken. Our vehicles were often shot at by the partisans in the woods. There were no casualties on our side, but we found six soldiers of the Lüttich security battalion murdered; they had been shot while resting. A reconnaissance patrol was fired on between Spontin and Dinant. The culprits were not found.

During the night of 5/6 September the Americans managed to cross the Meuse at Namur and repair a bridge that had not been properly demolished. The local commander of Namur fled to the east without informing his neighboring units, thus leaving the way into the heart of the Meuse defenses open to the Americans. A reconnaissance patrol of *SS-Aufklärungs-Abteilung 12* encountered an American advance-guard battalion on the Namur — Ciney road around 1100 hours.

I was on the way back from Siebken's command post when this bad news arrived. It seemed unbelievable to me, but the report was confirmed by another patrol at 1115 hours. The units were alerted at once and received orders to withdraw behind the Ourthe. The withdrawal could only be conducted at night. Speed was of the essence! We were pressed for time! The Americans would be at Durnal soon, and it would be only a matter of minutes before they reached the crossroads there. The operations staff would have to use those crossroads, if it wanted to escape from the Americans.

In an instant the operations staff was racing towards Durnal. I led the group down a steeply sloping field to reach Durnal through a patch of woods. Just before reaching Durnal, Hubert Meyer requested me to turn over the spearhead to *SS-Hauptsturmführer* Heinzelmann. I waved Heinzelmann past us and his vehicle overtook us just as we approached the first houses in

Durnal. The town lay in a deep defile; to the left of the road was a 1.5 meter wall around which the road curved to the east. As always, I was standing in the vehicle and trying to get a look "on the other side of the hill". As a result, I was able to see over the obtrusive wall to the main road to Namur. I shouted a warning to Heinzelmann, but it was too late! One round tore the lead vehicle apart, and the first American tank came around the corner firing.

The situation changed in the blink of an eye. It was no picnic attacking a tank column with a couple of *Volkswagen*. We could not turn round. I looked at the tank as it rolled slowly forward. From my own experience in similar situations, I assumed the tank commander would use this unique opportunity to overrun the operations staff or eliminate it by fire. There was nothing left to do but get off the road as soon as possible!

I leapt over a gate and a wire fence that separated the courtyard from the garden. What an unpleasant surprise, however! I was in a terrible trap. I could not escape behind the row of houses; the buildings had been built into the ascending high ground and also surrounded by a high wall. If I tried to climb it, I would present myself as a perfect target to the Americans.

The first thing I had to do was find a place that offered concealment. A chicken coop was the only option, so off I went! A body flew over the wire. Max Bornhöft saw me before I had disappeared. At that point, both of us were in the trap. At least the chicken coop offered some concealment from the enemy's view for the time being. We hoped to find our way back to our comrades after nightfall.

There was loud shouting from the road as the population cheered the Americans. The tanks rolled past. I heard an excited exchange of words in a nearby house and heard the name Köln. I never saw him again. *SS-Obersturmführer* Köln has been listed as missing in action ever since.

By then it had become 1400 hours; a light rain drizzled on to the chicken-coop roof. I could no longer take it. I had to know what was happening on the road. I crawled to the wire fence on my stomach. I had barely reached the corner of the coop when I experienced one of the most hair-raising moments of the war.

Some partisans came to the fence and talked to the farmer. They probably wanted to know if he had seen any German soldiers on the farm. The farmer shook his head. With gritted teeth I was only a couple of meters from the partisans. They leaned on the fence and were visually searching the hill. Would these be my last minutes on earth? I gripped my pistol firmly; they wouldn't get me without a fight. A nettle bush was my concealment.

Shouts and fired rounds attracted the fellows to the neighboring farmhouse. A comrade's life had met its end. We felt somewhat better. The farm had been searched, after all, and perhaps the rain would keep the curious away. The minutes became hours. We were pleased about the weather.

Suddenly, we were dumbfounded. The chickens were gathering in front of the coop and wanted to enter it. But they didn't want to share their quarters with us. They wanted us out. The story could not end well, so what had to happen happened. The little old farmer stayed at the fence, wondering what was going on. He then attempted to shoo his poultry into the coop. The critters were stubborn, however. They wanted to have their "empire" all to themselves.

Being curious, the farmer stuck his head into the coop. He should not have done that because, before he could open his mouth, he found himself sitting on an old barrel in the darkest corner. At that point, he had become the third man in the barrel. He looked at our pistols in terror. We could have done without our visitor. The situation had become more complicated. With our luck, the farmer's wife would soon be joining us. She would certainly soon miss her lord and master and go looking for him.

We decided to release the old man. He promised to keep his mouth shut and not to contact the partisans. He shuffled away quickly. Of course, we did not take his promise at face value. The old fellow had barely disappeared, when we climbed over the high wall and promptly landed outside the headquarters of the partisans.

I had not counted on that surprise. It couldn't get much worse than that. The partisans were housed in the church boiler room and a young fellow was standing in the cellar door enjoying his first American cigarette. Heavily armed partisans were coming up the cellar steps. We jumped, crawled, and dashed across the cemetery like weasels. The old graves and headstones protected us from being discovered.

We reached the compost heap in the corner of the cemetery. Because nothing else occurred to me at the moment, I covered Max with old wreaths and asked him to keep an eye out on the entrance to the church. I intended to hide behind some vegetation.

A shout echoed across the cemetery; it told us we were in a very dangerous situation. While still turning around I was looking down the muzzles of the weapons of two policemen on the church steps. The police were startled; they had not seen Max as yet. I raised my pistol like lightning and indicated I was about to fire. The police took cover. I had to get away! I ran to the southern edge of the cemetery and again found myself looking down the muzzle of a carbine. Its owner was standing in a doorway and took off when I ran directly towards him and threatened him with the pistol. We were surrounded. The old man had alerted everybody. I jumped over the cemetery wall, landing in the village street which was some 4 meters lower. Max was grasping for breath behind me.

Jesus! It was unbelievable how agile you could become if your life was at stake. The street inclined up the slope. My lungs seemed ready to burst. Rounds were whistling around our ears. I heard a scream from Max. I turned and fired a couple of rounds. Max was lying in the road. He had been shot.

My rounds had forced the "brave freedom fighters" to take cover. I turned towards the way out of the village. Just in time I saw two more partisans pulling guard there. Where could I go? I saw a small door held in place only by a large stone. I took cover behind it unnoticed.

I sat in the corner of a stable in complete exhaustion, peering through the cracks in the door. The partisans appeared in a few moments running excitedly up and down the road searching every bush. My disappearance could not be explained and they started blaming each other.

In a loud voice, one of the partisans demanded I leave my hiding place and surrender. He promised to hand me over to the Americans and respect "international law". I did not respond to his request.

My pistol seemed to become heavier and heavier in my hand. At one time we swore that we would never be captured alive. The grim experiences in Russia had made us do that. The time had come! There was a round in the chamber and a last one in the magazine. Should I fulfill my oath? Was it only valid on the Eastern Front? Weren't these completely different circumstances? Minutes passed. I looked at my pistol again and again.

I thought of my family and the unborn child. It was difficult, very difficult, to make a decision. The partisans were standing only a couple of meters away from my hiding place. I studied their faces. Some had embittered, brutal features, others seemed quite harmless, maybe they had only been given their weapons a few moments before.

The leader of the group asked me to give up again. A boy of about 14 stood next to him. It was obviously father and son. The little rascal had a carbine.

The boy suddenly pointed excitedly at the door and at the stone that had rolled to one side. He understood. It was dry where the stone used to be, so it must have been moved only a few minutes ago. The father asked me to give up again.

A round was fired through the door, and hand grenades were being called for. Two more rounds splintered the door and forced me further into the corner.

I called out to the father: "My gun is aimed at your boy! Will you keep your promise?" At once he pulled the boy to him and repeated his promise to treat me properly.

It was over. A counterattack by my comrades was my only hope. I threw the pistol's magazine into one corner and the pistol into another. What a horrible feeling it was being taken prisoner!

I slowly opened the door and walked toward the partisan leader. Some of the fellows moved to attack me; several pistols and carbines were pointed at me. Not a word was said. I took no notice of the threatening weapons. I looked into the father's eyes. With a wave of his hand he forced his companions to lower their weapons. They obeyed grudgingly and accompanied us to the church. The partisan leader told me he had been in Germany during World War I as a worker and had had only good experiences. He therefore had no reason to lead a gang of murderers. However, he said it was sometimes difficult to stop the young men from carrying out murder and manslaughter.

Max was still lying in the road; he had suffered a very unpleasant bullet wound in his thigh. We carried him into the police station where he immediately received an anti-tetanus shot. The village doctor was exceptionally friendly. He told us he hoped we would soon be able to return home.

The two policemen then took two pairs of handcuffs from their pockets and put them on my wrists. I almost fell to my knees in pain. The cuffs were practically pinching off my hands. They cut ever deeper into my flesh. They look at me expectantly; the partisans were also gaping at me. Those two must have often carried out this torture in the past. It was apparent they were waiting for me to cry out in pain. Max looked on and said: "The bastards!"

The ringleader returned to the room and gave the order to march me off. We stumbled across the cemetery and arrived back at the boiler room, the partisan's base of operations. Max was put on a straw mattress. In astonishment, I watched the two policemen open the boiler door and pull out their civilian clothes. They left soon thereafter clothed as partisans. I secretly cursed the German field-police unit that had been stationed next to the church. The fellows must have been asleep at the wheel; frontline troops were paying the price for their incompetence. Max was in severe pain. He asked me again and again to contact his father should I survive. He had little hope for himself.

The hours passed slowly in the cellar. The partisan leader brought some bread for us. He was worried; he knew there were German troops west of the village and they would probably pass through Durnal during the night. I listened to every noise. I understood from the partisans' conversation that the Americans had moved on towards Dinant and there were no Americans in Durnal at the time. A very quiet partisan guarded us. If only he wouldn't be so careless with his pistol. Every time I tried to make Max more comfortable, he shouted and waved his pistol at us. The young man was terrified. I found out why a few hours later.

The partisans departed suddenly at midnight and only left our guard with us. Before leaving, they moved a heavy table between the guard and us. It divided the room into two. The pistol was pointed at me all the time. Sometimes I got the feeling he wanted to kill me.

Vehicles moved through the village: Were they German or American? We did not have to wait long for an answer. After another hour I heard rounds being fired and the explosion of ammunition. Ammunition whizzed through the air. It was probably a German vehicle that was burning.

By dawn the noise was all around us. We could tell the distinct difference between German and American machine-gun fire. We listened anxiously to the noise of battle. Our guard became more and more restless; his pistol was permanently pointed at me. He even refused to give Max some water; he was afraid to turn his back on us.

The windowpane was suddenly shattered by machine-gun fire and the Americans asked us to surrender. That was an interesting turn of events! Terrorized, the guard threatened us with his pistol from a corner. I had to shout at him to open the door and stop the wild shooting of the Americans. The Americans continued to fire. Our partisan opened the door at last and called to them. Machine-gun fire slammed against the church wall, and the first American charged down the cellar steps.

In astonishment, I watched our guard receive a rude kick in the ass. He flew into the corner. The partisan turned out to be a deserter from Lorraine. He was upset because the Americans were treating him like a bum.

My astonishment vanished quickly; a submachine gun was pointed at my stomach. At the same time, a second American shouted: "Don't resist! My buddy wants your medals as a souvenir." Powerless with rage, I allowed my Knight's Cross to be stolen. It had been with me since April 1941.

The second American spoke fair German; his mother had been born in Germany. He said to me: "For God's sake, don't tell them who you are. Your soldiers are being treated badly in the rear!" I didn't understand the meaning of his words until 24 hours later.

We climbed the steps and ran into German machine-gun fire in the cemetery. I saw the firing position in the woods, barely 150 meters from the church. We lay between the graves and waited for a break in the firing. Before I really knew what was going on, my watch and rings were stolen. I had fallen into the hands of gangsters.

My money was stolen from me behind the church and another GI took custody over me. Angry that there was nothing left to steal, he hit me in the back with the butt of his weapon. After a few meters he had had enough. As we passed two frightened women standing in their front doors, I received another blow to my back. I stumbled a few steps and turned to receive a full blow to the side of my head with the stock of the weapon. I collapsed to the ground. As I fell I heard the protests of the women.

Further blows drove me to my feet and I staggered across the street. The cries of the women resounded in my ears. A few steps further on the gangster shoved me into a little garden. Blood clogged my eyes and poured out of my left ear. I couldn't think any more. I saw a large currant bush in front of me, into which I was shoved. So that was going to be the end of the road for me! An image of my family passed quickly in front of my eyes. The official report would be "missing in action". My real end was to be murdered and then dumped into an unmarked grave!

I looked at the man with burning disdain while he raised his carbine. I no longer saw him at all; I was already in the next world. In astonishment I saw the carbine drop; he left me with a "Damn" and took off.

My rescuer was the young lieutenant whose mother was German. He intervened at the last moment and stopped the soldier's act of violence.

The lieutenant tried to excuse the behavior of the soldier. He said it was the armchair warriors' irresponsible propaganda campaign that was responsible for the brutalities. So as not to stain his vehicle, I sat on a fender. The windscreen was red with my blood in a few moments. The blood was blown onto it as we moved. We stopped at an American supply column after a short ride. There I was handed over to the leader of the column with instructions to take me to a hospital.

The column was comprised of about 12 trucks and staff cars. Each was equipped with a machine gun. All the vehicles had a driver and two assistant drivers. I observed how well the soldiers were resupplied with an astonishment mixed with jealousy. An advance-guard task force composed of a tank battalion and an infantry battalion had assembled as if it were about to go on a parade. Tank after tank and vehicle after vehicle were staged closely together in an open area. The formation was resupplied without any ground or air defensive measures in place.

Five *Tigers* attacking out of the woods would have destroyed this entire group. But there were no *Tigers* between the Meuse and the German border. There were only battle-weary men in this zone. They were wandering around, smashed by fate. Would the Americans be halted at the *Westwall*? I didn't think it possible. I knew there were no more functional divisions and, furthermore, the *Westwall* was incomplete and had long been neglected.

In truth the Ruhr lay undefended in front of the Allied spearheads and nothing could prevent Montgomery from occupying Germany's weapons forge. One powerful drive by 10 to 15 Allied Divisions into the northwest area would break the backbone of German resistance and end the war in a few weeks. The conflict for Europe had been lost.

The groans of an injured comrade brought me back to reality. He was in a neighboring vehicle with an abdominal wound. I suddenly saw a hand waving in the third vehicle; it was Max in a fuel truck. The empty cans formed his sick bed. About 60 German prisoners had been gathered in the meantime. Some paratroopers, about 15 soldiers from my division and members of a security battalion were distributed among the trucks.

The column started off towards Namur late that afternoon. I carefully observed tracks through the grain fields. Would an escape attempt be successful? It was obvious that I had thoughts of escape. I nudged a young paratrooper and indicated my idea. He nodded his agreement and moved closer to the edge of the vehicle.

The Americans were careful. Each vehicle was covered by the loaded machine gun of the vehicle following and, in addition, an American with a submachine gun sat on each cab. As a result, only a road with numerous bends that passed through wooded terrain could be considered for an escape. We had no such luck; we arrived in Namur sooner than we thought. The best chance of escaping was gone, but the thought of escaping stayed with us. We wanted to be free again.

There was much activity in Namur. The bridge over the Meuse had been repaired by the American engineers without a great deal of difficulty. The population either watched us indifferently or had a threatening manner. Our column moved through the middle of the town and stopped in front of a big building, which I later discovered was the prison. It was close to the railway station, and we were surrounded by curious civilians. Women pointed to me. I was completely splattered with blood and made a pitiful impression.

I saw Max Bornhöft being lifted from the truck and carried into the building after a few minutes. Partisans and policemen received him at the entrance. The onlookers surrounded the group. Then the unbelievable happened; a shot rang out and turned the civilians into a raging mob. The road came alive with jeering, cheering and the clapping of approval. Without warning and in a fraction of a second, my companion of so many years — a brave soldier and good comrade — was murdered by a cowardly and mean animal. The accompanying Americans shook their heads over this murder lust and chased away the cowardly bullies.

The column moved off again. So, that was how imprisonment looked like after 24 hours as a prisoner! The brutal murder of bleeding, wounded soldiers! Oh, what naive fellows we had been a half an hour before we had just seen what had happened.

Our journey ended in a police-station yard. The station was in the center of Namur, and I clearly remember the entrance gate. Near the entrance there was an old Gothic church.

It was dark when we passed through the gate. Partisans guarded the entrance. That was a bad sign. The vehicles had hardly departed when we were shouted at by the young slavering fellows and ordered to fall in line. Those comrades who were the last to dismount were struck with a hail of blows from rifle butts. I stood to the right and saw an American talking to the partisans while pointing to me. The partisans nodded their heads and asked me to follow them. They took me to a *Feldwebel* who bandaged me. While that was going on, I heard cries of pain from my comrades. A partisan group was beating a group of soldiers. I asked: "What was going on? Why the beating?" The sergeant told me they were sorting out the *Waffen-SS* and *Fallschirmjäger* and intended to kill them! Then rounds rang out across the yard, and about 20 German soldiers were murdered on that 7 November 1944 in Namur. They did not die at the hands of Belgian soldiers but at the hands of fired-up adolescents who displayed their red scarves like important decorations.

At about 2200 hours I was taken through the empty streets between two partisans. Our steps sounded hollow in the dead of night. The sound resounded in my ears like the dull roll of drums of death. Once again I suspected my end was near but, this time, I was wrong. One of my companions suddenly started talking. He offered me an American cigarette and asked how much pain I was in. "You know," he said, "it is really amazing you can still walk with your fractured skull. We have orders to take you to a doctor, and we'll be there in a few minutes."

It looked like the Americans had told the Belgians I had a fractured skull. I was to be medically treated on the orders of an American officer. The "fractured skull" had to be traced back to the constant bleeding from my left ear. Well, I hoped the story would end well. I knew, of course, that I didn't have a fractured skull, but I did not know why I was bleeding. I was told later that the American had broken one of my blood vessels.

After a few hundred meters I was led into some sort of school. The partisans and youths shouted: "SS, SS...?" My escorts replied: "No, He is an *Oberst* of the *2. Panzer-Division*. The Americans wanted us to take him to a hospital." The youths mistrustfully pushed me into an ambulance and I was taken to a Catholic hospital. On the way, I was told by a seminarian and by one of my escorts that *SS* men and *Fallschirmjäger* were being shot at once.

I listened to that monstrous statement almost with indifference. Many a young soldier had died at the hands of the murderers, without himself having killed anybody. After all, many of my soldiers had only just been transferred to the division during last few days. Hardly 18 years old, they had become victims of an incited mob.

Once again I saw my end in front of me. It could only be a short time before I was identified. I feverishly thought how I could get rid of my *Soldbuch*. I could not leave it in the ambulance; it would have been found in a few hours at the most.

I was brought into an operating room and had to lie down on a plank bed. A very friendly, German-speaking nun administered to me. The seminarian and the nun were siblings. The partisans saw me for the first time in the candlelight and watched me with interest. They had their suspicions. The camouflage jacket seemed familiar to them and identified me as a member of the *Waffen-SS*.

It was high time for me to get rid of my treacherous *Soldbuch*. But how? Those fellows were watching me like hawks. At the last minute I asked the nurse to let me relieve myself. After a short hesitation, she gave me permission and I staggered a few doors on. One of partisans accompanied me. The pay book disappeared with lightning speed. I realized too late, however, that the pipes were broken and that my identification would only be delayed for a few hours. The doctor decided I should be put to bed first and X-rayed tomorrow. I staggered into a ward rather weak at the knees and was put to bed with the help of the partisans. The loss of blood had exhausted me. I really was at the end of my rope.

My filthy jacket was taken from me by the partisans, and I watched them going through the pockets looking for my papers. Would I continue to have luck in the future? It did not take long, and I was asked where my pay book was. It all depended on how convincing my reply was. I opened my eyes slowly and responded in a firm voice: "Americans". For a fraction of a second I didn't dare to breathe. Then I saw my guards were satisfied with my response. They shook my hand and disappeared. I was given a fresh pillow in the middle of the night because I was still losing a lot of blood. I fell asleep in exhaustion and dreamt of my part of Germany. I remained in the hospital for two weeks. The doctors treated me excellently. The nuns secretly gave me cigarettes and also bought me a snack occasionally. I felt better day by day and I sensed I would not be in the hospital much longer.

My thoughts turned feverishly to escape. I was on the third floor and the only escape was through the window. I would have to find some sheets and try to make a rope. I was moved to the Albert Barracks before I could carry that out, however.

The barracks were near the railway station and were used by paramilitary units. I did not like the fact I was the only prisoner and was kept alone in a deserted room in the corner of the barracks. My new lodgings were hardly suited to escape. High walls and stringent security obstructed my path to freedom. My solitary confinement ended after 48 hours. The partisans sent me a fellow sufferer in the afternoon. *Leutnant* Aumüller was caught north of Namur trying to reach the German border with a group of German infantry. They had fought their way through the woods and fields of northern France and Belgium for more than three weeks. Dependent solely on the assistance of the locals, they had marched hundreds of kilometers only to be caught just short of the border.

We worked together to try and improve our situation. Up to that point we had endured miserable conditions. The heating problem was swiftly solved. We took the chairs, tables and wardrobes apart and burned them in the fireplace. However, the rations could not be improved so simply. We were given the same soup every day.

An addition to our family arrived two days later. *Leutnant* Wagner, a platoon leader in an infantry division, was the third man to join us. Wagner, like Aumüller, had moved east for weeks only to be captured on the Meuse. He was what we had been lacking. He had somehow managed to hide a few hundred Francs which would help us improve the menu.

It was our intention to engage the guards in conversation so as to enlist them for our purposes. We were pleased that a former Belgian military cadet, who had been in a German prisoner of war camp until 1943 and who had been released on the intervention of the king, became our best ally. He had been a career Belgian soldier and acted accordingly. Showing disgust, he confirmed that German soldiers had been murdered. The red partisans had been responsible for that.

A Russian prisoner, who had been captured in 1942 and managed to escape from a Belgian mine in the spring of 1944, helped our friendly Belgian in supplying us. In due course two legitimate freedom fighters, who fought for their country out of true idealism and had also spent a long time imprisoned by the *Gestapo*, joined our helpers.

We could only maintain our physical strength with the help of these men. Of course, even they could only obtain some bread, potatoes, carrots and fruit. Indeed, they did not have much food themselves and were dependent on receiving rations. However, the decent, upright attitude of those Belgians helped us over many a difficult moment. There was heavy fighting around Aachen around that time and we thanked our caretakers for every bit of news with all our hearts.

The former Red Guard soldier was not very happy when he heard of Germany's critical military situation, especially on the Eastern Front. He did not view a Russian victory with eagerness. That simple Russian soldier seemed to understand more about Russian intent than the fathers of the Casablanca Treaty. In any event he had a sounder instinct than Sir Samuel Hoare, whose response to a warning letter from Franco, the Spanish Head of State, on 25 February 1943 was:

I cannot accept the theory that Russia will present a threat to Europe after the war. In the same way, I reject the thought that Russia could start a political campaign against Western Europe after the end of hostilities.

You argue that communism represents the greatest danger to our continent and that a Russian victory would enable them to triumph over the whole of Europe. We have a completely different view.

After all, would a nation be able to rule Europe completely on its own after this war? Russia will be preoccupied with its reconstruction and would be dependent to a large measure on the help of the United States and Great Britain. Russia does not hold the leading position in the fight for victory. All military efforts are completely equal in the achievement of the Allied victory. After the war, large American and British armies will occupy the continent. They will be composed of first-class soldiers. They will not be worn out and exhausted like the Russian armies. I dare predict the English will be the most powerful military power on the continent. British influence on the continent will then be equally as strong as in the days after Napoleon fell. Supported by our military strength, our influence over the whole of Europe will be felt and we will participate in the reconstruction of Europe.

Thus spoke Sir Samuel Hoare, one of the leading British politicians. I believe that the course of history has confirmed the fears of our Russian caretaker. His fears have been proven justified. Russia had become the dominant power in Europe, not England.

Each bomber squadron following its low-altitude course over the high ground along the Meuse carrying its deadly bomb load towards the burning homeland convinced us to hatch escape plans. However, no possible route to freedom could be found. We were guarded too closely.

One day the Belgians brought us pieces of German uniforms from old German stocks in order to add to the clothes we had. I received a field blouse and a coat. At that point we were more-or-less protected against the cold, but we looked more like a band of robbers than German soldiers.

At the beginning of October, two Americans appeared under the command of an MP major and we were put in a truck. Our guard was very strong; again, we could not consider trying to escape.

We arrived at Reims that evening at a police station that the MP's were using. The cells were filled with rampaging Negroes. They were drunk with victory. The next morning we moved across the battlefield of Reims further and further to the west. Up to that point we had been occupying ourselves with plans to escape and noting the key terrain features. Starting then, however, we sat depressed in the corner of the vehicle and observed the huge Allied supply depots in astonishment. There were unbelievable amounts of ammunition, fuel and stores. Supply depot after supply depot, kilometer after kilometer. Between them were airfields and additional depots with reserve artillery and tanks. The traffic on the roads and in the camps moved as if it were peacetime. No trace of camouflage or defense precautions could be seen. We were looking at a rich man's house. Did the Americans understand, as they stood on the borders of Germany, how great their superiority in weapons and equipment was?

We moved through Compiegne late in the afternoon. It was a long way from the Compiegne of 1940 to the Compiegne of 1944. The road in 1944 led to a large prison camp. Where would it lead from there?

The camp made an enormous impression on us. Barbed wire as far as we could see. Before we could reach the interior, we had to pass through two exterior gates which were guarded by bored American troops. We were immediately brought before the camp commander who questioned us closely and individually. I was registered as *Oberst* Meyer of the *2. Panzer-Division*. My pay book was treated as misplaced by the Americans in Namur. The camp commander turned out to be an old Berliner whose law office was on the Kurfürstendamm and who had emigrated to America in the 1930's.

After a wide-ranging discussion, I was designated as assistant to the camp commandant and asked to supervise the officer's camp. I was given a small room with Wagner and Aumüller, and we were happy to have a little area to ourselves. As a matter of top priority, we carried out a thorough reconnaissance the next morning. The camp was divided into three sections and held several thousand men. As a camp assistant I could move from section to section looking for comrades without difficulty.

I met a *Feldwebel* from the *1. Fallschirmjägerdivision* during my first tour of the enlisted men's camp. He lets me into the camp secrets. It was swarming with spies and traitors. Caution was the watchword!

It did not take us long before we had all parts of the camp infiltrated with our people. Food for escape attempts was organized systematically and distributed to different barracks. We were even the happy owners of a compass. *Leutnant* Wagner had managed to keep it despite all searches. Everything revolved around the method of escape. Even bandage material was obtained. A *Fallschirmjäger* doctor had joined our group and wanted to attempt an escape with us.

One day a few hundred prisoners arrived from the Aachen battlefield and brought us the latest news from Germany. Among the prisoners were some soldiers from the *Leibstandarte* who told me about what had happened to my division. I was shattered when I heard of the murder of my loyal comrade Waldmüller. *SS-Sturmbannführer* Waldmüller had become a victim of the Maquis on 9 September at Basse Bodeux, 10 kilometers northwest of Vielsalm. They had stretched a wire across the road. Wallmüller and his driver had been jerked off their motorcycle. Both were seriously injured and subsequently drowned like rats in a water-filled ditch. Brave *SS-Obersturmführer* Hauck had driven onto a mine which was detonated from ambush and had suffered serious burns.

According to further information from the soldiers, the *12. SS-Panzer-Division* was being refitted in an area around Plettenberg in Sauerland. At that point, our escape attempt went into high gear. We were aware that if our plan were to succeed we had to work with the woodcutting detail. Time was of the essence; we expected to be evacuated in a few days to make room for new arrivals.

I had a most interesting if quite unpleasant conversation with the camp commandant on 7 November at about 1700 hours. I met him in the corridor of our barracks and I was astonished when he pulled me into a corner just like a comrade and started the following conversation: *Herr Oberst*, I have a big favor to ask you. I'm in a very unpleasant situation." Completely surprised, I promised him my help, but I had no idea what had happened. "*Herr Oberst*, this is really incredible! According to the latest reports I had just received, a senior *SS* officer was in the camp. It would be a very big disgrace for me if I really did have this fellow in my camp!"

Well, it had happened! Thank God, the light was so bad that the recently coined American could not see the color of my face. Surprise was total. My tongue lay like lead in my mouth. I didn't want to answer right away. I had to win some time. I had to stay calm, very calm! After taking a deep breath, I promised the commandant my cooperation and asked the name of the *SS* officer and what he looked like. The reply was something like this: "Well, I don't know his name. I don't know what he looks like, either. I only know that he is smartly saluted and everybody grins when he goes through the camp."

Reassuring him that I would follow the matter up, I calmly said goodbye and disappeared to my room. There was no lack of surprise when I told Wagner and Aumüller that I had just promised to search for myself and hand myself over to the camp commander!

We quickly agreed to mount the escape attempt the next day. *Leutnants* Aumüller and Wagner and *Feldwebel* Müller would also take part. On 8 November we reported as woodcutters at the camp gate. We would initiate the long way back to Germany from the woods at Compiegne.

We were out of luck. Very few woodcutters were needed. We crept back to the barracks with disappointed faces. I was summoned to the camp commandant at about 1100 hours. I said a few words of farewell to my acquaintances and followed the MP's.

The commandant was sitting behind his desk playing with his nightstick. He had a determined expression on his face. The expression on his face told me everything. I immediately noticed that he was not absolutely certain, however. I had to keep calm!

Military policemen as tall as trees stood at my side. Their nightsticks were held loosely in their hands. It took a serious turn! The newly coined American shouted at me: "Take your jacket off! Remove your shirt!" I stood there with the upper part of my body naked. "Raise your arms!" My God! The color of the camp commandant's face changed. He gasped for breath and gripped his desk tightly. He angrily stammered a couple of curses in English before he asked me about the origins of my blood group symbol.

Only then did do I realize why I had to strip. I really had not thought about that treacherous tattoo. I really had to gain time to find a plausible answer. I only responded at the second request. I replied with a question. I politely asked the camp commandant for an explanation. This nearly drove the good man crazy. He continued to shout at me and claimed I was an *SS* pig and was the wanted senior *SS* officer.

The time had come for me to act. I looked at the camp commandant and said: "You are mistaken. Although the *Waffen-SS* introduced the blood-group markings, the army *Panzertruppe* started using it as well on its tank crews after it had proved so useful with the *Panzer-Divisionen* of the *Waffen-SS*. You must come to terms with the fact that all students at the *Panzertruppenschule* were sent back to their units with the tattoo."

This explanation worked like a bombshell on the MP's. I was allowed to put my clothes on again and leave the room. The camp commandant remained sitting in his chair, exhausted by the confrontation.

The whole camp was turned upside-down half an hour later. The MP's were looking for radio sets. The inmates had planned an escape for 9 November and had requested an airdrop of weapons for that evening. We had to laugh at the imagination of the Americans.

We were suddenly ordered to assemble in march order. The transport to evacuate us had arrived. Within an hour we had marched through the town and climbed on board a waiting freight train. Wagner and Aumüller were with me in the same car. Wagner immediately sat on the floor and started to cut a hole in it. He started cursing up a storm. The car had been constructed out of hard wood, and he could hardly make an impression on it. We would not be able to open up the floor. The two of them remained undeterred. They wanted to escape.

I heard my name being called just before departure. I was asked to leave the train and to climb into a waiting jeep. They had gotten me after all.

We quickly returned to the camp. I was really surprised when I recognized a *Leutnant* in tramp's clothing in front of the administration building. He was getting ready to go on a little outing in the vicinity of the camp with some Americans. He looked away shamefacedly but in vain. My shout of "Miserable bastard!" reached him. That was the first cowardly creature I had encountered in German uniform. He was the camp's spy and had, in fact, been a young *Leutnant* in an infantry division.

The camp commandant forced me to stand against the wall in such away that only my fingertips touched it and my feet were about 1.2 meters away. Every time I moved I received a light blow with a rifle butt in my back and was mockingly ordered to maintain discipline. Everything eventually comes to an end and I was soon taken to a special room. There I was constantly guarded by two Americans with drawn weapons.

I was taken to Paris via Compiegne at some ungodly hour the next morning and, from there, to an airport west of Paris. By 1400 hours I was sitting in a plane watching the old battlefields of 1940 passing below me. We left the continent at Dunkirk and only then did I accept my destiny and forget all thoughts of escaping. The war was over for me. Europe lay in ruins behind me, bleeding from countless wounds. The Channel passed below and misty England appeared in front of me — the England that wanted to liberate Europe but, in truth, tossed it in the clutches of the Communists.

Kurt Meyer in Antwerp on 14 November 1943. (Jost Schneider)

Right: The swearing in ceremony of the *12. SS-Panzer-Division "Hitlerjugend"*.

318

Normandy, first days of June 1944, just before the invasion. Kurt Meyer driving the divisional commander, *SS-Brigadeführer* Fritz Witt, in motorcycle with side car. Behind Meyer sits the regimental surgeon, *SS-Sturmbannführer Dr.* Erich Gatternigg. (Jost Schneider)

Left: A divisional exercise is observed from a prime mover by (from the right): *Feldmarschall* von Rundstedt, *SS-Sturmbannführer* Hubert Meyer, *SS-Obergruppenführer* Sepp Dietrich, *SS-Oberführer* Fritz Witt, *SS-Standartenführer* Kurt Meyer. (Hubert Meyer)

A delivery of food to the Invasion Front.

Left: Summer 1944, France. Left to right: Wilhelm Mohnke, Max Wünsche, Kurt Meyer, Gerd Bremer. (Jess Lukens)

House-to-house fighting in Normandy.

Normandy 1944, *Panzermeyer* in a tailor-made jacket and cap of Italian camouflage material. (Jost Schneider)

Normandy 1944, Kurt Meyer as commander of *SS-Panzer-Grenadier-Regiment 25*. (Jost Schneider)

From the left: *SS-Standartenführer* Kurt Meyer, *SS-Brigadeführer* Fritz Witt, *SS-Obersturmbannführer* Max Wünsche in the Ardenne Abbey, beginning of June 1944. (Hubert Meyer)

Left: 37th birthday celebrations for the divisional commander on 27 May 1944 in Tillieres. lst row, 3rd from left: Schürer (divisional ordnance officer); Springer (adjutant); Kurt Meyer, (commander, *SS-Panzer-Grenadier-Regiment 25*); Witt; Mohnke (commander, *SS-Panzer-Grenadier-Regiment 26*); Schröder (commander, *SS-Artillerie-Regiment 12*); Rothemund (adjutant). 2nd row: von Reitzenstein; Manthey (divisional engineer); Pandel (commander, *SS-Nachrichten-Abteilung 12*); Wünsche, (commander, *SS-Panzer-Regiment 12*); Hubert Meyer (division operations officer); Buchsein (divisional staff officer); Schuch (divisional headquarters commander). 3rd and 4th row: Weiser (corps adjutant); Krause (commander, *I./SS-Panzer-Grenadier-Regiment 26*); Urabl (commander, *SS-Feld-Ersatz-Bataillon 12*); Bremer (commander, *SS-Panzer-Aufklärungs-Abteilung 12*); Dr. Kos (divisional commissary officer); Müller (commander, *SS-Panzer-Pionier-Bataillon 12*); Waldmüller (commander, *I./SS-Panzer-Grenadier-Regiment 25*); Siebken (commander, *II./SS-Panzer-Grenadier-Regiment 26*); Kolitz (commander, divisional trains); Hanreich (commander, *SS-Panzerjäger-Abteilung 12*); Ritzert (commander, *15./SS-Panzer-Grenadier-Regiment 25*). (Hubert Meyer)

Summer 1944: Kurt Meyer as *SS-Standartenführer* and commander of *SS-Panzer-Grenadier-Regiment 25*. (Jess Lukens)

Right: A corps artillery battalion, schweres *SS-Artillerie-Bataillon 101*, supports *SS-Panzer-Grenadier-Regiment 25* at the beginning of June 1944. 3rd from left, the battalion commander, *SS-Sturmbannführer* Steineck; *SS-Sturmbannführer* Hubert Meyer; *Panzermeyer*; *SS-Obersturmführer* Bernhard Meitzel, *SS-Hauptsturmführer* Günter Reichenbach. (Hubert Meyer)

Commander conference at the divisional command post in Caen-Venoix on 13 or 14 June 1944: *SS-Obersturmbannführer* Max Wünsche, *SS-Brigadeführer* Fritz Witt and *SS-Standartenführer* Kurt Meyer. (Hubert Meyer)

A bunker is built at the divisional command post in Airan. Standing in the middle, from the left: *SS-Oberführer* Kurt Meyer, divisional commander; *General (Luftwaffe)* Peltz, corps commander of a *Luftwaffe* field corps; *SS–Sturmbannführer* Siegfried Müller, commander of *SS–Panzer–Pionier–Bataillon 12*; and, *SS–Sturmbannführer* Hubert Meyer, divisional operations officer. Normandy, July 1944. (Hubert Meyer)

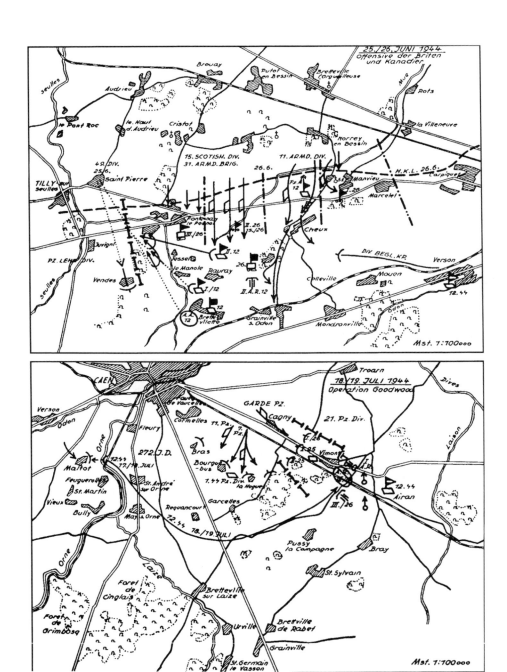

Allied ground offensives in June and July 1944, including the massive "Goodwood" operation.

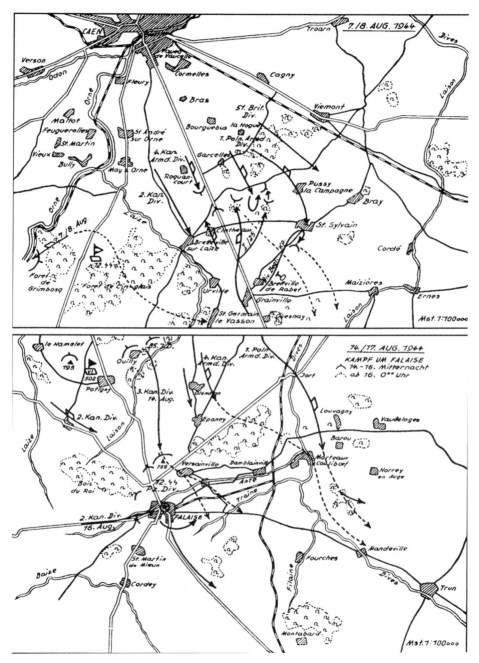

The end of the Normandy Campaign, the closing of the "Falaise Gap", August
1944.

The Employment of the 12. SS-Panzer-Division "Hitlerjugend" from the End of the Invasion to the End of the War

by Hubert Meyer, Former Divisional Chief of Staff

It was a heavy blow to the division to lose its commander, especially since his fate was uncertain. The men had not just heard-tell of him; instead, all of them had seen him at some point wherever there was fighting being conducted, particularly in difficult situations.

Those members of the divisional staff and the staff of *SS-Panzer-Regiment 12* who had survived the clash with the Americans at Spontin without being wounded hid in a patch of woods at the outskirts of the village until darkness. They were called on to surrender by the partisans who suspected that they were there. No one dared to enter the woods, however. The soldiers heard the American columns moving through the village and being enthusiastically welcomed by the civilians who came out of their hiding places. They also heard firing towards evening between their comrades hiding in the village and the partisans. They feared their divisional commander was in the village but, of course, did not know exactly where he was. They were not in a position to intervene with only two or three pistols among them.

The group of nine men marched east during the night. They didn't have a map; their only navigational aid was a compass. At first they went cross-country, then along a railway bed. They wanted to link up with their comrades.

They encountered an outpost of *Kampfgruppe Siebken* after midnight. It had been able to pull back under the cover of darkness. A patrol had been sent to Spontin; it reported the village clear of the enemy. A civilian said that a senior German officer with a decoration around his neck had lain injured in the street and was taken away by the police. We could only hope that the commander was still alive.

The operations officer assumed command of the remnants of the division, along with the staff of *SS-Panzer-Aufklärungs-Abteilung 12*. The *Kampfgruppe* of *SS-Panzer-Grenadier-Regiment 26* remained committed under army command. The rest of the division was pulled out and transferred to the Saar for refitting. The division was a shell of itself. The field replacement regiment under the command of *SS-Obersturmbannführer* Krause was brought up from Kaiserslautern. It had been pulled out of the lines earlier. The remnants of *SS-Panzer-Regiment 12* came from the Lüttich area. All the armored fighting vehicles were issued to the units employed along the *Westwall*.

The personnel roster showed that of an original total strength of 20,000 men (including reserves) in the division, about 10,000 men, including 21 unit commanders, were lost.

The artillery was practically without guns. There were virtually only light infantry weapons left, and the vehicle complement had shrunk to a quarter. The situation offered little hope, but nevertheless, training and refitting started at once. *Panzergrenadier* replacements arrived from their own replacement battalion and replacements for the other branches of service arrived from the respective replacement units. We received men from the *Kriegsmarine*, *Luftwaffe* ground personnel and even flight crews from the *Luftwaffe*. They had hardly any infantry training and their integration into the formations was not easy.

In place of the missing *II./SS-Panzer-Regiment 12*, which remained at the training area, *schwere Panzerjägerabteilung 560* was attached to the division from the Army. *SS-Brigadeführer* Kraemer assumed temporary command of the division, however, he transferred to the *6. SS-Panzer-Armee* at the end of November as its chief of staff. *SS-Oberführer* Kraas then assumed temporary command of the division in his place.

During the Allied airborne landings in the Arnhem area, a *Kampfgruppe* of company strength had to be detached temporarily. It returned to the division after a few weeks. In the meantime, the division had moved to the area around Sulingen on the lower Weser to be ready in case of an Allied landing on the coast. At the end of November it was moved to the area west of Cologne. It seemed the formation was being held as an operational reserve for the sector of the front east of Aachen; in reality, it was being trained and prepared for the planned offensive in the Eifel.

The division was moved to the Eifel under low cloud cover and in snow-storms in two night marches on 13/14 and 14/15 December. It was staged in the area east of Sistig and organized into three pursuit groups. Despite insufficient training and a shortage of vehicles and equipment, the division held a firm hope that it could help change the situation in the next operation. It was ready to give its utmost.

The division received the following order:

After the forward divisions break through the American lines, the division will move out on two separate march routes in two *Kampfgruppen* to pursue the enemy. On the first day it takes at least one crossing over the Meuse directly south of Lüttich. The *1. SS-Panzer-Division* is employed to the left of the division; the right flank is open.

The following information was briefed about the enemy:

The American 99th Infantry Division was deployed in the front line in a loose series of strongpoints across the division's axis of advance. Friendly reconnaissance had repeatedly managed to penetrate the enemy defensive system by several kilometers. The American 2nd Infantry Division, which had suffered considerable casualties on the Saar front, was in the Elsenborn training area. It was not fully combat ready. The American 1st Infantry Division was in the Verviers area for refitting. It was likewise not combat ready.

The offensive started on 16 December 1944. The *272. Volksgrenadier-Division* and the *3. Fallschirmjäger-Division* started the attack in the *6. SS-Panzerarmee* sector after an enormous surprise artillery barrage. The barrage hit few enemy positions because of inadequate reconnaissance. For the most part the *272. Volksgrenadier-Division* came into contact with the enemy only after the paralyzing impact of the artillery fire had ended. The attack soon stalled in the difficult wooded terrain east of Krinkelt. Part of the division had to pull back to its original positions.

To get the attack rolling again, The *I./SS-Panzer-Grenadier-Regiment 25* was placed under the operational control of the *272. Volksgrenadier-Division* but no progress worth mentioning was made. The enemy sent reinforcements to the *Schwerpunkt* of our attack. All of *SS-Panzer-grenadier-Regiment 25* and *SS-Panzerjäger-Abteilung 12* were attached to the *272. Volksgrenadier-Division* on 17 December. They reached the stream 5 kilometers east of Krinkelt by evening.

The reinforced *SS-Panzer-Grenadier-Regiment 25* was returned to the command of the 12. SS-Panzer-Division on 18 December. In the heavy fighting, during which 200 Americans of the 99th Infantry Division were taken prisoner, *SS-Panzer-Grenadier-Regiment 25* penetrated as far as the outskirts of the woods east of Krinkelt. The *I./SS-Panzer-Regiment 12* was ordered to attack Krinkelt. It entered the town towards evening despite fierce enemy resistance. Prisoners were taken from the American 2nd Infantry Division. In the meantime, the *1. SS-Panzer-Division* had managed to take Büllingen against weak enemy resistance by taking advantage of a breakthrough in the

3. Fallschirmjäger-Division sector. It had advanced further west towards Stavelot. As a result, the attack at Krinkelt was not continued and the division was moved up via Schmittheim, Manderfeld and Büllingen. When *SS-Panzer-Grenadier-Regiment 26* arrived south of Büllingen on 20 December, the enemy had reoccupied the town. The *I./SS-Panzer-Grenadier-Regiment 26* retook it.

On 21st December *SS-Panzer-Grenadier-Regiment 26* and *schwere Panzerjäger-Abteilung 560* started to attack Bütgenbach through the Bütgenbach estate. A *Panzerjäger* company entered the estate, but the attacked stalled in the murderous defensive fire. The company was pulled back. A new attack was launched on 21/22 December. After initial success, it too bogged down due to the artillery fire after it had become light. The estate was defended by elements of the 26th Infantry of the American 1st Infantry Division, which had been brought up from Verviers.

Another attack on Bütgenbach — down the western side of the estate — was made with and controlled by elements of the *272. Volksgrenadier-Division* on 23 December. At the same time, the remaining formations of the *12. SS-Panzer-Division* were assembled in the area around Amel. The *III. (gep.)/SS-Panzer-Grenadier-Regiment 26* entered Bütgenbach. The success could not be exploited, and the battalion was pulled back to its original position. There were heavy losses.

On the 26 December the entire division was moved to the Samrèe area, five kilometers northeast of Laroche. There it was placed under the operational control of the *II. SS-Panzer-Korps*. It was given the mission of breaking through the enemy-occupied forest north of Samrèe to Durbuy and then forcing a crossing over the Ourthe. The attack started after nightfall on 28 December. Elements of *SS-Panzer-Grenadier-Regiment 25* managed to advance as far as Sadzot, where they encountered units of the American 75th Infantry Division, which had just been deployed there.

Enemy tanks had already appeared; our own tanks could not be moved up through the woods. The attack, which had not been very promising from the start and which had been ordered despite our serious and energetically voiced misgivings, was called off and the units pulled back. For the third time the division had had the misfortune of being employed with too weak a force against a freshly deployed enemy and, once again, far too late.

The division was moved to the area north of Bastogne on New Year's Eve and came under the control of the *5. Panzerarmee*. Moving from Bastogne, the enemy had broken through the German positions northeast of the town. The *12. SS-Panzer-Division* was to reestablish the front with a counterattack. The enemy spearheads were thrown back by the *I./SS-Panzer-Regiment 12* south of Bourcy during the night of 1/2 January 1945. The division, whose combat power had shrunk to a quarter of its authorized strength, started an attack on Bastogne on 2 January. However, the neighboring formations did not participate in the attack in accordance with the operation orders. Despite that, Margerite was successfully taken on 5 January after heavy fighting and with heavy losses.

Hill 510 west of Margerite was taken in a difficult night attack by *SS-Panzer-Pionier-Bataillon 12* and the five remaining serviceable tanks of the *I./SS-Panzer-Regiment 12* on 3/4 January. It was soon lost, however. The enemy, which comprised the American 26th Infantry Division and the American 9th and 11th Armored Divisions, wanted to prevent us by all possible means from gaining observation points overlooking Bastogne. The operations south of Bourcy and at Margerite and Hill 510 could be considered as an excellent showing by the exhausted division, which had been bled to death. *Feldmarschall* Model paid high tribute to the division.

The Ardennes offensive had failed. The division was pulled out and moved to an area west of Cologne in a series of moves between 10 January and 6 February.

It is not the purpose of this chapter to analyze why the self-sacrificing operations of the units involved did not lead to the desired success. Doubtless they gave their best, but they were forced to operate out of character by conducting frontal attacks in bogged-down situations in order to attempt to change the outcome of the fighting. Starting from a disadvantage, they had the further bad luck of being committed every time against a fresh and numerically superior enemy.

The division was entrained between 2 and 6 February in the area west of Cologne and moved to Hungary. Replacements joined their units at the railheads. Some of them were wounded from military hospitals, some *Kriegsmarine* and *Luftwaffe* personnel who were insufficiently trained for ground combat. Final integration into their units could only be carried out after reaching Hungary. The men from the *Kriegsmarine* and the *Luftwaffe*, some of whom already had combat decorations, had to fight with the division without any training. It is not intended to demean those men when one states that the combat power of the division was not increased in proportion to the numbers received.

The division arrived in the Raab area between 7 and 16 February; it was made out to be training and construction units. Along with formations already positioned in Hungary, the *6. SS-Panzerarmee* was to eliminate the Russian and Bulgarian formations west of the Danube or throw them back across it. The establishment of a shorter and more easily defended front would have freed up forces for the Oder Front and secured the oilfields west of Lake Platten. The destruction of the Russian bridgehead across the lower Gran was a precondition for the planned operation.

The *I. SS-Panzer-Korps* was given that mission. The *12. SS-Panzer-Division*, elements of *SS-Panzer-Regiment 1* and *Grenadier-Division "Hoch und Deutschmeister"* were employed. Starting 14 February, the *12. SS-Panzer-Division* moved with organic means and army assistance to the area around Ersekujvar via Kisber and Komorn. The division moved up to its attack position southwest and south of Kolta during the night of 16/17 February. The division was employed as follows: On the right the armored *Kampfgruppe* and *SS-Panzer-Grenadier-Regiment 25*; on the left, *SS-Panzer-Grenadier-Regiment 26* (minus the *SPW-Bataillon*). The division started its attack around noon in the misty weather. The area north of Köbölkut was reached by nightfall; fierce enemy resistance was encountered there. The infantry division on our left made little progress; its attack bogged down north of Bart.

The enemy counterattacked with tanks from Köbölkut during the night of 17/18 February and tried to throw back *SS-Panzer-Grenadier-Regiment 26*. They were repulsed with heavy losses, however. Towards noon on 18 February the armored *Kampfgruppe* and *SS-Panzer-Grenadier-Regiment 25* attacked from the northwest. They advanced along the railway line, entered Köbölkut and took it. Meanwhile, the *Kampfgruppe* of *SS-Panzer-Regiment 1* had reached the area north of Muzsla.

The attack was continued along the road from Köbölkut to Muzsla on 19 February. Muzsla was taken at about 1100 hours. The commander of *SS-Panzer-Grenadier-Regiment 26*, *SS-Obersturmbannführer* Krause, was killed in action during an artillery barrage. During the course of the afternoon *SS-Panzer-Grenadier-Regiment 25* and elements of *Grenadier-Division "Hoch und Deutschmeister"* took Parkany; *SS-Panzer-Grenadier-Regiment 26* took Ebed on the Danube.

The division was moved to the area north of Bart on 20 February to eliminate the rest of the bridgehead in a night attack. The attack was stalled by heavy defensive fire from Bart and Beny and could not be continued successfully on either 22 or 23 February. *SS-Panzer-Grenadier-Regiment 26* was therefore ordered to attack Beny during the night of 23/24 February. After heavy fighting around the strongly fortified railway embankment west of the town, the regiment, supported by tanks and the *SPW-Bataillon*, was able to take Beny at 0730 hours on 24 February. It then advanced on the Gran. Bart was also taken on 24 February. By doing that, the Gran bridgehead was completely eliminated. It was the division's last successful major operation. Some units had suffered critical losses and the casualty ratio among the officers was again disproportionately high.

The division moved via Komorn and Bankesy into the area north of the junction between Lakes Platten and Velence between 25 February and 3 March. It assembled for an attack on the night of 4/5 March. The *1. Kavallerie-Division* was on the right and the *1. SS-Panzer-Division* on the left. The *II. SS-Panzer-Korps* assembled to the left of the *I. SS-Panzer-Korps*. Once again *SS-Panzer-Grenadier-Regiment 25* was on the right and *SS-Panzer-Grenadier-Regiment 26* on the left. The armored *Kampfgruppe* was supposed to follow *SS-Panzer-Grenadier-Regiment 26* as soon as the terrain conditions permitted it. The ground was still partly covered with slush. There had been a surface thaw, but the soil beneath the surface was still frozen.

SS-Panzer-Grenadier-Regiment 26 attacked Ödin-Puszta at 0445 hours. *SS-Untersturmführer* Rechers and a few men of the *II./SS-Panzer-Grenadier-Regiment 26* only penetrated the defensive system, which consisted of 5 lines of trenches, in the afternoon after heavy losses. Ödin-Puszta was taken with the support of tanks and a *SPW* on 6 March at 0500 hours. Around 1100 hours Major-Puszta was finally taken; a counterattack was repulsed. The area 4 kilometers north of Deg was reached on 7 March after breaking through an antitank belt. The commander of *SS-Panzer-Grenadier-Regiment 26*, *SS-Sturmbannführer* Kostenbader, was killed during an air attack.

Deg was taken during the night with the help of tanks and *Panzerjäger*. Several positions were overrun that night by tanks and *SPW*. The retreating enemy was pursued well into the night through Meszezilas and Igar. The enemy tried vainly to regain his positions on 10 March. All his attacks were driven off.

The division assembled on 11 March to attack Simontornya and across the Sio. Tanks of the *12.* and *1. SS-Panzer-Divisionen* entered the town at the same time. The *Panzergrenadiere* reached the canal at 1430 hours. The *I./SS-Panzer-Grenadier-Regiment 26* did not manage to cross the canal until the evening and then form a small bridgehead on the south bank, which was immediately subjected to energetic counterattacks. The bridgehead was enlarged on 12 March when Hill 503 was taken. All the counterattacks on 13 and 14 March were repulsed. The *1. SS-Panzer-Division* relieved *SS-Panzer-Grenadier-Regiment 26* in the bridgehead on 15 March.

The Russians had started their counter-offensive north of Lake Velence in the meantime. The division was pulled out of the Sio sector and moved to Stuhlweißenburg. The Russians had broken through the Hungarian sector of the front and penetrated the *3. SS-Panzer-Division "Totenkopf"* sector in depth. As soon as the divisional units arrived they were deployed in the *Totenkopf* sector to halt the enemy. The enemy had broken through at Moor and it took some effort to contain him.

It was extremely tough to move the division into the mountains when it was insufficiently equipped with vehicles and the road network was saturated. The division pulled back via Dudar to the Margaetethen Position at Zirc on 21/22 March, but it was impossible to fight there as the enemy had already overrun positions to the south and north of the division without meeting any resistance. The division fought its way back to the Raab Position by 27 March, being repeatedly outflanked and threatened in the rear by the enemy.

We reached the unoccupied *Reich* Defense Line on 30 March. The enemy had already overrun the neighboring sectors. With difficulty, we managed to break through the enemy encirclement at Ödenburg on 31 March. The division reached the Vienna Woods at Hirtenberg on 2 April and formed a defensive line under considerable adversity.

The combat-support units established combat units. With their help, the front was eventually stabilized. (*SS-Panzer-Grenadier-Regiment 25* had only about 60 *Panzergrenadiere* left in the front lines at the time.) The front east of Altenmarkt was held until 21 April, but we had to pull back through the mountains via Rohr to the Tradigist area after being outflanked by the enemy on our left. According to prisoner statements, the enemy was establishing defensive positions there to await the Americans.

The division disengaged itself from the Eastern Front on 5 May in order not to be exposed to the danger of surrendering to the Russians. It marched into captivity on 8 May. Two kilometers from the demarcation line formed by the Enns, the divisional commander, *SS-Brigadeführer* Hugo Kraas, had the division march past him for the last time. The dwindling remnants drove past him to an uncertain destiny. They were disciplined and maintained a soldierly bearing despite the attempt to humble them by disarming them. They ignored the order of the Americans to display a white flag on each of their vehicles.

*

An explanation concerning the question of cuff titles is perhaps in order. It was true that all *SS-Divisionen* of *6. SS-Panzerarmee* fighting at Lake Platten were instructed to remove their cuff titles on orders of Adolf Hitler. The commander-in-chief, *SS-Oberstgruppenführer* and *General der Waffen-SS*, Sepp Dietrich, did not pass on the order. In any event the units were not wearing cuff titles at the time for deception purposes. The order came into being in the following way:

During the Lake Platten offensive, the Army Group Commander, *General der Infanterie* Wöhler, encountered a *Nebelwerfer* crew from the *Leibstandarte* moving to the rear. He called them to account. The gun commander reported he had lost contact with his unit because his vehicle had broken down and he was now looking for it. During the course of a telephone conversation with the Chief of the Army General Staff, *General der Infanterie* Wöhler justified the failure of the offensive by stating even the *SS* could not hold the line and was pulling back. That conversation was briefed during the evening *Führer* conference, which led to the order to remove the cuff titles.

SS-Oberstgruppenführer Sepp Dietrich sent a staff officer to the *Führer* Headquarters to report the true situation, conditions and behavior of the units and to ask for the order to be rescinded. Adolf Hitler regretted having giving the order on the basis of an incorrect report and cancelled it.

That is what actually happened. All other accounts are based on false information or propagate an agenda. I am rendering this account of the facts only for the sake of the truth. There are still attempts to falsify true events to the disadvantage of the surviving soldiers of the former *Waffen-SS*, the dead and their surviving dependents.

*

Further material provided by Hubert Meyer in 1993 concerning the division's losses:

On 1 June 1944, the division had a strength of 20,540 officers, noncommissioned officers and enlisted personnel. Losses up to October 1944 officially amounted to 8,636 soldiers, but a total loss of 9,000 is closer to the truth inasmuch as some of the casualty lists were incomplete. The division's strength was about 11,500 men after the operations in the west. The bulk of the division was moved to Germany to be refitted at the beginning of September 1944. Only a *Kampfgruppe* remained committed. It was composed of the remnants of three *Panzer-Grenadier-Bataillone* with about 150-200 men each; the remainder of *SS-Panzer-Aufklärungs-Abteilung 12;* the Division's *Begleitkompanie;* two combat engineer platoons; a mixed battery with 200 rounds of ammunition; 10 *Nebelwerfer* with 251 rounds; a 75 mm antitank gun; and, a single 88 mm *Flak*. All the tanks and *Panzerjäger* were handed over to those *Panzer-Divisionen* still in action.

Imprisonment in England

The aircraft glided easily through the mist and crossed the English coastal cliffs at Dover. We were soon circling over an airfield south of London and floating down onto our designated landing strip. So, This was England! I stood freezing on the runway and waited for the next thing to happen. It seemed the Paris to London connection was not working well. It was only after a few telephone calls that a car came to collect me and take me to the interrogation camp. I was amazed during the ride that there was no bomb damage to be seen and the outskirts of London gave an impression of absolute tranquility. We moved quickly to the camp through well-maintained streets with brightly lit houses.

A strange silence hung over the complex as I was accompanied through the door and taken to an isolated room. A heavily armed Englishman stood guard. I was presented with a questionnaire which I filled in at once so as not to be bothered anymore. A fellow dashed through the door, shouting hysterically, and stood in front of me with his fists twitching. His eyes were hidden behind dark glasses. He shouted at me and flung my overseas cap into the corner in an attack of false rage. The young man was too keen to make an impression to seem genuine and played his assigned role very badly. I was moved to solitary confinement.

I was given some bread through the door the next morning and afterwards taken out for some fresh air for twenty minutes. It was impossible to identify my fellow prisoners during the exercise; large shelter halves had been hung to separate us and prevent our seeing each other. After a few days on my own I was glad to be taken to interrogation again.

A newly coined American (ex-German) awaited me and started a discussion about the German military situation and about the peace prospects for Europe after the German surrender. In his opinion, Europe was moving towards a rosy future under English leadership. When I tried to explain to him that I did not share his opinion and that the Allies were going the best way to destroy the old Europe forever and open the doors to communism, he stopped the discussion. He almost became angry about my comparison of Russian and Allied soldiers. He bade me farewell and informed me of my transfer to the so called "Generals Camp".

Up to that time I had never known about my departure until the moment before the time had come to go. This time I was told in advance and also that I would be meeting my commander-in-chief, *General der Panzertruppen* Eberbach. I spent my last hours in the interrogation camp eagerly anticipating meeting my former commander again.

I arrived at the Generals Camp at Enfield late that afternoon. It was a manor house surrounded by high barbed-wire fences. A dozen German generals and staff officers were already living there. *General Ritter* von Thoma was the camp leader. Von Thoma had been captured in North Africa and was highly thought of by the English. Relations between him and the guards were excellent.

After having reported to the camp leader, I was taken to my room and met *General* Eberding there. He commanded the *67. Infanterie-Division*, which had defended itself so fiercely on the islands of the Scheldt and was forced to surrender in the battle for Breskens. In gratitude, I shook hands with my old commander-in-chief, *General der Panzertruppen* Eberbach, and met *Generäle* von Schlieben, Ramke, von Choltitz and Elfeld (who had been reported as missing ever since Chambois).

Our discussions on the first evening not only touched on the conditions in the camp but also the new political developments in Germany. The last topic was brought up repeatedly by *Oberst* Wildermuth who later became a minister in the German government. I sensed that one group of gentlemen did not understand the reality of the change that had occurred. They held on to their dreams and thought that captivity was only something temporary. Others had lost all their illusions. Their world was in ruins and they did not know what was to replace it.

The most disagreeable thing during the entire time there was the constant drone of enormous bomber formations. We had to stand by helplessly as death flew overhead on its way to Germany.

The availability of some books was greeted with gratitude and we all took much pleasure in playing chess. It was played for hours and tournaments were organized. There were, of course, some newspapers in the camp; one in particular was recognized as being especially insidious with lies, inventions and propaganda against Germany. We were thus all the more pleased when that paper emphasized the "miracle" of the German Army during the Ardennes Offensive in December, especially after the enormous strains of the last few years and the continued good morale shown by the German soldier.

The offensive came as a complete surprise. We knew our Army's condition and the enemy's strength. It was incomprehensible that my magnificent division had once again fought as a division after losing more than 10,000 men, including 21 unit commanders. The outcome of the offensive under such circumstances was inevitable and could only end in a further bloodbath for the German Army, considering the enemy's large materiel superiority.

I was taken to London for several days of lengthy interrogation in the spring of 1945. It took place on the orders of the Canadians and under the supervision of Major General Barker. Besides Barker, the participants were: Lieutenant Colonel Boraston, Lieutenant Colonel Page and Lieutenant Colonel McDonald. Colonel Scotland from the London District Prisoner-of-War Cage was also present.

Colonel Scotland later gained a dubious reputation with regard to his treatment of German prisoners. Among other things, he was not ashamed of slapping the face of the brave *General der Fallschirmjäger* Ramcke. The propaganda ballyhoo concerning Scotland knew no limits and stamped him as a hero of the British Secret Service. It is still maintained today that Scotland served as a German General Staff Officer on the staff of *Feldmarschall* Kesselring during the war. As a result, he was able to deliver important material to the Allies. That was, of course, a fairy tale, but it is still believed throughout the world. As it happened, Colonel Scotland later once went to see the so-called War Criminals' Prison at Werl. On that occasion, he was promptly asked by the *Feldmarschall* about his experiences on his staff. An icy silence was the only response.

In contrast to *General der Fallschirmjäger* Ramcke, I was well treated by Colonel Scotland. He showed me some of the sights of London on the way back to the camp and also gave me some tobacco.

My interrogation first touched on purely military or political questions, and then turned to the events in Normandy. From the way the questions were formulated, I soon understood that these gentlemen wanted to make me responsible for all events on the battlefield. I was to be singled out as a scapegoat.

During the operations north of Caen the Allied propaganda machine ran at full speed and presented my division as a gang of fanatical young Nazis. The evidence for the press campaign had to be delivered and preparations for the "festival" of revenge made.

I decided to answer in the negative any question about my division, which was still fighting at the time, that could bring it even the slightest amount of grief.

I was returned to my camp on the third day of interrogation and transferred to a camp at Windermere two weeks later. We traveled through the countryside for many hours in an express train and only reached Windermere late in the afternoon. A car then took us to the idyllically situated Camp Number 7. It was surrounded by mountains and served as a vacation retreat for a large company before the outbreak of war. After going through the outer gate we were led through two more wire fences and then stood before the old manor house. We were registered individually. Our personal details were verified and a search of our belongings undertaken.

A newly coined Englishman who had spent a long time in Berlin rummaged in my baggage as if he had found a gun runner. My meager belongings were strewn across the room. He took my print, "The Knight, Death and the Devil", out of its frame and threw the totally destroyed picture back into the box.

After that process *Oberst* Bacherer took me to his room and put me in the picture as to camp procedures. *Oberst* Bacherer was made camp leader by the English until he was replaced by *Oberst* Wilk, the defender of Aachen. I was very pleased to hear from Bacherer that Max Wünsche was in the camp and expecting me. My impending arrival had been announced several days previously. I noted with great pleasure that the officers' spirit was unshaken and they viewed the future with optimism.

As I left Bacherer's room to look for my assigned area, I was completely surprised to find myself in front of a group of officers. They had gathered without Bacherer's knowledge to welcome me to the camp. Max Wünsche came towards me beaming with pleasure and told me of his experiences South of Falaise and his capture.

There were officers of all branches of service in the camp, and we constantly hatched pranks against our dear guards. Two comrades were sitting in a dark cell at that moment because they had made an escape attempt. Unfortunately, they were only able to enjoy their freedom for a few days. They were soon recaptured and subsequently put in confinement for two weeks.

I was summoned before the camp commandant at the end of April. The manner in which I was summoned didn't bid good. Comrades Wünsche and Lingner were also summoned with me. To my astonishment, the commander was not present; instead, I was shouted at by a uniformed figure who obviously had no idea how a soldier in uniform should behave if he did not wish to become a laughingstock.

Dr. Otto John, later President of the Office for the Protection of the Constitution, erstwhile wanderer between two worlds, gave himself the honor of being a henchmen for England. He was bursting with venom as he shouted at me: "Don't leave this camp without permission. If you should attempt to escape your dead body will be brought back to camp. Regardless, you will never see your family again."

Had that pitiful creature sensed with what disdain I listened to his hateful claptrap, he would surely not had paraded through the camp like a peacock spreading his tail. Instead, he would have sunk into the earth with shame.

I was told at the same time that I was no longer allowed to join the group walks, but I could only move around under escort. The same order applied to my two comrades, Wünsche and Lingner.

The English directive didn't affect me, nor did the rumors going around the camps. *Kapitänleutnant* Eck, who was a camp assistant, and the camp doctor, *Dr.* Weißpfennig, drew my attention to the fact that the English considered me a candidate for the death sentence. Neither comrade knew that he himself was listed as a "war criminal" and a few months later would pay with their lives in Hamburg.

We were isolated like three black sheep during our next walk and made to feel like common criminals. Two Tommies marched us out through the camp gates and we climbed up into the nearby mountains. As soon as we were out of sight of the camp they slung their submachine guns over their shoulders and barely paid attention to us.

We had a pleasant surprise the next time we were escorted out for a walk. *Admiral* Hüffmeier suddenly stepped forward from the main group of our comrades as asked to be taken out with us. He gave his reason: "I'm a German officer and wish to be treated in exactly the same way as my comrades." *Admiral* Hüffmeier's action gave us a lot of pleasure. Regrettably, he was only allowed to do that a few times. The Tommies did not like such proofs of comradeship.

One day I met my old commander, *Generalleutnant* Pflieger, in whose battalion I received my first training in 1929. It was both a pleasant and sad reunion. Most of our comrades were no longer with us.

We followed the fighting in our country with heavy hearts. It was incomprehensible that they still fought on and offered resistance against the Western Allies. We somehow believed in the possibility of the peoples of Europe changing their minds and preventing the military occupation of eastern and central Germany by the Red Army, but we were mistaken. Destiny took its course and let the Asiatics up to the Elbe and into the heart of Europe.

The complete collapse of Germany struck us to the core. We had been waiting for it to happen for weeks, but it still hit us all deeply. The reports about the horrible events in the Russian Zone practically drove us crazy. We had had no news from our families for months. None of us could say for certain where his family was and whether it was still alive.

Was our fifth child alive? Was it the boy, we had longed for, or did I now have five daughters? Our senses reached out to Germany day and night, but all the worrying was of no use. We could get no answers. Our lives were confined more and more; soon everything only took place behind barbed wire. The walks stopped. A new international law came into effect — the victors' justice.

Feldmarschalle von Kleist and Sperrle arrived at the camp and we were shattered to hear the first eyewitness reports from a dismembered Germany.

The debates about the end of the war, the political mistakes, the insufficient preparation and the lack of weapons on all the battle fronts went on endlessly. The failure of our leadership was discussed with much bitterness. Most of the officers rejected the assassination attempt of 20 July 1944. I didn't find a single one who was for the perpetrators.

I had an especially pleasant surprise at the end of May. An *Oberst* arrived from another camp in southern England. He had been with my brother-in-law, an *Oberleutnant* in the *Luftwaffe*, and therefore he had the last news of my family.

I had had no communication with Germany since August and I knew nothing of the whereabouts of my loved ones. The continual worry over the well-being of our wives and children was the hardest thing to live with. Our own misery hardly affected us; we had been too hardened by war, but our families' misery almost threatened to break us. It is therefore impossible to describe how much I was touched by the few words of the *Oberst*. A single sentence made me the camp's happiest inmate. It was: "I bring you greetings from your brother-in-law. He wanted me to tell you that your family escaped to western Germany and your son was born on 15 February."

I was transferred in June, with some one hundred comrades, to Camp Number 18 at Featherstone. The transfer was a pleasant interruption to our paralyzing captivity. We were put into buses and driven all through England. The new compound was on the Tyne and served as an encampment for the Americans. It consisted of countless barracks divided into compounds. They were surrounded by virtual walls of barbed wire.

The entire complex left a bleak impression. Our welcome did not please us either. The buses had barely halted when a gang of wildly gesticulating uniformed men met us. They ordered us to leave the buses and trot towards the reception blocks. The behavior of the newly naturalized Englishman — originally from Berlin — was anything but gentlemanly. The guards, armed with nightsticks, surrounded us like a pack of wolves waiting for their victims.

We were led through a barracks to be checked once again. After the "filtering" process, we didn't have to worry about our baggage any more. All those things we had acquired with difficulty and which we had paid for out of our own pay were gone. We were then allowed to go to our barracks.

We realized during the next 24 hours that we were no longer prisoners-of-war in our guards' eyes. We were treated like criminals. We were classified for the first time and from then on were considered to be white, gray, or black Nazis. I found myself in the best of company. *Fallschirmjäger*, *U-Boot* officers and about 20 *Waffen-SS* officers were classified along with me as black Nazis. Surrounded by that company, captivity became endurable. We immediately started organizing advanced training courses and quickly arranged educational activities. *Oberst* von Viebahn deserved especial praise for his service on our behalf.

346

We received a new commandant in the summer who instituted welcome changes. The old commandant had been replaced after our complaints to the Red Cross representative. Life behind barbed wire became easier to bear. Lieutenant Colonel Vickers not only allowed us certain liberties out of humanitarianism but also understood how to tame the restless spirits and keep alive hopes of an early return home.

I was surprised in the middle of September to receive orders to get ready to leave in half an hour. *General* Kroh and Max Wünsche went with me to the gate. The commandant told me that I had been summoned to the London District Prisoner-of-War Cage and had to start the trip at once.

It was a beautiful summer day as I traveled to London in a special compartment accompanied by a captain and two noncommissioned officers. The journey was very entertaining for me. It did not last long and my three musketeers were sound asleep. Everything went well until we reached London; we even held the same views on political matters. The good relationship changed very quickly when I was handed over in London. A sergeant who had once lived in Berlin greeted me with hate-filled eyes and forced me to give him my shoelaces and my belt. The fellow took obvious pleasure in watching me climb the stairs. I can still hear him saying: "He will soon be finished."

I was taken to the fourth floor, then down a long corridor to the room at the end. There was already a comrade from Austria there. The poor man had been in Italy until recently and was skin and bone. The treatment in the Italian prisoner-of-war camps must have cost the lives of a lot of comrades.

I came across more comrades next morning. We were allowed 20 minutes of fresh air. My neighbors were *Generale* von Manteuffel and Schimpf. After a few days, Schimpf had figured out a way so we could communicate between the cells.

I experienced such a surprise one day I thought I was hallucinating. Glancing through an open door I saw Meitzel, who had been reported missing, and *Hauptmann* Steger, who had been reported killed in action. I made contact with all my comrades in less than 24 hours and they told me that the Canadians intended to put me on trial as a war criminal. After hearing that my last doubts as to the nature of my interrogations were removed. I prepared myself for the expected trial.

Hundreds of questions rained down upon me in a short time, all of which I answered most readily. The chief interrogator was Lieutenant Colonel B. J. S. MacDonald; he was assisted by C. S. Campbell. I got to know Major J. J. Stonborough as an able translator who underpinned the work of his superiors with his clever questioning. The latter supposedly came from Vienna and, perhaps precisely because of that, was one of MacDonald's most eager assistants. I knew from Meitzel that Major Stonborough had already condemned me to death and made his view public in the Canadian POW camps. My death had thus been decided before the first day of trial.

When no doubts remained that I would be tried,. I asked for permission to talk to *Oberst* von der Heydte and my old commander-in-chief, *General der Panzertruppen* Eberbach. Permission was granted after a few days.

Up to that point I had answered any question which might have incriminated my division in even the slightest way in the negative. I didn't see any reason to the give the "victors" an opportunity to try German soldiers. The victor had no right to pass sentence on the defeated! Even a neutral country was barely in position to be able to do that. Was there any human being who understood the reality of the battlefield and believed that only beasts fought on one side and angels on the other? All of us had stood in the melting pot of murderous battle, and men on both sides failed under the pressure of events.

The defeated were tried by special laws and special courts, not because the crimes with which they were charged were proven, but because they "were considered to be proven." These judgments were only the result of the victors' feelings of hatred and revenge. The "war crimes trials" were one sided and only carried out against Germans. They thus ran counter to any normal sense of justice.

Back To Germany

One nasty November morning, I was handcuffed without prior warning and taken to an airfield. We drove through waking London in silence and flew over the Channel to Germany without a word being said. I had no idea where we were heading but the lay of the land fairly quickly showed where we were. We flew above Ostfriesland at an altitude of a hundred meters. Straight-as-an-arrow roads, calm canals and lonely windmills greeted us. An airfield suddenly appeared out of the mist. A big crowd of soldiers, journalists and photographers awaited the "beast" of Caen. Had the story not been so terribly serious I would have loved to have a hearty laugh. That assembled mass of journalists, security personnel and planted spectators spoke highly for the prosecutor, Lieutenant Colonel B. J. S. MacDonald, as the propaganda chief for the planned "circus".

I was the last to leave the aircraft, accompanied by the whirring of recording equipment. I was met by two officers on the steps. I saluted these two gentlemen in the traditional fashion. It was not returned. As fast as lightning, I was chained to the taller of the two when my hand came down. Major Arthur Russell assumed command of the "escort party," and I climbed into an armored vehicle that was standing by with Captain W. H. J. Stutt. We were accompanied by armored cars and motorcycle riders on our journey down the blocked-off and barricaded roads towards Aurich. There was something operetta-like in the entire procedure. That came home when I discovered that Captain Stutt kept his hand in his right-hand pocket for the entire trip. I could see the outline of a service revolver.

We soon reached Aurich where I was taken to the former Naval Signals School barracks and a body search was carried out at once. That procedure left nothing to be desired in terms of unpleasantness. The "visitation" was carried out in front of the prosecutor and some officers.

Special measures had been taken for my accommodations. Two cells at the end of a long row were separated by a thick iron door, making them an especially secure section. One cell was equipped for interrogation; the other was to be my "home". The cell had been specially "modernized" for me. The "bed" was constructed from such heavy wood that it was impossible to move. It had just been completed by a German master carpenter. It was made without the use of nails and brackets. Two blankets completed the furnishings. A large square hole had been cut into the door and a guard had his head permanently stuck through it so as to keep me constantly under watch.

There could be no doubt at that point I was not only in captivity but "locked up". I had hardly made myself familiar with my cell when I was again chained to Captain Stutt and taken to the regimental headquarters of the Royal Winnipeg Rifles. I was guarded by three men. The staff was housed in the training center. The reasons for my arrest were finally given to me at that point. Lieutenant Colonel R. P. Clark stood in front of me and read out the charges and had them translated by an interpreter.

The bill of indictment contained two main points. They were:

The accused, *Generalmajor* Kurt Meyer, an officer of the former *Waffen-SS*, part of the armed forces of Germany, now in the custody of the 4th Battalion of the Royal Winnipeg Rifles, Canadian Army, Occupation Force, Canadian Army Overseas, is accused as follows:

1. That Meyer committed war crimes in the Kingdom of Belgium and the Republic of France, during the course of the year 1943 and prior to 7 June 1944, as commander of *SS-Panzer-Grenadier-Regiment 25*, in violation of the laws and customs of war, by inciting soldiers under his command and advising them to refuse pardon to Allied soldiers.

2. That Meyer had committed war crimes in Normandy by being responsible, as commander of *SS-Panzer-Grenadier-Regiment 25*, for the killing of seven prisoners-of-war in the proximity of his command post in the monastery of Ancienne Abbaye Ardenne.

I then knew what I was accused of and why I was supposed to be tried. I felt a certain relief having received that information despite the hardly pleasant situation. I then knew how I could prepare myself for the accusers' charges and refute them with corresponding evidence.

The handcuffs were snapped around my wrists again and, as we marched away, our steps echoed hollowly from the empty building's long walls. I was sized up by a few curious Germans who were working for the Canadians in the barracks complex.

I was back in my cell in a few minutes, trying to make friends with the soldiers guarding me. I was successful. The young soldiers had fought against my division and behaved in a truly comradely manner towards me. At his point I would like to say that the Canadian soldiers and officers always treated me properly. I was never once mistreated. The difference between front-line soldiers and armchair warriors was obvious. Unfortunately, one of the young soldiers was replaced after a few days because he constantly supplied me with newspapers.

Captain F. Plourde from the North Shore Regiment and Captain Wady Lehmann, visited me late that afternoon. Captain Lehmann introduced himself to me as the defendant's interpreter. Captain Plourde had been assigned as deputy defense counsel. It was not known who would be the chief defense council. Colonel Peter Wright, who was to defend me initially, refused the task. He thought that I had been prejudged, and a fair defense was not possible. Colonel Maurice W. Andrew from the Perth Regiment was later designated as defense counsel. The last named three gentlemen stood up for my interests in a very soldierly and conscientious manner.

I discussed the charges and the prosecution evidence with Captains Plourde and Lehmann on the first evening of my detention in Aurich. I also told them who I wanted as defense witnesses. Based on the available material — especially the witness statements for the prosecution — neither of the gentlemen thought I would be convicted.

That opinion, by the way, was also shared by my guards. I often heard, "If you are going to be sentenced then so must our officers. The same things happened on our side and it's not right just to hold the loser responsible." But in 1945, no value was placed on the opinions of front-line soldiers; they had done their duty and had to keep their mouths shut. Armchair strategists, paid re-educators and other parasites had the floor.

Close confinement in the cell and the permanent guard gradually got on my nerves. I had to perform even the most intimate functions under the eyes of other people. Thank God the soldiers had different ideas about human dignity than their superiors obviously had. I felt especially disturbed by the continuous strong light. But remedial measures were taken in due course, and the light was directed into a corner.

I finally had the opportunity to make contact with my wife. I had no idea at all where my family was and how they were surviving. I asked Captain Lehmann to send a few words to my wife. He had my family traced to Offleben near Helmstedt where they were living with my mother.

I waited restlessly for Lehmann's return so I could finally find out how my loved ones were. He stood in front of me two days later. With great pleasure he extended greetings from my wife and showed me a photograph of my first son. A small blond boy was laughing out at me; tears ran down my face. Captain Lehmann had brought my wife and my second daughter with him and quartered them in a hotel. The Canadian Army had ordered that to be done. I was very thankful to those responsible for that example of sincere humanity.

Captain Lehmann gave me a disturbing report about my family's living conditions. My wife and her five children struggled from Ludwigslust to Heide during the first days of May. She was subjected to several attacks by low-flying aircraft near Lübeck and two children were lost from sight. The family was only reunited by chance when she reached Heide. Heide was overcrowded, and my wife and children lived in a single room with straw on the floor until September, when my mother was able to take them into her home.

Meanwhile, stupidity had joined heartlessness. In Offleben some "democrat elites" had risen to the top who were busily engaged in trying to send my children to glory. As a first step, those "heroes" refused my children ration cards and the so-called residence permit. I owe it to acquaintances, many miners among them, that my family survived.

Yesterday's enemy, Captain Lehmann, cleared up the situation through his personal intervention. The office of the District President in Helmstedt was obliged to issue the necessary instructions to the "gentlemen" in Offleben.

Captain Lehmann told me that I could see my wife the next day. Sleep was out of the questions — forbidden newspapers helped me pass the night.

During the following morning's exercise period I was able to greet my comrades. They stood at the windows and waved at me. I also become aware of the presence of *Generale* von Schweppenburg and Eberbach. The fact that proven comrades were nearby was a great reassurance.

The great moment came in the afternoon. The minutes passed slowly. I paced up and down in my cell, waiting for Captain Stutt, who was supposed to take me to the visitor's room. Suddenly, while turning around in front of my barred window I saw a jeep and a woman and a small girl climbing out of it. They become indistinct as my eyes misted over. I recognized my wife and daughter. They were led into the training center.

Captain Stutt came for me a short while later. The cold metal of handcuffs was felt around my wrists again and we marched off. I walked quickly, my pulse racing. Stutt opened the door and pushed me in front of him but that politeness was for naught. After all, we were chained together. I nodded towards my wife with a helpless gesture and waited impatiently for my chains to be loosened. Finally, I was free!

We walked towards each other silently, speechlessly and forgot about our less-than-enjoyable surroundings. Our daughter looked at me, laughing and crying at the same time. Time had lost its power; yearning, fear and desperation disappeared. We had been united again and felt strong. Our faith was greater than we dared admit. Unfortunately, time passed too quickly. The allotted 20 minutes were far too short, of course, to discuss anything more than the most pressing of matters. Only the knowledge that, we were henceforth allowed a visit every other day, made the parting easier. I laughingly observed the snap of the lock on my handcuffs. Our first visit had ended.

Colonel Andrew finally reported as my defense counsel. I was happy to know that at least one Canadian soldier was on my side as a defender after having been told that no German lawyer was prepared to defend me. I only found out later that, in fact, no German had been asked.

Colonel Andrew was a lawyer and had served in the Canadian Army as a reserve officer. (Although his desire to defend me properly was genuine, he nevertheless said goodbye quickly after the sentence was pronounced and left me alone with Captains Plourde and Lehmann.)

We went through the charges point by point and determined which counter-arguments and witnesses for the defense would be used. Unfortunately, we came off second best. My most important witnesses were only "found" long after I had been sentenced, even though they were in Western custody and could have been reached easily. I only saw witnesses at Aurich whose whereabouts I had personally provided to the Canadians. The main witnesses for the defense, *Dr.* G. and *Dr.* St., were only interrogated by the prosecution in March 1946.

The Trial

The proceedings against me were opened at exactly 1030 hours on 10 December 1945 in the former Naval Signals School. I virtually had to run a gauntlet of journalists before reaching the courtroom. They were trying to take a photograph of the branded beast to satisfy people's curiosity. Whether they succeeded in presenting the desired picture to their readers is beyond my knowledge. I was no longer affected by those scribes' impertinent curiosity. I was already standing in front of my judges. My whole existence rested on the duel between me and the prosecutor.

The chains were removed from me shortly before entering the courtroom. My hands were free, as free as my heart. I entered the room as a soldier and not a depressed defendant. Determined to prove myself in front of the tribunal and also demonstrate proper bearing as an example to my soldiers, I approached the bench through the courtroom audience.

My judges — five generals — were sitting in front of me. I looked for and found the eyes of General Foster, who had been my opponent on the battlefield in 1944. He had been designated President of the Court to pass judgment on me. What a strange encounter between two soldiers! The victor was now chosen to administer "justice" over the vanquished after they had fought each other with every fiber of their being for months on end.

The selection of the President and his co-judges was an impossibility according to international law. All of the gentlemen had fought against me and were thus involved in the case. I thought I saw understanding and sympathy in Foster's eyes. In any event, I felt I was standing in front of a soldier and not a civilian in uniform. After the usual formalities had been dealt with, the individual charges of the prosecution were read out. After that the prosecution was given the floor.

The prosecution tried to prove that I had issued written orders for murder in the autumn of 1943 during training in Belgium. As proof of that assertion, he gave the court a photocopy of the infamous "order". When that "exhibit" was put on the judges table, I did not know what to think. Was it stupidity or impudence which made the prosecution regard that "order" as an "item of evidence"? The presentation of that scrap of paper was an insult to the judges who were to decide the authenticity of the "order". The following nonsense was presented to the court.

Exhibit T3

12. SS-Panzer-Division "Hitlerjugend"

Secret Orders

1. Behavior towards the civilian population in occupied territories: If the member of the populace looks at a *SS* soldier in a disdainful way or expectorates on him, the person in question can be beaten and arrested. If interrogation leaves the impression that the person arrested is hostile to Germans, he should be executed in secret.

2. If somebody tries to get information about weapons or ammunition, he will be arrested and exposed to severe interrogation. If the interrogation leaves the impression that the person arrested is hostile to the Germans, he will be executed for espionage. Soldiers who give information about security will receive the same sentence.

3. Guards will not leave their posts, nor are they allowed to eat, drink, sleep, smoke, lie down or put their weapons down while standing guard. The soldier who quits his post before being properly relieved or who reveals the challenge and password to the civilian population will be sentenced to death. The challenge and password is the most important part of guard duties.

4. Behavior at the front: *SS* units will not take any prisoners. Prisoners will be executed after interrogation. SS soldiers will not surrender and must commit suicide if they have no other choice. Officers have stated that the English do not take any *SS* soldiers prisoner.

5. Information about enemy troop movements will be transmitted as quickly as possible. Written information will be learned by heart at the same time. As soon as a soldier gets into danger, all papers have to be burnt or eaten. He will carry nothing but his dog tag. The strictest silence has to be kept in all matters. Traitors will be executed, even after the war.

6. Observers who return from the front with information and their accompanying officers will not use the same route they took to the front.

Where had that absurd nonsense come from? The prosecution gave the following explanation:

SS-Schütze F. Tobanisch was a member of the *15./SS-Panzer-Grenadier-Regiment 25* and deserted from his unit in April 1944. He later joined the Belgian resistance movement. Tobanisch is a Czechoslovak. He stated that the order was read aloud in front of the assembled company and all soldiers had to confirm what they had heard with their signature. Furthermore, the company first sergeant warned everyone against revealing the contents and directed the company to memorize what had been read. The witness dictated the order from memory to the Belgian resistance movement. The Belgians then wrote it down in Flemish, and I had the "order" translated into English.

The "proof" was a photo copy of an English translation based on the Flemish transcription of a German/Czech oral statement.

It was only with difficulty that I managed to keep a straight face when that heavily weighted "evidence" was presented. How could one call that scrap of paper "evidence"? But it looked official. A large sheet of paper had been photocopied and indeed gave the impression of a document.

The "order" was a mixture of undigested memories of training provided by platoon leaders and company commanders and malicious defamations. It is absurd to assume that during training it was said "*SS* Soldiers will not surrender and will commit suicide in the most extreme situation." Such instructions would have undermined the unit's morale. Furthermore, it must be mentioned that the unit was not allowed to carry out the interrogation of prisoners. It had not been instructed on how to do that.

With regard to the signature, I must say that every German soldier had to confirm in writing that he had been comprehensively instructed by his company commander about desertion and spying. That signature was also obtained from Tobanisch. No other confirmation was asked of unit members. I had the impression that the prosecution had no doubts as to the true character of Exhibit 3 and knew pretty well that the nonsense written had emanated from the fantasy of a deserter and murderer. (Tobanisch had murdered a German officer.) But what importance did MacDonald attach to the truth? He had conducted interrogations in Europe and America for a whole year to bring about my downfall.

I now had to go down. Proof of his efficiency had to be shown. He put Exhibit 3 forcefully on the table and, in so doing, could not help throwing a triumphant look at me. "Witness" Tobanisch was not available to the court and could "not be found".

I had to say that I had little hope for a fair trial after the first act of the "hearing of the evidence". I waited tensely for the second act. The next witness for the prosecution was *SS-Panzergrenadier* Alfred Hazel of the *15./SS-Panzer-Grenadier-Regiment 25*. He also came originally from Czechoslovakia and became a soldier in 1943 as an ethnic German. In contrast to Tobanisch, Hazel was a brave soldier who was wounded in the attack on Bretteville in June 1944 and captured three days later. Hazel was a tall, solidly built fellow who had turned 19 during the trial.

As I already indicated, MacDonald had driven or flown from camp to camp to collect evidence against me. MacDonald had found Hazel in Hull (Quebec) and sent him to Aurich as an interesting witness against me. Immediately after Hazel's arrival in Aurich, he was visited by MacDonald and once again his attention was drawn to the importance of his statements. MacDonald had gained such a good impression of Hazel that he decided to use him as the first witness against me, to condemn me from the start. Hazel had explained in his statements in Canada that I had said in Belgium in 1943: "There are no prisoners from my regiment!"

MacDonald described the appearance and examination of Hazel in court as follows:

When I asked him about his earlier statements, Hazel was sitting on the witness chair. He replied in an evasive manner; the color of his face changed and he slipped further and further into his chair. He made excuses; suddenly, he couldn't remember anything. In the end he denied that Meyer had ever said such words.

I was completely speechless about the behavior of that witness, especially as he gave evidence so readily in Canada. I envisioned other witnesses retracting their statements.

Dismayed by that start, I looked around the room and suddenly I saw the reason which had led to the failure of the witness. Meyer, who sat diagonally opposite the witness, fixed Hazel with such a penetrating stare, the likes of which I had never seen before. One could virtually see sparks shooting across the room. The unfortunate witness sat on the chair in panic, like a bird caught by the hypnotic stare of a venomous snake. The physically powerful figure had lost all strength; he had evaporated like dew in the heat of the sun. It was an astonishing demonstration of the enormous disciplinary power which that officer still had over his former soldier and of the fear which his presence created. I decided to use an old tactic and positioned myself between Meyer and Hazel. But it was of no use; the court was not inclined anymore to attach much credibility to his statements.

I had a surprise the next day which did much to enlighten me. My defense counsel asked to be allowed to speak and informed the court that Major Stonborough had been caught the previous day trying to influence Hazel in his statements. Stonborough had talked insistently to Hazel for more than ten minutes and had shouted at him angrily. The court asked Stonborough to leave the courtroom and allocated another interpreter for a few days.

What had I really said and what really happened between Hazel and me in the courtroom?

During the exercises on the training area at Beverloo in Belgium we officers naturally used every opportunity to impart our experiences to the soldiers that were entrusted to us. The unit was being prepared for fighting on the Eastern Front. It would have been a crime to leave these young soldiers guessing about what was waiting for them on the Russian steppes. It was simply an act of comradeship to impart our bitterly won experiences to them. We had to tell them about the horribly butchered soldiers who had had the misfortune to fall into Russian hands. My words to these young soldiers at Beverloo were: "Men, no one from my regiment will be captured. Believe me, it is better to fight on to the bitter end!"

MacDonald's courtroom observations do not need to be commented on. I can, however, tell Mr. MacDonald one thing: I have never met any former soldiers from units I led before or during the war who was afraid of me or who ever crawled away in fear. No, Mr. MacDonald is mistaken! I was a soldier among soldiers, and I never trained or led by fear. My soldiers respected me, but they were never afraid. I wish Lieutenant Colonel MacDonald could see the comradeship which I am permitted to experience again and again today, ten years after the end of the war.

In the next few days former prisoners presented evidence about their experiences during the fighting and after having been taken prisoner. Lieutenant Colonel Charles Petch, commander of the North Nova Scotia Highlanders, stated, among other things, that A and C Companies had been overrun by the *III./SS-Panzer-Grenadier-Regiment 25* and that only one officer and 23 men and one officer and 25 men respectively had come back. All other soldiers had either been killed in action or been captured. Major Learmont, along with 20 survivors of C Company, had been taken prisoner by the *III./SS-Panzer-Grenadier-Regiment 25* in Buron. He had seen Private Mexcalfe, who belonged to his company, shot after being captured because he still had a hand grenade in his pocket. On the way to the rear, the prisoners came under Canadian artillery fire and Private Hargraves received a leg injury. He was not able to walk and was shot by a German soldier.

All those events occurred right at the front and during the young soldiers' first days in combat. They had just gone through hell; heavy naval shells had thinned their ranks and fighter-bombers had continuously strafed the battlefield. The irregularities mentioned above happened during the first minutes of that bloody struggle and were committed by individual soldiers. They did not happen because their commanders wanted it that way or ordered it, but because the enormous burden was simply too much for the young soldiers. That's why they simply cannot be condemned according to the legal code alone. It is of course easy to condemn the perpetrators, but only real combat soldiers had that right. They knew from their own experiences how long it was until somebody was "normal" again after the fighting. The members of A Company also had similar experiences. For example, Private Richards was fired at, then bandaged by a medic and taken to rear. He also admitted in court that he was supplied with water by the German soldiers.

In the end, I was accused of having ordered the shooting of seven Canadian soldiers on 7 June at our forward command post. That accusation was based on the statements by a Pole named Jesionek made in the spring of 1945 in Chartres so as to be able to join the Polish army in Italy. Jesionek originally came from Upper Silesia and entered the *Waffen-SS* in 1943. His father was in a German prison at the time. Jesionek stated the following:

On the morning of 8 June I was near the forward command post of *SS-Panzer-Grenadier-Regiment 25* because my vehicle had hit a mine. My comrades and I were not wounded. At about 1000 hours a German led seven prisoners into the monastery yard.

Jesionek followed the prisoners into the inner yard and claimed to have heard the soldier report to the regimental commander. In reply, I am supposed to have shouted at the soldiers and have said:

What are we supposed to do with these prisoners? They only eat our rations.

Then Jesionek claimed I had spoken to one officer and loudly ordered:

No more prisoners are to be brought here in the future!

He claimed to have seen the prisoners individually shot by a German noncommissioned officer.

The Pole's infamous accusations hit me hard, of course. At that point I had to prove my innocence; it was no longer incumbent on the court to prove me guilty. I was considered guilty from the start. It was up to me to prove the contrary.

The witnesses I requested were not available. It was said at the time: "The witnesses cannot be found; we do not know whether they are still alive." It was interesting, however, that these men emerged immediately after the sentence was pronounced and were at the prosecution's disposal.

I do not want to bore the reader with the question and answer game. In total, more than 4,000 questions were asked. Instead I will let a Canadian journalist,who later wrote an extensive report on the trial, speak. Among other things, he wrote:

Although Jesionek was never unsure about in the main points — he had been interrogated countless times and therefore knew his story word for word — he did not make identical statements in different versions of his story. For example, he said in Chartres he had heard Meyer giving the direct order to shoot the seven Canadians. In another version, Jesionek failed to mention the explicit order. His evidence did not match that of other witnesses in many details.

He said he saw Meyer in the monastery church at about 1000 hours in the morning. Meyer stated that he had been out touring the front at that time and he only returned towards midday. At that point, he climbed the tower at once to observe the combat zone. Two German officers were able to confirm that.

Jesionek said that Meyer was wearing a long vulcanized rubber coat in the monastery church, but Meyer said that he wore only his usual camouflage uniform his division wore at the time.

Jesionek's statements about his platoon's and his vehicle's positions on the day in question contradicted other witnesses statements. The entire contents of Jesionek's story was questionable. Everything hinged on the door through which he claimed to have seen the seven Canadians marching to their deaths. A 16-year-old French youth, Daniel le Chevre, who lived in the abbey, stated that the door was obstructed by an air-raid shelter.

There, where Jesionek claimed to had seen seven corpses in a pool of blood, and where they were found about ten months later, Le Chevre and two or three of his friends had seen nothing which might have attracted their interest.

Jesionek had mentioned a number of steps beside the doorway, but Monsieur Jean-Maris Vico, who had also previously lived at the abbey, said that he had built these steps himself in July 1944, an entire month after Jesionek claims to have seen them.

Furthermore, Jesionek claimed that he had seen my loyal Cossack with the prisoners. That assertion could be refuted immediately. The Russian was not in France at all during the time in question but was on leave in Germany. Jesionek's statement that my driver, Bornhöft, was present was also an easily proved lie. Bornhöft only joined the *12. SS-Panzer-Division* after I had assumed command of the division.

It was equally nonsensical to claim that he and his *Schwimmwagen* crew got into one of our minefields and was, therefore, a witness to the events in the monastery. Every soldier knew that there would not have been much left of a *Volkswagen* after hitting am antitank mine. Jesionek tried to explain that the consequence of hitting that mine was neither the wounding of the passengers nor the total destruction of the vehicle. Jesionek's wild story was even too much for the judges. His statements were examined by the court and dismissed.

Jesionek's examination concluded the case for the prosecution. My defense counsel, Lieutenant Colonel Andrew was allowed to speak. I decided to make my statement under oath. That same day the Canadian Forces magazine "The Maple Leaf" published the following article:

For Canadian troops.
Tuesday, 18 December 1945
Volume 5, Number 32

The Maple Leaf

The atrocity stories did not worry *Panzer-Meyer. Panzer-Meyer,* the former *SS* Major General, appeared calm in front of the court, although his cold, blue eyes twitched nervously. He sat erect as the prosecution finished its case against him.

He was attentive during the first week of his trial but showed little emotion as he followed the statements which tried to prove that his former regiment was a gang of young, fanatical murderers who took pleasure in shooting Canadian prisoners-of-war. But even witness accounts of the atrocities which his men had committed were unable to change his iron countenance. He arrives in court every day accompanied by two officers of the "Winnipeg Rifles." He stops close in front of Major General Foster, the president of the court, bows briefly and is then requested to take a seat.

That act is not without drama if one remembers that not much longer than a year ago Major General Foster and Major General Meyer faced each other in combat as commanders. At the time General Foster commanded the 7th Infantry Brigade of the 3rd Division. It was defending Bretteville while under attack from Major General Meyer's units.

With the exception of General Foster, whom he respects, Meyer pays little attention to members of the court.

Frau Meyer, who brought her daughter, Ursula, with her to Aurich, attended some of the hearings last week. Her husband usually looks to see if his wife is there as he enters in the morning and afternoon.

Although Meyer wears a simple soldier's gray-green uniform, it is hardly necessary to see his golden general's epaulettes to realize that he was a senior officer. It is incorrect to describe him as arrogant or a despot; instead, he is a man used to giving orders but not receiving them. It therefore comes as no surprise to hear that he was known as a daredevil, a hard-fighting commander and a strict superior in all theaters of the war.

This is, in short, the man accused of bearing the responsibility for the shooting of Canadian prisoners-of-war. On the eve of the opening of the defense, he faces the prosecution coldly and with apparently unshakeable confidence.

Colonel Andrew opened my defense at 1400 hours on 18 December. He asked me to take the stand. I had to describe the entire course of the fighting on 7 and 8 June once more and establish my actual whereabouts and activities at each moment.

Colonel Andrew made an enormous effort to refute the far-fetched "evidence". He was able to substantiate a great deal but could not do away with the "documents". The "murder order" was on the table and that which was on the judge's bench in black and white indeed had to be true!

Colonel MacDonald's cross-examination started after a day-long examination by Colonel Andrew. The cross-examination lasted a day and a half. Hour after hour passed in uninterrupted cross-examination. Several sergeants had to write down each question and answer in short hand. I sat alone. Hundreds of eyes watched my every movement and every expression on my face. I was unable to relax for a moment.

A group of men with total authority, spurred by honor and professional ambition, against one man deprived of his rights. Questions and more questions: The unit's training program; the content and meaning of talks I had given years ago; about the character of conversations held in front of French fireplaces etc. An annihilating barrage designed to bring about one person's downfall, to break his spirit and destroy him. Annihilating fire from behind the front line — no grenades or violent combat, but destruction simply using malicious words, distorted statements and questions, confusing allusions and distorted pictures from the past. That was how MacDonald tried to force a victory over me, but I refused to capitulate.

I was not to be defeated by the advocate's war. MacDonald then tried to catch me in a lie: During the initial interrogations I had denied finding the dead Canadians in the monastery but that I later confirmed they had been found. He tried with real gusto to present me as a liar and even denounce me as a conspirator. That armchair strategist could not understand that during the last months of the war and right up to the trial, I had no cause to incriminate any German soldier. I had never accepted the arbitrary Allied military tribunals as legal courts of justice. To me, they were and remained "victors tribunals".

MacDonald wrote at the time:

Meyer's behavior was polite most of the time and of straightforward demeanor. However, after the first day's cross-examination, I insisted on an explanation of his statement in which he denied all knowledge of finding dead Canadians. When I placed that question Meyer lost his composure and stared at me intently. I then felt the hypnotic influence that the first witness must have also felt.

I do not know whether Meyer learned that technique as a Prussian officer or as a civilian but I do know that I became dizzy under the astonishingly intense and fearful fire of his gaze. I had no experience in such matters and had to act quickly to overcome it. Meyer asked to have a portion of his earlier statement repeated. I pointed out that he only wanted to gain time. His look had reached its most intense stage when he answered (answer 3190): "I do not need any time to give you the answer." I replied: "And your intense looks cannot influence me, you can be assured of that." I then read out the requested statement. Meyer then looked out at the courtroom as if a veil had fallen over his eyes.

I don't want to waste words on a grown man's childlike fantasies — perhaps the good Lieutenant Colonel was overworked — but I can confirm one thing: I was so surprised by his reproach that I truly doubted his sanity. That's why I threw a questioning glance at the courtroom.

Further witnesses were interrogated following my cross-examination to support my statements or to provide a different perspective. Several officers were able to shake a lot of MacDonald's explanations. Above all, they were able to confirm that I stayed with the forward-most units most of the time and, as was self-evident, had no time to occupy myself with interrogating captured prisoners.

SS-Hauptsturnführer Steger tried all means available to him to prove that it was highly unlikely I told people not to take prisoners or even take revenge on unarmed soldiers. Steger had been a *Wehrmacht* officer and a company commander in *SS-Panzer-Grenadier-Regiment 25* since October 1943. He made the following statement to describe my character and to negate allegations of atrocities against me:

An American bomber was shot down by German fighters during our training period in Belgium. Some of the crew escaped by parachute and landed on the training area at Beverloo. The then *Oberst* Meyer at once ordered that the injured Americans be given medical attention and then taken to the military hospital. One unwounded American got the opportunity to warm up in the mess and to have a cup of coffee with some of the regiment's officers. Meyer talked with the American for about an hour. The prisoner was later handed over to the military police.

The captured American's attitude so impressed Meyer that he immediately dictated a special order presenting the American as a splendid soldier and a model for his own regiment. The special order was read out in front of the assembled companies of *SS-Panzer-Grenadier-Regiment 25*. I had learned a lot from Meyer as an officer. I respected him for his qualities of leadership. Meyer was respected by his soldiers and was popular with them. He never demanded more than he was himself prepared to give.

General der Panzertruppen Eberbach was the next witness. He said:

I must emphasize that as a soldier I deeply regret all atrocities committed and I condemn them sharply. Whoever the perpetrators may be, they have stained our soldiers honor and I feel no pity towards them. I consider wholesale condemnation wrong, however. The *12. SS-Panzer-Division "Hitlerjugend"* was, on the whole, no worse than any division of the Army. In my opinion, the atrocities committed in its sector may be linked to just a few perpetrators. Incidentally, a quarter of the division's officers were from the Army. The *Waffen-SS* did not have enough of its own officers available to staff that formation. Those Army officers were men of impeccable character who had the choice of leaving the division and who would not have stayed had the division's behavior been dishonorable.

I knew the *Hitlerjugend* Division from an inspection tour I made in March 1944 and, furthermore, it was attached to me in Normandy from 6 July 1944 to the end of that month. That unit's performance was always above average for the entire period. I had a particularly good impression of the men, as well as the officers.

My predecessor in Normandy was *General der Panzertruppen* Geyr von Schweppenburg. Being a Catholic and a career army officer, he was anything but a friend of the *SS*. He also judged the division's performance, attitude and discipline to be above all other divisions. Its success ratio was threefold that of the *21. Panzer-Division*. In my experience, a division that was not solid to its core could not continue such a performance for long.

It is for that reason that I'm convinced that the atrocities were only the actions of individuals. The Canadians were brave soldiers but rough fellows. It was reported to me repeatedly that they took no prisoners or shot captured German prisoners. I also had a written report with regard to that. Incidentally, I read similar statements in an Allied brochure while in captivity.

Meyer had been a member of the Mecklenburg state police from 1929-34 and subsequently transferred to the *Leibstandarte*. He was primarily a career soldier. Meyer is married to a wife whose character makes a perfect impression. He has four daughters and a son. Meyer was an unusually good soldier. He was most caring towards his men, who respected and loved him. He was also an excellent trainer and good tactician. His personal bravery was beyond praise. He far from hated the Canadians. It would have been entirely out of character for him to kill prisoners or issue such an order. *General der Panzertruppen* Geyr von Schweppenburg and I concur in that judgment.

Concerning the Facts of the Case

First of all I wish to emphasize that the *12. SS-Panzer-Division "Hitlerjugend"* captured and turned over the most prisoners of any formation while under my operational control. I could call *Generale* Geyr, Schack, Obstfelder, Schimpf and Sievers to confirm that the *"Hitlerjugend"* Division had a good reputation for discipline. Furthermore, there were no rumors about atrocities or unnecessary harshness.

During the fighting there was an appalling case of rape involving a member of the division's signals battalion and the girl concerned died. Meyer had the man court-martialed and the perpetrator was sentenced to death. The sentence was carried out in the presence of the local mayor and a French priest. It would have been nice if the

Allies had exercised such strong measures against their own soldiers in Germany in such cases.

One has to admit that a commander who wields such a strong hand over a subordinate in such circumstances could not possibly be the man who either encourages or even tolerates atrocities on other occasions. Meyer was always level-headed in even the most difficult of situations. Based on my experiences in war, that was the measure of a good character.

The shooting of Canadian prisoners is supposed to have taken place during the first 24 hours of the invasion. At that time Meyer was still a regimental commander. The execution of the prisoners is supposed to have happened in the vicinity of the command post. Meyer was hardly ever at the command post; he stayed at the front with his units.

On hearing of the incident he started an investigation and transferred the regimental adjutant to the front where he was killed in the next few days. Although there was no direct evidence against the regimental adjutant, he was still incriminated as the shootings occurred when he had most likely been at the command post.

The investigation was further hampered by combat operations and the heavy losses that had taken place in the meantime. No clear evidence as to the culprit arose. Shortly thereafter Meyer was appointed divisional commander in place of the fallen *General* Witt and was so occupied by the continual heavy fighting that he was unable to take an active part in the investigation, much as he would have liked to.

I do not believe Meyer to be capable of such a deed or of complicity in it as it was not in his character. I especially want to emphasize that Meyer was too good a tactician to allow such an event to occur. As an officer fully trained in tactics Meyer knew with certainty that with our total lack of air support or any other means of intelligence gathering, we were entirely dependent on prisoners' statements for information about the enemy. He knew how decisively important it was to furnish higher headquarters with the latest enemy reports by sending prisoners quickly to the rear. Meyer was much too intelligent not to have realized that. According to Canadian statements, no further shootings occurred after Meyer took over command of the division. Had he been guilty of the shootings which had already taken place, one would assume that such incidents would have occurred more often during the time he led the division.

The trial was adjourned until 27 December because of the Christmas holidays. I was taken to my cell in handcuffs to experience my 35th birthday in a miserable environment of searchlights and iron bars. It was announced that my brave wife, who had witnessed the trial day after day, was to visit me on 24 December.

I was afforded an unforgettable experience on 23 December. Canadian soldiers brought me birthday greetings from my comrades, at least those who were in Aurich as witnesses. I was even given a small picture as a birthday present. That little pen and ink drawing of a North German landscape could not, of course, remain in my cell. A Canadian soldier passed it on to my daughter.

On the morning of 23 December I suffered the odious chains around my wrists and greeted God's wonderful world with profound joy. Beneath the tall trees and in the fresh air I forgot my impending fate. I recalled my 35 years with pleasure, happy that I was able to experience them as a fulfilled man. Each guard post made the effort to impart a little pleasantry, but I had my greatest surprise that evening.

After darkness fell there was an entirely unscheduled rattling of keys in the cell corridor. The sound of doors being opened echoed through the gloomy building. Suddenly some officers appeared in front of my cell and harshly requested me to step into the corridor. I was once again in chains before I was completely awake and being trotted off into the darkness. Two noncommissioned officers walked ahead with loaded submachine guns and the same "guard of honor" followed me.

Rather than going the usual way, we walked across the yard and suddenly found ourselves in front of the officers' quarters. My accompanying guards were stationed at the doors and windows to guard them from the outside. Not a single word had yet been spoken, and I had no idea what lay ahead or what this melodrama meant.

Sensing nothing good I allowed myself to be led into a well-lit room. An officer came towards me unlocked the cuffs at my wrist and let the cold steel fall to the tiled floor. Speechless with surprise I tried to get to the bottom of the officer's mysterious behavior without success. Two other officers present introduced themselves and asked for my word of honor not to attempt an escape, to which I agreed.

Next I was asked to use the bathroom to wash up. After I had done that the gentlemen led me to a door, opened it and propelled me into the room. I thought I was dreaming. Before me was a festively decorated table that could hardly bear the weight of the food and drink it was carrying. Candlelight gave the room a truly festive touch. I took one hesitant step towards about half a dozen officers who were standing in front of me.

Suddenly they broke into "Happy Birthday". My enemies had arranged a birthday party for me. I allowed the sight to touch me in silence and was unable to prevent the tears streaming down my face. The change from prison to birthday party simply overwhelmed me. I was especially deeply touched when the officers introduced themselves as unit commanders of the Canadian 3rd Division. These were the same men against whom I had fought from June to August 1944 and whose comrades I had allegedly been guilty of shooting.

We spoke freely that night, extensively discussing the whole problem of war crimes. The result of the conversation was what these gentlemen told me: "If you are found guilty and the sentence is endorsed by the world and by history, then the Canadian Army will have no generals tomorrow. They would all have to follow in your footsteps."

It was interesting to hear about my prosecutor's remarks when he drowned his anger in whisky at the bar after the daily proceedings in court. The field officers did not have a good opinion of Colonel MacDonald. We decided to end the "party" at midnight. Shortly afterwards my cell door slummed shut again. I was a prisoner once more.

On 24 December I had the great fortune to be allowed a visit from my wife and my daughter Ursula. We were allowed to sit hand in hand for almost 30 minutes and talk about the children. My wife said goodbye to me with dry eyes. My daughter was looking forward to a Christmas party to which the Canadians had invited her.

The trial continued on 27 December and at 1615 hours the judges rose to announce their first verdict. After exactly three hours of deliberation Major General Foster announced the following verdict: "Major General Meyer, the court has found you not guilty of the second and third indictments. Please resume your seat."

That meant that the court did not hold me responsible for events during the fighting or for the related shootings. Furthermore, it absolved me of Jesionek's accusations. The court was convinced that I had not given the orders for the shootings on 7 or 8 June. The trial continued on 28 December with statements from *General der Panzertruppen* Eberbach and my wife. The "Maple Leaf" wrote the following:

The most dignified of all defense witnesses appearing for Meyer on the last day was his dainty, blonde wife Käte. She fought to control her emotions. Her slim, pale face showed the strain of the fourteen-day trial. She told the court about her happy marriage with the former General, what a good husband and father he was and how her five small children loved their father. "I can say no more than that we were very happy," she said at one point, forcing a nervous smile to play round her lips. She left the picture of a proud and brave woman behind her.

Captain I. A. Renwick of the 28th Tank Regiment, who was captured on 9 August 1944, made the following statement:

I was captured by soldiers under command of the accused and taken to his command post. Meyer interrogated me and showed me a copy of the "Maple Leaf" which contained General Crear's accusations that Canadian prisoners-of-war had been shot. Meyer said that that was not the way to wage a war. Nor could he understand why Canadians were fighting Germans. I had the impression that he was a very capable officer who knew what he was fighting for. I was not threatened or bullied while I was there. His behavior was extremely proper and always that of an officer and a gentleman.

Captain Renwick's examination concluded the hearing of evidence and I received permission to say a few words in conclusion:

I heard many things during these proceedings that were a complete surprise to me. These excesses were in no way perpetrated by young soldiers. I am convinced that the culprits may only be found among soldiers who had been brutalized in many aspects by the experiences of five years of war. I made the effort, as a commander, to train my young soldiers and mould them into worthwhile people. To that end I did not shy away from passing the most severe sentences against officers and men. I take overall responsibility for everything that I ordered, tolerated or encouraged. Whether a commander can be held responsible for the actions of individuals will be your decision. I await your judgment.

The court withdrew to deliberate at 1120 hours and I was led into a side room. The officers accompanying me expected me to be sentenced to a year of prison to satisfy the outraged public. My counsel was of the same opinion and did not deem it necessary to ask my wife to leave the courtroom. He felt that she could listen to the judgment without a problem. Thank God, I succeeded in getting my wife to leave the courtroom.

At exactly 1145 hours, that is, only 25 minutes later, the court returned. You could hear a pin drop. The president, having obvious difficulty in controlling his emotions, delivered the judgment:

General Kurt Meyer, the court has found you guilty of the first, fourth and fifth points of the indictment. You are hereby sentenced to death by firing squad. That sentence will only become binding on confirmation. These proceedings are closed.

My prosecutor described that moment as follows:

Meyer stood erect as the judgment was passed. With gritted teeth and a grim look, but showing no other emotion, he bowed slightly, turned sharply and marched out of the room.

The trial was over.

Only after I had left the courtroom did it become clear that I had been sentenced on the basis of the so-called "Secret Orders" of the Czech deserter held responsible for the shootings in the monastery. With impotent fury — despising all Pharisees — I stalked through the waiting reporters and inquisitive crowd into my cell.

On Death Row

The bars on the windows threw long shadows on the wall as I looked at the photographs of my children and tried to imagine their life without me. At the same time my wife was hearing about the death sentence. Captain Lehmann had promised me that he would break the bad news himself and not to leave it to chance.

So that was how my life was to end. A volley would crack out in a sand-pit somewhere and my body would disappear into a nameless grave! The knowledge of my impending death did not depress me. I was no longer alone. I stood once more in the midst of the fight and surrounded by comrades. It was too dark to see the bars when a voice called to me: "General, don't be afraid, we are fighting for your life!"

What an irony of fate — my enemies had become my friends! I thought about God the most that night. We communed with one another, and I await-ed the bright morning with renewed strength. Death was not without its importance. On the contrary, the flame of life burned very brightly and the knowledge that my hours were numbered was not easy to bear. However, there was no abyss to be crossed. I was living in the presence of the Creator with the knowledge that death was part of His creation. I prayed during the long night for the strength to meet death as a man.

I was handcuffed again early in the afternoon of 29 December and led to the visitors' room. My wife was waiting for me. I was terrified of that meet-ing. How much had that fateful blow affected my wife? She had worried about me for years on end only to greet me at that point as one marked for death. I showed gratitude to the Canadian officer who removed the handcuffs so I could enter the room unshackled.

My wife came forward without tears, but the facade dropped as our hands touched. Her tears fell on my awards. I gave them to her to give to my son. Our daughter Ursula smiled at me through her tears and told me of my son's arrival. I had never been so proud of my wife as at that time. Despite her boundless grief, she knew how to give me the strength to return to my cell.

Colonel Andrews bade farewell and asked me to petition Major General Chris Vokes, the commander of the Canadian 3rd Infantry Division, for par-don. I rejected that proposal. Colonel Andrews and I separated as soldiers; for us the war was now in the past.

As a result of pressure from Captains Lehmann and Plourde and the request of my wife, I handed in a petition for clemency. I did it to do my duty to my family. The petition was handed in on 31 December and had the following text:

Sir,

As I write this petition, I think with pride about the reputation of my units and on my reputation as a soldier. I think about my wife, who has been a faithful companion to me and a good mother to our children. She will now have to bear the responsibility for rearing our five children by herself.

I think about her above all else as I use this last opportunity to write to you — not on my account, but instead for my children who will be deprived of a father. I am lobbying for the reputation of my units and for my reputation as a soldier.

In this last appeal to you I wish to leave aside all of the formalities that are otherwise found in the courtroom and speak to you man to man and as soldier to soldier.

Two soldiers have accused me of having encouraged my men not to take prisoners. That accusation is false. The statements were made by two deserters of Polish and Czech nationality. One of them did not appear before the court, because he cannot now be found. Another witness, likewise a former member of my regiment and a Czechoslovak citizen, renounced his earlier testimony and emphatically the contents of the so-called "Secret Order."

Of all the men in my division who are now prisoners-of-war in England, Canada or the United States only three were prepared to appear against me as a witness. If one were to ask the hundreds who had belonged to my unit about the accusations against me, I am convinced that their statements would contradict what was said by a few, so much so that one could not believe them.

The facts prove that from the first day forward my unit took in large quantities of prisoners. A former officer, who had returned home and had been released from the army, voluntarily came back to Germany to testify on my behalf.

The former Commander-in-Chief West told the court that my division had brought in three times as many prisoners as any other combat formation.

I wish to emphasize that my soldiers were trained in accordance with international agreements and I fulfilled my duties as an officer to the best of my abilities. I tried to accomplish my mission as well as could be expected from a soldier, but my mission was difficult and my responsibilities went well beyond the norm.

The fighting during the invasion in Normandy demonstrated that the morale of my regiment was good. They had been committed to battle in a short period of time and were bombed heavily on the way to the front by low-flying fighter-bombers. We were the lead elements of the division. Our sector was unusually wide. Our left flank was exposed and we had to count on the employment of paratroopers or air-landed troops at any moment.

My young soldiers — 17- and 18-years old — fought for three entire months without being relieved or having the hope of being relieved and without having had a single night of good sleep. After the initial phase of the fighting, the division had between 3,000 and 4,000 casualties. The sector assigned to it became larger instead of smaller.

If my soldiers could withstand such ground and aerial attacks for a quarter of a year, then the majority of them must have been good and well-trained men.

I can only reiterate that I never encouraged my soldiers not to take prisoners! That happened neither during the training period before the invasion —when we still didn't know where we were going to be employed — nor during the fighting after the invasion. No oral, secret or any other type of order nor the example provided by my behavior or attitude gave any impression whatsoever that prisoners could be killed.

I cannot understand why I am being made responsible for the actions of some members of my command. Responsibility can be direct or indirect, and it follows that a commander cannot be held responsible to the same extent for individual actions of one of his soldiers acting on a momentary impulse, as perhaps for the actions of one of his staff officers in his immediate vicinity. Does that responsibility also not depend on the degree to which the situation and the general environment are normal or abnormal?

Can a commander of young, inexperienced troops — who are being employed for the first time under unexpected and abnormal conditions in far too large a sector with exposed flanks, under permanent artillery fire and air attacks, with inadequate supplies and no hope of reinforcement or support and continuously threatened by paratroopers and air-landed troops — be made as responsible for individual actions of his troops as a commander of experienced troops on a quieter front under normal conditions who has complete control over his units and has the opportunity of being able to observe, lead and control them better?

I want to stress once again that, if abuses happened, they happened during the first days of the fighting, when conditions were chaotic and remnants of coastal units were still in my sector. They happened at a time when I was a regimental commander and not authorized to control troops besides my own.

Following my appointment as a divisional commander, when I received full command authority over the divisional sector, the situation became more bearable and no further abuses occurred.

I have turned to you, Sir, a soldier with considerable combat experience, in the hope that you will understand under what conditions I fought and the situation in which I consequently find myself.

Although I have been found guilty as charged, I only consider myself responsible to a certain degree. I have never felt myself to be so guilty and responsible that the death penalty would be justified. That is still my opinion.

I therefore request a revision of the judgment passed on me to bring it in line with the degree of responsibility and guilt which actually still applies today.

Aurich
31 December 1945

Captains Lehmann and Plourde came to see me late in the evening and informed me of the results of my petition. General Vokes rejected the petition with the following comment: "I have considered this appeal and cannot see my way clear to mitigate the punishment awarded by the court."

Both officers expressed their astonishment that the general had made the decision so quickly. He had read the petition and immediately wrote the rejection. In all probability there was no longer any chance of a rescue; execution seemed inevitable.

By questioning the guards I gradually became aware that I would probably be shot during the next eight days. From another quarter I was told that my execution squad was already training hard in Oldenburg and was shooting at beer coasters.

A couple of days later I was told that the time had come for me to take my final farewell of my family and that my dependants were already waiting for me. Despite the utter bitterness of the hour I felt a profound pleasure. I was looking forward to meeting my boy. I was finally going to get to know my son. Would I ever see him again after that first encounter? My wife held my son out to me. The little fellow grabbed my golden epaulettes and held them tightly, shouting with joy. I was not a stranger. He came willingly into my arms. My daughter Irmtraud sang a Christmas carol to me and our oldest daughter Inge comforted me with the words: "Father, you can't really die. Look, our boy will be just like you some day. I will always look up to him."

Oh, how happy these children made me! Our eldest had just turned 10-years-old, but what a reliable friend she already was for her mother. I was also allowed to say farewell to my mother. She was going to bear the entire responsibility for caring for my loved ones. My wife and my mother had grown in stature during those last few minutes. Time passed enormously quickly. A last word, a short embrace and a trembling smile ended our last meeting. My life had come to its end.

I waited for my execution during the next few days. I noticed, with astonishment, that death held no fear for me. Had the horrible experiences of war brought that about? Every morning, when the doors rattled and I heard the guards' voices, I quickly glanced again at the photographs of my loved ones — each hour could be my last.

I found out I was supposed to be shot on 7 January 1946. The English wanted to interrogate me once again on 6 January about the fighting at Dunkirk in June 1940. The English interrogator greeted me with the words: "You can finally tell us everything. You're going to be shot, after all!" The good man was most distressed he had to go home having achieved nothing.

While I was awaiting my execution the fight to prevent my death got underway outside. Captain Lehmann advised my wife to look for a German lawyer who might do something, perhaps organize a petition. I was unaware of their efforts. My wife was a stranger in Aurich. She did not know a lawyer and was penniless.

She went from place to place and was met by closed doors. In her helplessness, she chanced on Emdener Straße 11. That chance discovery was our good fortune. In the same house in which Bismarck once resided was the lawyer, *Dr.* Schapp. He was from the Friesian district of Germany and as tall as a tree. Like all his fellow countrymen, he had an almost fanatical sense of justice. *Dr.* Schapps' first step was a personal petition for clemency addressed to the Commander-in-Chief of the Canadian occupation forces, Major General Vokes. General Vokes answered *Dr.* Schapps' private petition on 6 January and a first, if admittedly faint, ray of hope began to shine. Major General Vokes' letter was as follows:

Dear Sir,

Your letter dated 3 January 1946, addressed to the district president of Aurich, concerning the matter of Major General Kurt Meyer, who has been sentenced to death by the military government, has been forwarded to me and is gratefully acknowledged.

I have taken notice of every detail but am compelled to inform you that I have the duty and responsibility for ensuring the relevant Canadian laws in the matter are respected and applied. I am sure you will fully understand my situation. I can therefore do nothing else but express my acknowledgement for the moderate nature of your petition and to thank you for having written to me so extensively on the matter.

Dr. Schapp made the following note on that correspondence: "Answer polite; position not clear, but neither is it a clear rejection."

The running battle with death continued. The Aurich Red Cross collected signatures. President Hollweg, acting as the highest member of the Reformed Church of the diocese of Hanover, forwarded a petition of clemency to General Vokes with a copy to the military chaplain at the headquarters of the forces of occupation, the Reverend Wilson.

At that moment Captain Lehman arrived with another message of doom. My petition had been rejected. He only told my wife at first as he was afraid that *Dr.* Schapp might be discouraged and give up all further attempts. However, to the contrary, our fanatic for justice really got going then. He tried to persuade Marahrens, the Protestant Bishop of Hanover, and the Catholic Bishop of Münster, Clemens August *Graf* Galen, to support my case.

My mother immediately traveled to see the Bishop of Hanover with a letter from *Dr.* Schapp, but the bishop regretted he was not in a position to support a former *SS* Officer. One of Dr. Schapp's assistants went to Münster. The aged *Graf* pointed at the wastepaper basket which was already overflowing with all manner of petitions, but he allowed my case to be presented. Suddenly, he pulled out the typewriter. A few minutes later Dr. Schapp's assistant was on his way back to Aurich bearing an important document. *Graf* Galen wrote:

According to information provided to me, *Generalmajor* Kurt Meyer has been sentenced to death because his subordinates committed crimes which he neither ordered nor tolerated. As a representative for the Christian understanding of justice, according to which each human being is responsible only for his own actions and can only be punished for the same, I accordingly support *Generalmajor* Meyer's petition for clemency and ask the sentence be commuted.

Clemens August *Graf* Galen

Meanwhile, General Vokes decided to fly to London to get the advice of his colleagues. He spoke on 9 January with General J. C. Murchie, Brigadier Orde, Lieutenant Colonel Bredin and Mr. John Reid of the Foreign Office who happened to be in London. General Vokes expressed his dislike of the death sentence and was of the opinion the sentence should be altered. He then flew back to his German headquarters and, on 13 January, commuted the death penalty to a life sentence.

He argued his decision as follows: "I did not feel the "degree of responsibility" established at the trial warranted the extreme penalty."

The doors rattled especially loudly on the morning of 13 January and there was a certain unease in the corridor. Several people stopped in front of my cell door. The door was opened and two Canadian officers entered the cell. I no longer had any doubts; my last hour had arrived. I listened completely unemotionally as one of the officers started to read something out to me. I presumed he was telling me the sentence had been approved. My eyes wandered over my family photos again and again. The text being read no longer interested me. Suddenly I listened more carefully as the words "life sentence" and "Canada" resounded in my ears. It took me a long time to understand fully what I had heard and assimilate the new situation.

My reaction was extreme. I was barely alone when I sank onto my plank bed a totally defeated man. I did not expect that turn of events. The thought that I was to spend my whole life behind bars was a crushing prospect. Hours had passed before I was able to come to terms with the new situation. The will to live started to make itself felt again. By afternoon I was already making plans for the period after my eventual release. I simply could not imagine finishing my life behind bars. I was allowed to see my wife once again on 14 January. As a token of farewell she brought me the first spring flowers from home.

On 15 January I was sitting in a "Dakota" in chains flying from Bad Zwischenahn to Odiham, Hants, England. My escort, Major L. M. Fourney, took me directly to Reading from the airfield. I entered the musty buildings of a civil prison for the first time. It was the same penal institution in which Oscar Wilde had been imprisoned decades ago and where he wrote his famous ballads. It was serving the Canadians as a military prison and housed some hundreds of deserters and other perpetrators. The discipline matched the conditions. Loud commands echoed through the building the whole day. Marching feet made even the heavy cell doors tremble. During my first exercise period behind the prison's red walls I moved back and forth beside the walls in the company of a sergeant. It did not take long to find out that a near-by building contained the gallows and that we were walking over the remains of the executed. In accordance with English custom, the executed were buried under the prison pathways. (The same custom had been observed in Hameln).

After a short time the institution was returned to civil use, and I was taken to a barracks near Aldershot. Hundreds of deserters were also housed there. My accommodation in the camp was good; I was even allowed to have my own small garden and enjoy the sun. The guard was composed of veteran soldiers who had all fought against my division. Their behavior was proper, and they always comported themselves in a military manner. Those fellows even had so much understanding of my situation as to set up a telephone call to the Irish Republic so as to give me an opportunity to hear a German voice and talk a bit with a young lady. They had done that without my asking for it. It was also there in the camp that I heard about the reaction of the Canadian public to my pardon. The news reports were simply incredible. The demand to drown me in Halifax Harbor was nowhere near the worst proposal for sending me to the other world.

My transfer to Canada was announced at the end of April 1946, and I was put into a Canadian uniform at the same time. As night fell, I left the camp for Southampton, where I was led onto the troop transport "Aquitania". The "Aquitania" was the sister ship of the "Titanic".

I was led into the bowels of the ship and locked in a cell. But that didn't last too long. Not even the ship's officers had any idea who their prisoner really was. I still remember a humorous event. We were playing cards in the salon when an older officer came over and said: "Men, don't make such a racket; let the German sleep a bit." Everyone quieted down in accordance with his request; the poor fellow didn't even realize that he was looking at the German General's cards.

The journey was very interesting and full of change. The ship was a so-called "love boat". It was taking a large number of wives and children of Canadian soldiers who had married in England. Nobody noticed the "Nazi beast" walking around on deck instead of sitting in his dark cell.

Some miles from the Canadian coast I was locked in the cell and warned of the attitude of the incited population. Thousands of people were on the pier as the ship docked. The joy of the women at seeing their husbands again was indescribable. Someone gave a welcoming speech and music droned through the enormous arrival halls.

General Foster, my counterpart in Normandy and my judge at Aurich, was standing at the gangway as I was led on deck. In order not to cause any commotion I was asked to leave the ship without escort and get into the car waiting nearby. And so it happened I stepped onto the soil of the American continent disguised as a discharged Canadian soldier. The civilians did not notice the "Nazi beast" leave the ship although all eyes were directed to the gangway. After a short evening drive through the streets of Halifax we arrived at an old fort, part of the former defenses. We were quickly issued with road rations and the seating plan was arranged. I sat with two officers and one sergeant while three sergeants followed in a second car. We drove off quickly and left Halifax behind us.

Dark forests engulfed us. I noticed with astonishment that intense snowstorms raging in the countryside had caused heavy damage. Many a telegraph post was lying by the ground. It was probably still very cold there even by 1 May. I was told during the trip that our destination was a prison near Moncton in New Brunswick. We drove for hours on end through the dark night across the Nova Scotia peninsula. The night was so cold that the convoy commander ordered a halt, and we jogged up and down the road to warm ourselves up. It would have taken little effort to sneak off into the bushes and disappear. But why should I risk an escape? In all actuality, I had become stateless. Where could a German soldier turn in 1946? My escort's behavior was proper in every respect. I was treated as an officer.

The bright lights of Dorchester prison appeared at about 0400 hours. The building's massive walls and towers dominated the Bay of Fundy. The enormous mass of stone and steel created a forbidding atmosphere. No one spoke a word and we drove up the hill in silence. I felt like a spectator at my own funeral.

The cars ground to a halt in front of the main gate. A bell alerted the guard who opened the outer gate. Between the outer gate and the interior one of the complex we had to pass through a couple more gates. In front of the last one, the sergeants said goodbye in a traditional soldierly manner. The two officers also saluted in parade-ground fashion. Having taken leave of my three escorts with a handshake, I stood face-to-face with my jailers. The difference was obvious. I realized within the first few seconds that, from then on, I was only to be a number in an army of nameless men. Neither the warden of the institution nor his senior assistants uttered a single word to me. As far as they were concerned, I no longer had the right to be treated as a human being. I was led to a special room and a couple of old rags were thrust into my hands for clothes. I was only able to keep my orthopedic shoes with difficulty.

After that spectacle I went for a long walk through the prison and got an initial crushing impression of degrading imprisonment. Cell after cell; they looked like cages for beasts of prey but, instead, they contained men. There was no longer any private life. The cells were not closed by doors but by bars. The prisoners were never allowed to experience refreshing darkness; their cells were always illuminated. A repulsive smell wafted towards me. We passed through some grilled gates and then I was given to understand that the hole in front of me was my cell. "Here's your new home!" With the ironic words of the assistant warden ringing in my ears, I was suddenly standing behind bars. I fell down on my plank bed in exhaustion. However, having made the decision not to throw in the towel under any circumstances, I stretched out my limbs.

I was awakened by my neighbors early in the morning. They were disappointed I couldn't give them any news from Montreal. It was only later that I was informed that my fellow prisoners were told I was transferred from Montreal to Dorchester. Before long I was taken to a big washing and changing room filled with a great number of prisoners where I had to shower and then sit down on a barber's chair. I innocently took my seat. The barber, a prisoner of course, made painstaking inquiries about the duration of my sentence without sensing, however, to whom he was talking. The duration of the sentence made no impression on him. He had been sentenced to twenty years because he had killed his uncle. I was glad when I was rid of my beard and no longer felt the blade of his razor at my throat. I awaited my haircut with my eyes closed; suddenly, however, it was as if I were electrified. The fellow had gone right over my scalp with a large pair of scissors and very calmly cut all my hair off. Two officials sneered at me. My hair fell in the dirt, and I was trembling with rage. After that, the number 2265 was painted on my chest. My name had been replaced with a number.

My left-hand cell neighbor turned out to be a professional criminal who had been "lodged" in penal institutions in the USA and Canada with short interruptions since 1917. He had been sentenced to life imprisonment because he burned seven people alive. My right-hand neighbor was a sex offender. The bastard had raped his own daughter.

I will never forget the disgrace inflicted on me behind Dorchester's gray walls. I was forced to note with horror that schools of crime flourished even in highly developed western countries. Young offenders who, of course, never normally shared a roof with hard-boiled criminals, were only educated in prison to become criminals.

I was given a job in the well-equipped library after four weeks. I got to know an excellent man, the department head, Mr. J. E. L. Papineau, who was an Air Force officer and came to understand the war from all sides. His gentlemanly behavior was exemplary. Newspaper propaganda against me was still running at full course and babbling complete nonsense. Tub thumpers from the Canadian servicemen associations claimed to be speaking for frontline Canadian soldiers. I had it confirmed once again to me that the common soldier holds a completely different opinion from that propagated by official spokesmen. The less people knew about the real events, the louder they asked for my head.

I received my first greetings from home in the summer. My family was making it by hook or by crook. It was supposed to exist on 119 *Reichsmark* in social security. The thought of their misery was almost impossible to bear. Once more I must express my deeply felt thanks to the chivalry of *Dr.* Wilhelm Schapp. In pure charity he did not shy from any sacrifice to preserve my family for me. Among other things, he took on all my wife's hospital costs. She had collapsed completely and was suffering from a serious heart problem. My family's well being rested on my mother's shoulders.

One day followed the other with no change in my situation. They took their course with mechanical precision. I want to spare my reader descriptions of the horror of captivity as there are hardly any Germans of my generation who don't know about that misery or who did not experience it first hand.

I had some revealing conversations with church representatives over the years of my imprisonment. Whether bishop or simple priest, they all spoke freely and fell on their swords for me. I still remain in contact with some of these gentlemen even today. Representatives of the Catholic Church made the strongest impression. They presented their views without fear and understood how to command respect.

I gained a friend in Major James R. Miller from the Eastern Command in Halifax. Miller was a military chaplain on General Foster's staff. He came to visit me every fourth week and also stayed in contact with my family. I don't know to what extent I am indebted to him, but I do know that he did not avoid any effort to improve my situation. He was often called to Ottawa to report on my situation. He conducted himself as a true Christian, his Christianity consisted of deeds. He was not a man for unctuous speeches.

A couple of years had passed in the meantime and feelings had calmed down. The newspapers were publishing the other side's words by then and allowed an open discussion. I therefore read with astonishment one day that Army and Air Force officers were petitioning for my freedom and objecting to the sentence passed on me. General Foster frequently sent his regards, while other officers tried to give my children some pleasure.

I got to know a truly wonderful person in Fritz Lichtenberg from Moncton. *Herr* Lichtenberg emigrated in 1911 and made a new life in Canada as a building contractor. His home was only a couple of kilometers away from the prison. He persistently fought for permission to visit me. One day the time came. After a long wait we stood face-to-face and blurted out words of welcome in German. The joyful welcome threatened to overwhelm us. *Herr* Lichtenberg showed me photographs of my children and also told me about the latest goings on in Germany. That man's loyalty to Germany touched me deeply. He proudly declared his love for Germany, just as we had learned as children.

I was allowed to see that magnificent man every fourth week in the company of Major Miller. It was not long before he brought me news that a lawyer's office in Halifax had taken over the task of working for my release. I hasten to add that Canadian officers made available a not inconsiderable amount of money for that purpose.

The Canadian lawyers made tremendous efforts to convince the government of the intolerability of the sentence, but the political atmosphere was still stronger than justice.

The arguments in the press continued. All parties and organizations participated in that discussion. The discussion didn't center around the human interests of a prisoner and his family, but about the "holy principles" of Nazi prosecution. Ralph Allen, editor of "Maclean's Magazine", joined that "media war" on 1 January 1950 and published the following article:

Has Canada given a just hearing to her only war criminal? A war correspondent who was present says no, but *Panzermeyer* is still serving a life sentence.

The man who sat in the middle of the judge's bank spoke very slowly, almost regretfully, as if he were thankful for the breaks in what he was saying: "The judgement of this court is death by firing squad. The sentence becomes final once approved. These proceedings are now closed."

The man standing in the middle of the room was as rigid as in a vice. He took one step backwards, bowed slightly, waited a second for the officers accompanying him and then turned to the left. As he stalked out of the room he held his head a fraction of an inch higher than usual and his tense face was a fraction of a shade paler than usual. His steps echoed in the silence like a distant, ghostly roll call, directing our attention towards an awful past and an unknown future.

The death penalty has not been carried out in Kurt Meyer's case.

And now, four years after the sentence had been passed down by a Canadian military tribunal, neither the extent of Meyer's responsibility for the past nor the significance of its meaning for the future has been clearly established.

The inconsistencies of justice, injustice and doubt in that highly bizarre, disputed and possibly most important trial in the history of Canadian law are not as loud today as they were when Major General Vokes, then Commander of the Canadian occupation forces in Germany, unleashed an international storm by commuting the sentence to life imprisonment. However, disagreement with that decision is no less deeply rooted. There are hundreds of thousands of Canadians who believe that Meyer should not be in prison.

Summarized briefly, the facts of the case are as follows:

As the commander of *SS-Panzer-Grenadier-Regiment 25* of the *12. SS-Panzer-Division "Hitlerjugend"*, Meyer commanded German soldiers who shot prisoners-of-war during the first bloody days after the Allied landing in Normandy.

He is the only enemy officer who was interrogated according to Canadian war crimes statutes of August 1945. Those statutes attempted to lay down guidelines under which enemy war criminals could be sentenced for their actions between 1939 and 1945.

Working through the United Nations War Crimes Commission, all Allied nations passed similar statutes soon after the war in Europe had ended. Meyer's trial in December 1945, presided over by a Major General and four Brigadiers in Aurich, Germany, was the first trial of a German combat soldier or commander under such statutes, Canadian or otherwise. That was why it had such importance attached to it beyond the fate of individuals or the laws of a nation.

What are the main facts? The one fact at the heart of the matter is the question of guilt. Was Kurt Meyer guilty?

Was he found guilty according to a process and a legal code which could serve future generations of Canadians as a useful and practical guide without challenging its legacy of Anglo-Saxon justice? Did he have a just hearing? Was he interrogated according to the rules of impartiality? Were these rules applied impartially by the courts which examined him?

As a reporter who witnessed Meyer's trial, I say: No! I returned to Canada and said no to all these questions and to everyone who would listen. It was not, perhaps, a miracle that there were not yet many who wanted to listen at all, considering the mood of the population 4 years ago, when memories hurt like open wounds and revenge had not been taken for all offences.

I have just come from Ottawa where I checked the files of the trial and that's why I still say: No! Under the accumulated dust, the protocol still seems to say what it seemed to say four years ago in the courtroom in Northwest Germany in between bouts of uproar and silence in the courtroom. Among other things it seems to say:

That, under cover of Canadian laws, Meyer was interrogated according to rules which contradict the most fundamental and precious principles of Canadian law.

That, although the words of the laws were in line with the fundamental principle of the innocence of the accused, they nevertheless contained a flexible clause which presumed him guilty until he could prove his innocence if it was once proven that crimes were committed (not necessarily by him, on his orders or with his knowledge or approval).

That the court assumed itself to be authorized in such cases and held it for permissible, to "construct" its own guidelines during the course of the trial without restriction and often to the disadvantage of the accused.

That the main witness against Meyer, without whose evidence the case for the prosecution would have been weakened to the point of non-existence, completed his first statement against Meyer under threat of death.

That the same witness had been interrogated at least eight times outside the proceedings and that, under the elastic guidelines of the trial, the prosecution lawyers were allowed to present parts of each of his eight statements and bring him into the witness box twice.

That the same witness very often contradicted himself despite months of preparation and checking of minor as well as major points of the evidence.

That most of the incriminating details of his evidence could not be substantiated by any other witness but could, at least partially, be disproved by half a dozen witnesses.

Should somebody ask the question whether justice has been done to Meyer? I do not believe the answer would be yes.

In the small, paneled courtroom Kurt Meyer showed up under the gloomy floodlights of examination like a documentary film which told the story of National

379

Socialism from its birth to its ruin.

The dead legions marched again. The supermen stalked the face of Europe with undiminished glory, and an old comrade warmed and praised them with a voice fall of pride — at times close to tears — for what might have been.

Kurt Meyer, an unknown 18-year-old police cadet, had hitched his wagon to the star of National Socialism. When he was transferred to the *Leibstandarte Adolf Hitler* in 1939, he had party number 316714 and *SS* number 17559.

When the *Wehrmacht* struck in 1939, Meyer was at the sharp end of the spear. His reputation as a soldier was unique and, without question, deserved. There was a special name for the new type of warfare — *Blitzkrieg*. There was also a special name for the young commander who rode with his dashing troops: *Schneller Meyer*. The places he visited still remember that *schneller Meyer*.

After operations in Poland, Holland, Belgium, France, Rumania, Bulgaria, Greece and Russia, *schneller Meyer* became the commander of a *Panzer* division in Normandy. He received his third wound and eleventh award.

He spoke well and readily in court about his campaigns. Like a lonely voice calling over a wide, unspanned abyss he spoke up again and again for his vanished comrades and the forbidden creed of "blood and iron."

Once he bent forward insistently and repeated parts of the address he once delivered to the young fanatics of the *12. SS-Panzer-Division:*

"We are now in Normandy. In our fists are the weapons of the German people. German workers forged these weapons with the sweat of their brows, not so that we may throw them away in cowardice but that we may fight. Should the Allies ever cross the Channel, their elimination won't be accomplished by the much vaunted *V-*weapons. Do not swallow the nonsense in the newspapers! Know the facts! Our enemy's elimination can only be forced through our attack."

And on another occasion:

"At the end of the war each prisonor should be in a position to prove he did not go into captivity willingly."

The trial had no scene more dramatic than the moment when Meyer, in a touching and yet somehow frightening manner, suddenly forgot his surroundings and directly addressed his wife. *Frau* Meyer had come to Aurich with one of her five small children to be close to her husband. Most of the time she sat alone at the back of the room. A silent character, surrounded by unending grief. She had been allowed to visit her husband briefly but always with a guard close by and with instructions that Meyer was not allowed to say anything which could be considered as guidelines for the rearing of his children.

One day the prisoner stood ready to explain the principles he applied to the training of his soldiers. When he started, his blue eyes appeared to alternate between the hard and the affectionate, whenever they caught hold of the eyes in the tenth row.

At times, he would speak with pride and certainty, at other times with tenderness and solicitation. The tortured, helpless face of his wife would then beam and come alive.

"The discipline of my young soldiers was good," he said. "It was based on the principles of a healthy family life. Due to their youth, our leaders had to find new ways to train these soldiers. There was a close comradeship between officers and men. The parents were involved to a large extent in their training. Alcohol and tobacco were prohibited for the younger soldiers."

Meyer paused, then he started again in a milder fashion. "Behaviorally, the ideal of motherhood was held high for these soldiers: A mother fights, lives, sacrifices and suffers for her children. She lays the foundations for a proper life in society.

"My soldiers had complete religious freedom. I told them: 'God cannot be proven, but we have to believe in Him. A human only becomes a human if his conscience makes him feel responsible to God. A soldier who does not believe in God cannot fight; he lacks the last reservoir of strength which can only be gained by the deepest of faith. Their guidance for fighting was the ideal perspective of the military profession: I am nothing; we are everything.'"

Meyer was prosecuted by Lieutenant Colonel MacDonald and defended by Lieutenant Colonel Andrew, both lawyers in civilian life. Each one pursued the case with cunning and diligence. Both of them as well as the officers of the court found themselves considerably hampered in working with laws which were foreign to their experience and Canadian tradition. They were the laws of the victor; freshly coined and never tested by a single precedent.

It was inevitable that the prosecutor asked for and received advantages over the defender he would never have asked for or expected if the new laws had been in accordance with the basic tenets of Anglo-Saxon law.

The statutes state that the court may consider all testimony or each deposition which appeared to be authentic, provided the testimony or deposition appeared to the court to be of use to either prove or disprove the prosecution's case, irrespective of the fact that such testimony or disposition would not permissible as evidence in a field court martial.

The statutes state that the testimony of a defendant or a witness who is present at the proceedings, regardless of whether the testimony has been sworn or has been rendered without a prior warning, is permissible at any time as evidence.

The statutes state that if it has already been proven that personnel of a unit have committed war crimes, the commander of the unit or formation can be found responsible for these crimes, that is, unless he can prove he is not guilty.

Furthermore, these special extensions to the traditional Canadian legal code and the war crimes statutes conclude with a clause which in 21 words says that anything is possible.

Section 17 states: "In cases not foreseen by these statutes, a course must be taken which promises to show that justice has been done."

During his cross-examination, Meyer made a passing comment that, at the beginning of the Normandy campaign, he had come across the bodies of half a dozen German soldiers laying at the side of the road under conditions which gave reason to believe that they had been captured and then shot by enemy troops.

The Canadian Brigade commander in the area in question at that time was the president of the court, Major General Foster.

The court listened attentively to Meyer's proven testimony about the shooting of German prisoners by Canadian troops under Foster's command, but when the prosecutor announced he would call witnesses to disprove that, the court said that would not be necessary.

General Foster said: "I do not believe the court has any questions in connection with that particular incident."

Officially, Kurt Meyer's case is closed. Meyer is serving a life sentence in Dorchester.

There is no procedure whereby the death penalty can be imposed a second time. The one procedure by which the sentence of life imprisonment may be further mitigated is a petition to the throne. Why not just leave the matter alone, the way it is — over and forgotten?

The laws by which Kurt Meyer was condemned are not based — as all such laws must be — on the solid foundation of rules and clearly articulated principles, but on the unobtainable prerequisite of human omniscience.

If the Kurt Meyer trial was unjust, then the greatest injustice was not against Meyer himself. It was the ultimate, all-surpassing injustice against a whole series of precepts which, if we deny them to our enemies, they can also be denied against us.

That article, of course, stirred up a hornets' nest. The arguments for and against me raged for weeks in the Canadian press. Canadian lawyers worked on extensive legal briefs during that period and came to the conclusion I was sentenced unjustly. The full text of those briefs has never been published.

I was informed in a roundabout way in the spring of 1951 that my release was close and that my stay in Canada would only last a few more days. I simply could not believe it. The news came as too much of a surprise. Only when I was sworn to secrecy did I dare believe I was about to be released or at least transferred to Germany. I waited for the words of relief day after day, week after week, but I waited in vain. As chance would have it, a special report by a major Canadian newspaper fell into my hands some time later. The report read:

Meyer's Transfer Planned

The petition to pardon German general Kurt Meyer , who is being held at Dorchester prison, has been temporarily deferred, as we discovered today.

The petition for release, reversal of the judgment and return to Germany had been drawn up by his lawyers.

The government had been favorably disposed to granting the petition until an avalanche of protests reached Ottawa from veterans and other organizations and individual persons from all parts of Canada. It was said that the government had been prepared to transfer Meyer to Germany and reduce his sentence, on the grounds that the act for which he was accused — responsibility for the death of Canadian prisoners in France — had also be done by Allied soldiers and the United States had already released similarly accused persons.

It looks as though he will remain in Dorchester for some time to come. "The case has obviously come to a dead end" said one government spokesman.

From Dorchester to Werl

When I made my way to my cell on 17 October 1951, I was totally surprised to hear the chief warder ask me for my hat size. I was completely puzzled and unable to tell him. Shortly thereafter I was taken to the warden who informed me, in the presence of several civil servants, that I would be leaving the institution at dawn and flown to Germany.

I had a sleepless night, my mind filled with mixed feelings. My past experiences gave me little cause to be happy. I no longer had faith in anything. The hours passed slowly. Thank God, I was able to read. Our cells were always lit up. Winston Churchill's memoirs, the very volume in which he explained why he issued the order to destroy German air-sea rescue aircraft, shortened my night. The hands of my watch indicated 0500 hours when I had been pacing up and down my cell for more than an hour. Nobody showed up. The other inmates and I received our morning coffee at 0630 hours. Following that, I was led to my work place. With a feeling of boundless disappointment, I looked out over the wide bay and dreamed of thundering engines. White sails gleamed in the distance and disappeared towards Nova Scotia. Their majestic appearance renewed my dreams of freedom.

Relief came at midday. I was briefly told that thick fog had prevented the planned takeoff from Moncton and that I would be driven to Nova Scotia to fly to Germany from there.

Strict security was observed due to the expected "media frenzy". I was led to the waiting car through a side door. I had been put into Canadian uniform once again and was driven away in the company of two officers. The warden and other gentlemen took their leave with a handshake — in contrast to my arrival.

Our route crossed through huge, undeveloped country for hours on end until we reached the airport at 2000 hours and swept across the runway shortly afterwards. The aircraft was a North Star, used as a supply plane to Korea. The heavy bird rose after a short takeoff and rushed away to the north. Canada's last farewell was a stream of glittering lights. It was only then that I believed I was returning home. I regretted I was unable to say goodbye to my friends. I would have liked have said farewell to Fritz Lichtenberg and James Miller. I will always remember their kindness.

The only other passengers in the gigantic plane were two officers and some sergeants. We had a large dinner as soon as we had reached the open sea. That was a pleasant surprise that did me a lot of good. We had to stop over in Newfoundland because of the danger of icing. The flight continued after about two hours.

It did not take long, and everybody was fast asleep. My escort behaved properly in all respects. They were combat soldiers who had fought against me. I was told confidentially at midnight that the bomb had already gone off in Canada and that a variety of spokesmen had demanded my return. The press did not have to worry about sensational headlines any more.

I was slightly disappointed by the flight as I could hardly see anything except clouds and fog. On our approach to England we saw some troop transports taking the Canadian 27th Infantry Brigade to Germany. When I had been taken to Canada in 1946, the Canadian forces followed soon after and were demobilized. It was different at that point. People wanted to defend the freedom of the Western World in Germany and incorporate German troop formations into NATO.

Did the planners really believe that Europe could be defended by force of arms? There was no defending Germany nor the rest of Europe, only the destruction of Europa.

The flight to Bückeburg continued after a short layover in England. We got into thick fog over Holland and received instructions to land in Wunstorf instead of Bückeburg. Wunstorf was easier for landing and better equipped. So that was how I saw Germany again! My homeland's first greetings were patches of fog, the odd light and flashing searchlights.

The plane began its approach in complete darkness and seemed to hover in the fog. We were unable to make out a single thing. Suddenly the plane received a stiff jolt, reared up and then crashed its full weight down on its landing gear. All the loose objects in the plane flew around our heads. The plane slid forward a couple more meters and then came to a halt with a worrying list. The bird had gone beyond the runway and promptly landed in the neighboring potato field. The proud bird was now lying in the furrows without an undercarriage.

I was glad to have the soil of Germany under my feet again. The fog had upset the apple cart. The gentlemen who had been expected to receive me at Bückeburg had to be directed to Wunstorf.

The RAF officers there asked me into their mess, and it was not long before our conversation turned to the fighting in Normandy. They had all operated over Caen, and I found their reports of great interest. For example, I heard they had an enormous respect for the infantry's defensive fire and that it had destroyed many a plane.

By the time the British officials who were to escort me to Werl reached Wunstorf around midnight, a few bottles were standing in a corner. They had been emptied in a toast to my forthcoming release. The transition from my cell to an officer's mess had taken its toll.

All these impressions had to be assimilated. It was not at all easy to get used to normal life again, especially for someone like me who had kept his own company for six years.

Not knowing what would be awaiting me, I walked to the waiting vehicle in order to finish the last leg of the trip to Werl as soon as possible. The driver opened the door, and I was virtually struck dumb — a former *SS-Oberscharführer* from my old division, who then earned his living by driving the English commandant at Werl, was standing in front of me. Well, things were certainly looking up!

We drove slowly down the *Autobahn* to Werl. I had read a lot about Landsberg in Canada, but I knew nothing about Werl. I was immediately taken to the cell block and led to my cell on the fourth floor once the heavy gates had closed behind us. Out of curiosity, I quickly glanced at the names on the cell doors. The first name I saw was Kesselring; that was followed by von Falkenhorst, von Mackensen, Gallenkamp, Simon and von Manstein. I breathed a sigh of relief to be among soldiers once more.

Werl's commandant was the former commandant of Camp No. 18 in England. He had sent me to London in 1945. In contrast to his successor, Colonel Meech, Vickers was an understanding commandant.

I received my first word of welcome from my lawyer, *Dr.* W. Schapp, in Aurich. I was extremely pleased about his telegram; it went beyond the usual call of duty. I was allowed to meet my wife a few days later. We had not seen each other for six years and had only lived on sparse information in censored letters. Those six years had not passed without leaving their mark, but the time had not drawn us apart. I was happy to hear of our children's' development and that a healthy family was waiting for me.

Colonel Vickers told my wife that he was prepared to grant me 10 days leave of absence on my word of honor as soon as the usual formalities could be completed.

He kept his word. I was allowed ten days leave at the end of November. I speedily donned a borrowed suit to leave that unfriendly building as quickly as possible. I was worried the special leave might be withdrawn through unusual circumstances. So, off I went! There was no time to lose.

For the first time, I marched out without chains or escort and with my old uniform under my arm onto the soil of Germany.

I arrived home late that evening and stood without a word in front of my grown-up daughters. They were taller than I. Those few days passed quickly, especially since many old comrades visited me unexpectedly and also because the local people made a great effort to make my leave pleasant.

Canadian officers unexpectedly descended on our apartment one day. They had been visiting the inner-German border and wanted to say hello to my family. My leave struck their accompanying reporter, Douglas How, like a bombshell. Parliamentary inquiries and a considerable storm throughout the Canadian press resulted from his report. It took a long time for the Canadian public to calm down concerning the incident.

Meanwhile, efforts were being made in Canada as well as in Germany to petition for my freedom. In Canada, it was my good friend Fritz Lichtenberg, well supported by Canadian officers. In Germany, *Dr.* Schapp and his wife worked relentlessly for my release.

In June 1953 *Dr.* Adenauer participated in a Silesian pilgrimage to Werl which commemorated a similar event in Annaberg, Upper Silesia. *Dr.* Adenauer had received permission to visit war criminals at the same time. I suddenly and unexpectedly found him standing in my cell, shaking my hand and promising to do everything possible to put an end to my captivity.

Following a proposal by Brigadier Sherwood Lett, the Canadian government reduced my life sentence to 14 years. That decision was announced to the Canadian citizenry on 15 January 1954. I was only officially informed of that measure 14 days later. The Canadians had forgotten to inform the English of that act of leniency. Campbell MacDonald made the following announcement on radio station CFRA in Ottawa:

Former *General* Kurt Meyer will leave prison a free man some time this year, possibly on 7 December. He is presently behind bars in a German penal institution to which he had been sent from Canada.

As long as he was in Canada, he lived as a prisoner in Dorchester Prison in the province of New Brunswick. His transfer from Canada to West Germany two years ago was the first step towards a possible release.

Today's announcement was no real surprise to anyone familiar with the case. *Generalmajor* Kurt Meyer owes his freedom to the relentless efforts of a German-born Canadian citizen who became convinced a number of years ago that Meyer had been unjustly imprisoned. The name of that man is Fritz Lichtenberg. He is an entrepreneur and leads a quiet life in Moncton.

He gave all that up to provide the necessary financial means to get Kurt Meyer out of Canada, back to Germany and finally out of prison. He had thrown down the gauntlet. He started by proving that the judgment was a perversion of justice and that Kurt Meyer had become a victim of circumstance because of the very fact that he was on the losing side in the last World War. You will remember that *Generalmajor* Meyer had been tried by a Canadian military tribunal and sentenced to death as a war criminal.

Why had he been sentenced to death?

Because he was the commanding general of the troops in Normandy at a time when Canadian prisoners were shot. That was in 1944.

We want make the details of his charges absolutely clear. He was in no way accused of having given his men the order to shoot the Canadians. The evidence during his trial showed that he had no knowledge of these shootings. He was only charged with having been the commander in the area where the Canadians had been shot.

The sentence was death by firing squad!

Here in Canada the sentence was in accordance with the will of the people. The war was still fresh in Canadian memory. They were incensed that captured and defenseless Canadian soldiers had been shot in cold blood. The sentence satisfied their need for revenge. Kurt Meyer served as a scapegoat.

Then came the shock!

A Canadian general, red-haired Chris Vokes, reviewed the evidence and commuted the death penalty to life imprisonment.

The result was a cry of rage and indignation from the Canadian public.

What the civilian population did not understand and probably still does not understand today is the fact that the shooting of prisoners had not been exclusive to the Germans. I'm firmly convinced that among those men who are listening to me this evening, Canadian veterans of the First and Second World Wars, there are some who either participated in the shooting of German prisoners themselves or are certain in the knowledge that their fellow soldiers shot one or several prisoners between the forward and rear areas.

German prisoners were shot by Canadians during the last war in Italy, France, Belgium, Holland and certainly in Germany. They were killed because, at the time, they were a hindrance. They were in the way. They were only more mouths to feed, more bodies to be dealt with. That's why they were killed.

The soldiers felt justified because somehow, in some confused way, they believed that the fighting endowed them with a God-given right to take such action.

They believed that in that way they were putting paid to the enemy for the bombing of London and the deaths of their comrades on the battlefield.

It is neither pleasant nor easy to talk about these things. They are a nightmare now that the sun is shining. We were happy we had peace. But one man is still in prison — *Generalmajor* Kurt Meyer. And his crime was no greater than that of some Canadian General, whose men did to Germans what Kurt Meyer's troops did — without permission, without his orders and certainly wholly without his knowledge

Neither am I trying to excuse what some Canadian troops did, but these things happened and we would be hypocrites if we did not admit it.

General Chris Vokes, commander of the Canadian Forces of Occupation, knew exactly what he was doing when he commuted Meyer's death sentence to life imprisonment. It is highly probable that he knew fully that the reverse could have occurred if we had lost the war instead of the Germans.

And so Kurt Meyer came to Canada and the Dorchester penal institution. Life was difficult. The guards put pressure on him. They made that proud general scrub the floor again and again.

All that while the German-born Canadian citizen, Fritz Lichtenberg from Moncton, was trying to accomplish Meyer's release. He gave up his life of retirement and went back into business. He visited Meyer in prison, spoke to him, collected evidence and flew here to Ottawa to present the case to Canadian officials.

He finally achieved success two years ago.

The Canadian government decided to transfer Meyer to an Allied penal institution in West Germany. It was then only a question of time before Meyer's sentence was commuted. Today, the Minister of Defense, Mr. Claxton, announced in the House of Commons that the life sentence had been commuted to 14 years imprisonment

With time off for good behavior, the general could expect to be a free man again late in the summer of 1954.

I have nothing to add to MacDonald's comments; he hit the nail right on the head. Humans, not angels, fought on both sides.

Meanwhile, conditions in Werl had changed fundamentally. Colonel Meech believed that he had to treat us like criminals and that he could ignore his assistants' experience and well-intended advice. His orders, which proved he had no experience, made him the most-hated man in the prison within a short period. He tried to reintroduce the tone of the immediate post-war years and failed to recognize that the times had changed. The consequences of his work were several suicide attempts. We found our best allies in his assistants. Our English guards showed themselves to be much more understanding than their German colleagues or superiors.

Among the German personnel, the unsophisticated men — the former enlisted men and frontline soldiers — stood head and shoulders above the rest. They stood up for their imprisoned comrades with all their heart and soul. Regrettably, I can not say the same of their unknown German superiors. They seemed to care less about the well being of their fellow countrymen.

Colonel Meech was no soldier. He was a civilian in uniform who had experienced the war as a home-front warrior. He finally managed to create even greater isolation for the so-called war criminals. He practically established a prison within a prison for us. A building complex within the penal institution became our "home" and completely separated us from the outside world. Even the German guards were removed. They were replaced solely by British officials.

Obviously, we were tempted to play tricks on Mr. Meech and to get out of his satanically devised isolation. My comrades were therefore afforded great pleasure when they heard I had been able to see my wife for more than two hours almost under Mr. Meech's nose.

Mr. Meech is still lord and master of Werl today, in the summer of 1956. My comrades still long for freedom behind its drab walls. There are no longer any important men buried alive there. Corporals, noncommissioned officers and minor officers cry out for freedom, but their appeal to reason and justice has continued to remain unheeded. Time passes them by remorselessly.

My release was fixed for 7 September, 1954. I could hardly imagine what freedom would be like and waited impatiently for my hour of release. How would I be able to adapt to the changed environment? That thought occupied my thoughts day und night. The 7th of September was the 10th anniversary of my capture. I had been separated from my family for 15 years.

My oldest daughter was then four years old. A young woman of 19 was expecting me at that point. I wondered whether I would succeed in winning the hearts of my children or whether I would first have to struggle a long time for their love? Those thoughts tortured me during the last few weeks of my ten-year-long captivity. My hunger for life was dampened by the fear of freedom.

The last months behind barbed wire were especially bitter. Time seemed to stand still. I had the greatest inhibitions about taking leave of my comrades. It was not pleasant to say goodbye to comrades whom you knew were bound by law to sit in their lonely cells for many more years.

I will never forget the handshake of a former corporal who was still awaiting his freedom. His children's big eyes always seemed to ask: "Why did you come without our father?"

Then the time came on the evening of 6 September. The time of my release had arrived and my parcel lay ready on the table.

A number of comrades had assembled in front of the main gate of the penal institution, despite the early morning hour, to welcome me and accompany me home. The British found the demonstration of loyalty unwelcome. They therefore preferred to release me in secrecy through a side door and take me to a hotel in Werl.

I was overwhelmed by the impressions which surrounded me at that point. I can only to describe them in part here. Comrades had also been expecting my arrival at the hotel. My wife came towards me, beaming with joy and led me to my fatherly friend, Fritz Lichtenberg. Good old "Uncle Fritz" had come to Werl from Canada so that he could personally take me into the circle of my waiting children.

I was especially touched by the bouquet of flowers from the Vice President of the North-Rhine Westphalia Red Cross association, *Frau* Else Weecks. She eased the misery of the inmates and their families out of pure charity and the desire to help. She was justifiably called the "Mother of Werl" by the war criminals.

Heinz Trapp, a comrade from the *1. SS-Panzer-Division* quickly drove me to Friedland, where the release formalities were completed. I was able to meet my mother and friends in the repatriation camp before I continued the journey to my children.

I was only a few minutes away from my children by nightfall and fidgeted restlessly back and forth in the car. We reached the small village of Niederkrüchten, where my family was staying, shortly after 2200 hours. But what was going on? We could suddenly go no further. It was black with people ahead. The fields on both sides of the road were packed with cars. A triumphal arch, as was customary for festivities in the *Rheinland*, welcomed me. The police tried to explain what awaited me, but I could only understand a couple of words. The sound of drums made comprehension impossible.

I could hardly take in what was happening. Only the sight of the familiar faces of my old soldiers and officers made me realize I was being welcomed ceremoniously. The local population had formed a long lane of torchbearers to light the way to my children.

Deeply moved, I listened to a neighbor's words of welcome and those of a representative of the homecoming organization. When they were finished, I finally embraced my girls and my boy.

Not even the thought of the long journey from Saarbrücken had dissuaded the only survivor of my command tank, the one-armed Albert Andres, from coming to shake my hand. I was deeply grateful to my comrades who had rushed to Werl for the welcome. Those were men with whom I had marched across Russia's icy steppes and through Normandy's hail of steel. We had been separated for ten long years, but we had still remained comrades. The misery of that terrible war had bonded us for ever. I speak out in that circle of comrades — the manual laborers, mechanics, white-collar workers and farmers of today: "Believe me, the years which lay behind me have left no hatred towards our erstwhile enemies. Let's not talk about the past but work for the future. We have to secure the future for our children and build a strong, vital Europe."

I was home at last after fifteen years in the wilderness.

Afterword

This book was written following the grim experience of war and not intended to glorify either individuals or units. My account was written to advance the cause of both the living and dead soldiers of the former *Waffen-SS*.

Those men were soldiers! They fought at the side of all branches of the *Wehrmacht* and were continuously employed at all the hot spots of the front. Of the 38 *Waffen-SS* divisions comprising 900,000 soldiers, 300,000 were killed in action and 42,000 are considered to be missing.

In contrast to the propaganda advanced by certain parties, there was a totally comradely relationship between the men of the *Waffen-SS* and the Army.

The claim that units of the *Waffen-SS* divisions were assigned as extermination units is misleading and defamatory.

Nothing can or should be glossed over in the interests of accurate historical reporting. Things happened during the war which were a disgrace to the German nation. The former soldiers of the *Waffen-SS* are man enough to know what were war crimes and to detest them as such.

It would be silly to reject all the events with which we were charged by our former enemies as propaganda inventions. They obviously made propaganda out of them and exaggerated them in a grotesque and totally incredible manner. They were the victors, after all, and we had no rights as the losers. But crimes did happen. It is irrelevant to discuss the number of victims; the fact it happened is incriminating in and of itself.

Nor does it help to present a counterclaim against our former enemies. We know that they committed a long list of inhumane actions, causing enormous grief and the death of millions with their expulsion of our people from our eastern provinces and with the carpet-bombing of our cities. The Allies must cope with that evil themselves. We cannot take on the task of weighing the grief or judging. The final judgment of what happened in our time rests on the truth. We do not know whether that truth will ever be known.

The *Waffen-SS* is now incriminated with events in the concentration camps because leading individuals of the government have placed special formations in the same category as frontline troops. By doing so, they have rendered a bad service to our units and have thrown thousands of excellent young people into indescribable misery.

The world took revenge on the soldiers of the *Waffen-SS* after 1945 in a way which does the victor no credit. Our young soldiers had to suffer inhumanely for events for which they were neither responsible nor had any opportunity to prevent.

The soldiers had neither more nor less knowledge of the events in Germany than the majority of the German people. For example, I never saw a concentration camp either from the inside or the outside. Soldiers who were constantly engaged in fighting and who received 14 days of home leave a year — if they were lucky — could not possibly be held responsible for political crimes. Especially if one considers that these men were, for the most part, between 18 and 22 years old.

The young men volunteered because they were convinced that there was an especially good esprit de corps within *Waffen-SS* divisions and they were led in accordance with that attitude. There was no question the divisions consisted mainly of volunteers. The volunteers were idealists who, true to their upbringing and the traditions of their fathers, were prepared to sacrifice themselves for the greater good. They did not volunteer for easy duty but to fight for their homeland.

The units were subject to hard discipline and educated along good Prussian military lines. The *Waffen-SS* was taught by officers of the old Imperial Army, the *Reichswehr* and the provincial police up to the outbreak of war.

However, we all made the effort to set old-fashioned ideas aside. That applied especially to combat training. New tactics were relentlessly sought and found. The men pressed for a revolutionary change in training and leadership, above all in the relationship between officers and men! In doing that they followed a development which had already started during the last years of World War I under the pressure of increasing total war. It had taken form in special frontline detachments and, later, the *Freikorps*. It is clearly the path being taken in future military establishments.

The *Bundeswehr* confirms that that type of leadership, indoctrination and training were right. The reforms undertaken by the *Bundeswehr* that one can take seriously already belonged, for the most part, to the fundamental principles of training in the *Waffen-SS*.

The *Wehrmacht's* performance should indeed be reason enough to properly judge its training methods and understand that blind obedience and lock-step methods of instruction were the exception. Could any army have held out against the demonstrably proven Allied superiority for six years had the training methods and the interpersonal relationships between officers and men not been the best possible?

393

The authority of the officers in the *Waffen-SS* wasn't based solely on the obligatory and omnipresent iron discipline. Nor was it in any way founded on individual commanders being supported by their immediate superiors. It was the true authority of real leadership and was thus based on the fact that these men were exemplary human beings and soldiers. Something emanated from them which, on its own, created unconditional obedience. They were, first and foremost, comrades to their men. They were bound to them, for better or worse, unto death. By doing that, they gained the complete confidence of every single soldier.

It was a matter of course that before each decisive combat operation the troops were familiarized with the situation and the importance of their mission within the context of that situation (as much as time would allow). As a result, even the youngest soldier could contribute to the operation, aware of his own responsibility for mission success. He could see a clear meaning for his own role, even if it had to be his own sacrifice. Often enough that pronounced sense of responsibility reflected back to the unit leader and helped him overcome any inner crisis. A look full of expectation and encouragement, a bit of gallows humor and manly composure was, without doubt, a wonderful comfort.

That is precisely the essence of a true warrior community — that the factor which binds the men to their leader was also effective in binding the leader to his men. If a nation has to fight a total war against half the world for its mere existence, then such a community comes about on its own.

All that glistened in the *Waffen-SS* was not gold. However, the fact that there was such an exceptionally high loss rate in officers of all ranks confirms the opinion of a young soldier who said: "Our officers were not superiors but warriors." The heavy toll of blood which the *Waffen-SS* had to pay during the war still does not seem to be enough today to concede that the members of that formation were soldiers. That recognition has been denied them by civilians.

Our erstwhile enemies emphasize our men's fighting spirit in their literature and pay high tribute to the divisions of the *Waffen-SS*. No soldier of the former *Wehrmacht* who had fought alongside *Waffen-SS* divisions would condemn them. It was never my experience that *Waffen-SS* divisions were unwelcome at the front. They were greeted with open arms.

The creation of the *Waffen-SS* and the *Luftwaffe* and Navy Field Divisions was a mistake from a military point of view. There can and must only be one armed force. The existence of several specialized armies means a dangerous weakening of military strength.

It is, however, illogical to make the men themselves responsible for the existence of the former *Waffen-SS*. That responsibility must be carried by those politicians who failed during those years of internal political strife and who are using the strongest of language against our former soldiers today.

Perhaps these gentlemen may one day remember that, at the time when the political dice were being thrown, the future soldiers of the *Waffen-SS* were not even old enough to vote and that the youngest boy destined to enter the armed forces was then only six years old.

The soldiers of the former *Waffen-SS* do not desire a return to the past. The wheels of history cannot and must not be reversed. They support the government and are represented in all the Federal Republic's political parties. There is no uniform political opinion among the former soldiers of the *Waffen-SS*. They reject all radical, short-lived and erratic groups. They stand for democracy. This is exactly why they are fighting for their rights.

Any man who knows war — and we really did get to know it — does not wish for another war! Furthermore, wars are rarely started by soldiers but almost always by politicians. It is the soldiers who then have to make the greatest sacrifices. We know, better than anybody else, that our nation cannot bear to make any further blood sacrifices, that its children must mature in peace and tranquillity to become a strong new generation!

*

At this point I must also acknowledge and thank those Canadian officers and men who left no stone unturned during my long captivity in fighting for my freedom. I also thank my brave soldiers and officers who took care of my family in time of need and who are continuing to forge those bonds today.

And finally, above and beyond all the worries arising from our membership in our former units, a last word about that which lights a flame in all our hearts: Germany!

We, who, when all is said and done, sacrificed ourselves in battle not for a political party but for our homeland, know that our Germany existed before 1933 and also continues to exist after 1945. As the fatherland's poorest sons are also the most loyal at the hour of danger — as the blue-collar poet Karl Bröger once said — so do we also feel an obligation to our nation, even though the state has not yet fulfilled its commitment to us.

We must stand united to fight for the emancipation and freedom of our comrades convicted of war crimes and put an end to the slander. Those in prison cannot defend themselves. Those who lie in countless graves no longer have a voice. However, their dependents, especially their children, still live. Their future is our responsibility.

Let us therefore lift our voices again and again, moderate in tone but conscious of the responsibility. I am convinced that the hour will come when we will be heard!

If that should happen and if the history of warfare has been enriched by the campaigns described herein and by the spirit and vitality of our unforgettable army, then this book has fulfilled its purpose:

A memorial to the grenadiers of the *Waffen-SS*

In Memoriam

By Hubert Meyer
former chief of staff of the
12. SS-Panzer-Division "Hitlerjugend"

Panzermeyer was an exceptional military leader, a personality with a unique character and strong charisma. Any reader of his account will testify to that even without encountering that man in wartime or having known him at all. The question automatically arises as to how that personality developed and how his comrades perceived him.

My description of *Panzermeyer's* life until 1937 is based on information provided by other people. After 1937, it is based primarily on own experiences. We were together in the kind of situations in which any man reveals his innermost soul. I therefore know how *Panzermeyer* really was as a leader and as a man. I want to relate it here.

I started my service on 2 May 1937 as a young platoon leader in the *10./SS-Infanterie-Regiment "Leibstandarte"* after having been graduated from the *Junkerschule* at Bad Tölz and finishing the platoon leader's course. Having previously served in *SS-Infanterie-Regiment "Deutschland"*, I knew almost none of the roughly 80 officers of that 17-company regiment.

One of the first officers I noticed was the commander of the *14./SS-Infanterie-Regiment "Leibstandarte"*, *SS-Hauptsturmführer* Kurt Meyer. He appeared to me to be a man of exceptional vitality who inspired his company with his unique spirit.

Much emphasis was placed on sports. The men were not meant to become athletes but agile, skillful, enduring and fearless. Among other things, that involved the entire company jumping into the swimming pool from the 10-meter board with the company commander in the lead. One or two of the other platoon leaders sometimes complained under the pressure of such demanding service. At the same time, they were all proud to belong to that company.

When that company was formed it was known as the *Panzerabwehrkompanie* (Antitank Defense Company). When that name was later changed to *Panzerjägerkompanie* (Tank Hunter Company) it was a much more appropriate name for that which had already existed for some time. In Meyer's company, you didn't defend, you hunted. He would have much preferred to have been commander of the *Kradschützenkompanie* but, seeing how that was not possible, Meyer made hunters of the men entrusted to him. Men who pushed ahead, who sought and engaged the enemy rather than waited for him.

Meyer took over the *Kradschützenkompanie* after the Polish campaign and was allowed to form the new *SS-Aufklärungs-Abteilung 1* in Metz after the Campaign in the West. He was in his element. With the young company commanders and platoon leaders he had selected, he created an instrument second to none in terms of mobility and which justifiably earned him the name of "*schneller* Meyer".

What was it that drew him on, that made him always seek and find those special opportunities? Was it unusual ambition or even lust for glory? Kurt Meyer certainly had a healthy ambition. Of course, he was proud of the successes of the troops led by him and of his well-deserved awards but what really ruled his character was a burning love for his nation and fatherland. It was the wish to serve with all his might.

He did not have to be ordered into action. His motivation came from within. When the enemy situation was being briefed during the issuance of an operations order, his inner motor was already warming up, and his intellect was searching for the enemy's weak points, feeling for opportunities to multiply the combat power of his limited forces through skillful deployment and surprise and so leading to a big success.

Once on the march, he searched the terrain like radar for any sign of the enemy. He assimilated all reports with unbelievable speed and in all their implications. As a result, he never held on to preconceived plans, come what may. Everything was fluid, and he was ready to adapt to any new circumstances or changes and exploit them.

Likewise, he was not to be found in a rearward march serial of a column during an advance or at his command post during an attack. He was at the front where he could see and scent the changes, where the reports were fresh, where the situation could be checked and clarified directly. His decisions were made at the front; his orders were issued there. His dashing spirit, his daring and his will to win carried over to his company commanders and his patrol leaders and inspired them. And whenever the extraordinary was demanded, he set the example at the head of his men, who followed him without hesitation.

As a typical example, here is a report by Gerd Bremer, then commander of the *1. (Kradschützen)/SS-Aufklärungs-Abteilung 1*:

It was at the beginning of the Russian campaign when we were advancing rapidly through the Ukraine. My company and I were given the mission of taking an enemy occupied forest south of Zhitomir and to then push on to the north. We successfully penetrated the enemy's main line of resistance, and the enemy conducted a fighting withdrawal into the extensive, impenetrable woods.

Not yet being completely familiar with the fighting tactics of motorcycle troops and, more importantly, with our battalion commander's swift style of operations, I deployed my company down both sides of the road and combed the terrain to clear it of pockets of resistance. An armored car suddenly appeared on the road, with the commander standing up in it. He called for me.

I made a quick report. Kurt Meyer smiled at me, his eyes gleaming, and ordered me to have my company remount and continue the advance north by moving down the road and firing off to the sides at the same time. On my order, my men rushed from all directions to their vehicles on the road. The commander called out to them: "Men, where is your fury at the Bolsheviks?"

The company had hardly started to move out in pursuit of the enemy in its usual reconnaissance formation — my command vehicle behind the lead elements — when Kurt Meyer overtook me and spurred us on to move faster. A race almost started between us. I couldn't stand the thought that the battalion commander might be ahead of me. By using that method, we actually managed to push through that large section of woods despite a strong enemy presence without taking serious casualties.

Why did his men risk their lives to follow Kurt Meyer? It was primarily his inspirational personality, not something hypnotic or even frightening, as the prosecution in Kurt Meyer's trial thought. The picture of such a personality is composed of many details. What probably had the greatest effect was that Kurt Meyer was in the middle of things; he did not shy away from going forward. He was always at the hot spot in critical moments. Was it possible for unrealistic orders to be issued under such circumstances? No one needed to think twice about his orders. Did anyone remain behind his commander? No one would have exposed himself to such shame.

Of course, success also bred confidence. Without doubt, Kurt Meyer was a man whom the gods of war liked. The cleverest and bravest leader was useless without a bit of luck, as Frederick the Great and Napoleon graphically demonstrated. But fortune only favors the competent in the long run. Everybody in the unit was confident that Kurt Meyer would not demand anything impossible or unnecessary; that, in the end, one could rely on his good fortune and that he would never fail anyone in a pinch. He would always try to help with all means possible and at the risk of his own life.

I want to illustrate that with an example from my own experiences. Shortly after the beginning of the chapter on the start of the Russian campaign, Kurt Meyer briefly described how he came to the assistance of a patrol from his own battalion and the *12./SS-Infanterie-Regiment 1,* which I commanded at the time.

It sounds so mundane and insignificant, but it saved the lives of a lot of men. It was our first engagement in Russia. In fact, the enemy was everywhere. The advance of *Panzergruppe von Kleist* had thrown the enemy into total confusion, and he was trying desperately hard to evade encirclement. My *company* was the advance guard of the *III./SS-Infanterie-Regiment 1.*

There was a great distance between neighboring battalions, and we had come into contact with dispersed Russian infantry. The battalion did not want to be delayed and created a new advance-guard detachment. My company had the mission of guarding the trains and leading them to the main supply route.

After staging, we took the trains down a hard-surface road that led to the main supply route and continued the march. We soon came upon disabled Russian vehicles, destroyed guns and the bodies of soldiers killed in action. One could see traces of fighting everywhere but no sign of the enemy. *SS-Obersturmführer* Bremer and a patrol from his *2./SS-Aufklärungs-Abteilung 1* caught up with us towards evening. He informed me that the enemy still occupied Klewan and Olyka and that the area was crawling with dispersed Russians. He had been in contact with the enemy a short time previously.

I decided to wait for morning before continuing the march and assembled most of the vehicles on high ground next to the road. Outposts were set up in all directions. Gerd Bremer and his reconnaissance patrol stayed with us along with one of the 5 cm antitank guns from the *III./SS-Infanterie-Regiment 1* which had lost its way.

Russians who were marching north down the same road behind us attacked us before the onset of darkness. Because the road was well below us, their fire went over our heads. Our antitank gun set fire to some trucks. Fortunately, they obstructed the road to the south. We heard the rattling of tank tracks on the road but could not make out a thing. Our antitank gun was positioned in such a way that it commanded both the road to the south and the plateau. It was only dark for a few hours around the summer solstice. What would the morning bring? The many trains vehicles were a great burden. We could not get any radio contact with the *III./SS-Infanterie-Regiment 1* on our field radio. To our good fortune, Gerd Bremer was able to establish radio contact with his battalion by using his 30-watt apparatus and was able to report our situation.

At first light we spotted a long column of motor vehicles approaching from the south. Initially, we could not tell whether they were friendly or Russian troops. Our antitank gun opened fire as they reached the Russian trucks that had been knocked out the previous evening and were obstructing the road. The crews and personnel on board immediately jumped out and began firing.

Our antitank crew had obviously made the right choice. There was no mistaking that those were Russian machine guns firing. The Russian infantry deployed quickly and started to attack down both sides of the road. Two Russian tanks also attacked unexpectedly across the plateau from the southwest; fortunately, they attacked right where our antitank gun had been emplaced. One tank was quickly knocked out and the other was hit but was still mobile. The mobile tank even succeeded in putting our antitank gun out of commission. It then turned around, disappeared into a depression and was not seen again. What luck, because *Panzerfäuste* and hollow charges did not yet exist. At best we could only have defended ourselves poorly with satchel charges against tanks.

The enemy infantry worked its way closer and closer through the grain fields. We were under fire from all directions and had already suffered several casualties. My driver, *SS-Sturmmann* Wolf, who had taken cover behind his car, was shot in the head. Although unconscious, he was still breathing. He was bandaged in the roadside ditch. How much longer would we be able to hold out against superior enemy forces?

Using Gerd Bremer's radio, we asked *SS-Aufklärungs-Abteilung 1* for help as soon as possible; superior enemy forces had encircled us. The battalion radioed back soon afterwards, promising help. The situation became increasingly critical. In the meantime, dawn had broken fully and the rising sun in the northeast behind the attacking enemy blinded us. The Russians quickly approached the road under cover of the grain fields. We were expecting them to break through our defensive lines at any moment.

We suddenly heard intense rifle and machine-gun fire coming down the road from the north. Was the main attack coming from that direction? Then we thought we had identified the firing of German machine guns. I was making my way along the roadside ditch to see for myself when Kurt Meyer himself came towards me down the same ditch with a pistol in his hand. I shook his hand in thanks. He really had arrived to help us at the very last moment. It was very typical of him: He had not charged one of his officers with the task but had himself taken the lead of a small *Kampfgruppe* to dig out his reconnaissance patrol, my company and the battalion's trains.

We then switched over to the attack and had soon chased off the enemy. They left behind a great number of trucks, guns and other weapons. They had obviously suffered considerable casualties. Our injured were taken back to the main dressing station as quickly as possible. The dead were buried, and we then continued our march towards Klewan.

Other formations and, especially, Kurt Meyer's own men often experienced such incidents, sometimes even daily. It was no wonder that they trusted him blindly. But how modest is the description he himself gives of the role he played. You should consider that throughout the book.

Assisting in the formation of *12. SS-Panzer-Division "Hitlerjugend"* again brought Kurt Meyer new tasks. Although he would have preferred to take over *SS-Panzer-Regiment 12*, Sepp Dietrich preferred, for understandable reasons, to charge *SS-Sturmbannführer* Max Wünsche with that task. Wünsche had commanded one of the tank battalions of the *1. SS-Panzer-Division "Leibstandarte"* with great success at Kharkov in the spring of 1943.

It is obvious that Kurt Meyer would certainly have commanded that regiment with success. He had already earned the nickname of *Panzermeyer*. Kurt

Meyer approached the formation of the new regiment with fiery eagerness and a lot of new and unconventional ideas. To his regret, the regiment did not have a single *SPW-Bataillon*. He understood only too well how to inspire these young men from the *Hitlerjugend*, how to maintain their interest and how best to prepare them methodically for their difficult tasks.

The basic principles of modern engagement techniques for *Panzergrenadiere* were tried and tested in his *SS-Panzer-Grenadier-Regiment 25*. From the very beginning, there were only simulated combat conditions and realistic targets. Moreover, all the other training was under combat conditions and conducted in the spirit of the *Panzertruppe*.

During his first operation during the invasion, *Panzermeyer* proved that he was not capable of leading just a reinforced regiment. The division was initially ordered by higher headquarters to stage in the area around Lysieux. That order was nonsense and eventually recognized as such even by our higher headquarters. We were then ordered into the area west of Caen. As a result, *Panzermeyer* and his reinforced regiment reached the new assembly area first. Without the remainder of the division and acting on his own initiative, *Panzermeyer* took charge and conducted operations in accordance with the divisional commander's intent. It was as if the divisional commander had been at his side. The loss of time as a consequence of assembling at Lysieux was, in my opinion, the reason for *Panzermeyer* being unable to start the attack on 6 June. Post-war publications confirmed our opinion at the time that a breakthrough to the coast might have succeeded and there was no one better qualified to do it than *Panzermeyer*. The situation in the British-Canadian invasion sector would have been completely different.

Less than 14 days after the start of the invasion Panzermeyer had to assume acting command of the division because *SS-Brigadeführer* Witt had been killed. There was no solemn handing over of command, but I will never forget our first meeting in that new situation. It was a mutual pledge to work together as friends. Although *SS-Brigadeführer* Witt had been an outstanding personality, who had the trust and love of his entire division, nothing really changed. *Panzermeyer* was a commander known and admired by all members of the division. He was trusted in the same way as *SS-Brigadeführer* Witt.

But both men were significantly different from each other in temper and style of command. When our sector of the front line also came into flux after the loss of Caen, the ability of the divisional staff to work together was admirably demonstrated. Even at division level *Panzermeyer* personally led at the hot spots, always mindful of the intent of higher headquarters. Despite that, he never lost contact with his operations officer. Thus he was always able to keep himself informed about the situation in the rest of the division's sector, to exchange ideas and to issue orders, direct measures and provide guidance. During all operations there was one fixed and one very mobile command post. The mobile command post was *Panzermeyer* and a *Volkswagen*. But both of the command posts comprised a whole. One may regard that as our divisional commander's exceptional achievement.

The division had meanwhile shrunk to the size of a small *Kampfgruppe* and was not much stronger than a reinforced battalion in terms of men and weapons. *SS-Brigadeführer* Meyer who, by then, had already earned the nickname *Panzermeyer*, showed all his masterly skills of command during the defensive fighting from 7-9 August north of Falaise. Through his skill and personal effort at the focal points of the fighting, he gave that little gang of exhausted but stalwart men the efficiency of a *Panzer-Division*, considerably weakened but still with considerable combat power.

In view of the overwhelming enemy superiority in the desperate situation at Cintheaux at midday on 7 August who would have even considered the possibility that we might attack? *Panzermeyer* did and had the courage to order it. His men believed that he was right although they would have thought it to be madness coming from any other man. That bold decision and its admirable execution so confused the enemy that the attack of the Canadian 4th Armoured and Polish 1st Armored Division were stopped in their tracks despite the enemy's considerable combat power.

The huge power of *Panzermeyer's* personality became apparent to me once again during our breakout from the Falaise pocket on 20 August 1944. Our small group, consisting of the remnants of the divisional staff and carrying only hand weapons, was just on its way towards the high ground east of the Dives. It had passed Chambois in the light of the rising sun. We moved forward from hedgerow to hedgerow — always on the lookout and covering each other. At the same time, groups of unarmed German soldiers — both large and small — were carrying white flags and making their way towards the enemy. The enemy was firing into the pocket with everything he had; it was causing casualties everywhere.

Individual unarmed soldiers who were wandering around joined up with us; there was nowhere else for them to gravitate to. Their number became increasingly larger; those few of us who had weapons were in danger of losing our freedom to act without endangering them. *Panzermeyer* stopped and shouted: "If you don't have a weapon, you're a coward and we don't want to have anything to do with you. You can only join up with us if you carry a weapon and are prepared to use it." Most of the soldiers became shame-faced and followed that demand from a man who no longer had a division let alone a *Kampfgruppe*. They didn't know who he was, but they knew he was a real leader. They disappeared for a short time and came back with any weapon they could find. Only a few — perhaps those who were experiencing battle for the first time — disappeared silently and tried to join the columns with little white flags.

One day *Panzermeyer's* luck failed him and he fell into captivity. That was a terrible fate, especially for a man who loved his freedom. But it was also especially terrible because the fight then had to be continued on without him and because he could no longer be with his men. But he never capitulated in captivity and played many a trick on his captors. He gained the respect and even affection of his former enemies when he stood before the Canadian tribunal after the collapse. He knew what could happen to him, but he had a clear conscience.

When they awarded the death penalty completely against their own convictions, it was accepted by an outwardly unshaken man who had looked death in the face a thousand times on the battlefield. But a wound came into being in his proud heart which never healed and which took him from us far too soon. The distress of his tortured fatherland, which he loved so much, especially in its hour of need, preyed upon his mind. What indescribable good fortune it therefore was when Fritz Lichtenberg, an unknown Canadian of German origin, visited him in prison in Moncton, took him in his arms and said: "Where there is a German heart, there is the German homeland."

After two world wars during which his old and new countries had been set against each other, that infinitely diligent and excellent man felt irresistibly drawn to the German behind bars. He made many sacrifices to alleviate the lot of that needy man and his family from his home country and help him to be free. It was no coincidence that *Panzermeyer* gained that noble man's love and the deep respect of his former enemies on the battlefield who, undeterred by the propaganda smear campaign and without regard to their future, stood up for his rehabilitation and his release. His brave, military bearing and his outstanding personality gained him these friends and helpers. They were truly worthy of each other.

Panzermeyer's first thoughts were never for himself but always for the greater good of the community. He considered it to be a lifelong responsibility that he had formerly been a *Waffen-SS* divisional commander and a leader of a formation of very young soldiers. He felt that not only behind prison walls after the collapse but also as a free man in the "Economic Miracle" of West Germany. The offensive defamation of *Waffen-SS* — more so in our own country than in other, formerly enemy, countries — drove him to help establish the truth.

Together, we wrote a report about the deployment of our division during the invasion while he was still imprisoned in Werl. He had hardly returned home and was being kept very busy with his new profession when he wrote this book within a few months, mostly in the evening and during the night. It was intended to show the general public both here and abroad that the *Waffen-SS* was not a gang of unscrupulous fanatics but an elite formation of soldiers. The men who fought in that formation, whether they were killed or survived, were not detestable members of a "criminal organization", as the victorious powers had proclaimed during the Nuremberg trials. He wanted to rectify the false impression created by the "reeducated" with all the power of the mass media at their disposal and which they intended to stamp upon our own children.

The surviving dependents of the killed and missing members of the *Waffen-SS* and those survivors who were disabled were not treated like former soldiers by the authorities but like second-class citizens. Not only was it intolerable that many of these people fell into poverty through no fault of their own. Much worse was the effect that the new state created. It used two tiers of laws and so undermined its own foundations.

The *Hilfsgemeinschaft auf Gegenseitigkeit der Angehörigen der ehemaligen Waffen-SS* (Mutual-Aid Society for Members of the Former *Waffen-SS*) was founded in 1950 to fight defamation and injustice, to help those in need and to clarify the fate of the missing. *Panzermeyer* had hardly returned home when he devoted himself to that aid organization. He became its national spokesman in 1956. In innumerable speeches to his comrades he admonished them not to become embittered by injustice but to assist in the reconstruction of the new state and help strengthen it for its difficult new tasks. At the same time, he continued negotiations with important politicians, which had been going on for years, to overcome the defamation and finally achieve equal rights.

The goal and spirit of his work with HIAG were probably best illustrated by the speech he made in Karlburg in July 1957. In it, he said, among other things:

…As with no other generation, we are challenged by destiny and confronted with the task of finding the right answers to those great questions that we, the living, are asked today. Where are those answers? We do not find them in mimicking the past, either before or after 1933. We can only look to the future.

…A human being will not survive these troubled times without some sort of a compass, which at least gives him the general direction in which to proceed. That's why we first have to reestablish absolute values and enter into binding obligations. I believe that is the task to which we have to dedicate ourselves above all other things.

WE — that is, everyone who believes they can still help their nation and humanity in these times.

WE — that is, those who approach a matter not because it appears to offer success from the outset but, instead, because it is the right thing to do.

WE — that is, all those who have not dedicated their lives to the satisfaction of material goods but, instead, have dedicated it to a meaningful cause.

It is not our task to propose the organizational form under which these thoughts may be realized. The possibilities exist for all to assist, however. It is our duty to fulfil our responsibilities in professional organizations, political parties, in every field where society has provided any opportunity for its citizenry.

All of us, no matter what our political backgrounds, are seeking today a goal and the way to achieve it. He who first finds a valid answer will also have our support. My comrades, we must participate in this work if we are to be more than just a club.

The intellectual breakthrough will happen one day, and we must not stand around like camp followers. We already know what obstacles are in front of us; but the object of that breakthrough is not yet quite clear. One day, however, it will stand clearly before us all. For our people to be able one day to participate in that breakthrough with dignity we must first become a nation once more. It is the task of each one of us to assist in that, even if only in a minor way.

We must build fortresses within our families, within our professional environment, within the circle of our comrades. These must be fortresses of morality, of strength of character and of decency…

Panzermeyer believed he had reached his goal of equal rights by the beginning of 1961. His negotiating partners in the *Bundestag* promised him they would be sponsoring an appropriate decision by the *Bundestag* and that it was forthcoming. The *Bundestag's* decision of 29 June 1961 was a big disappointment. In view of the foreign situation — marked by the Eichmann trial and the Berlin crisis — the attempt to make the *SS-Verfügungstruppe* and, later, the *Waffen-SS*, a military force the same as the former *Wehrmacht* was not realized. The Federal Minister of the Interior only promised that he was prepared to support those affected in case of need.

Panzermeyer was called from us by death in the middle of his struggle on 23 December 1961, his 51st birthday. The news of his death hit his comrades and friends like a bolt from the blue. It was incomprehensible to us all. All those who had heard the news and who somehow found it possible assembled in Hagen to say goodbye to their *Panzermeyer* on 28 December 1961. More than 4,000 loyal friends attended. One may seldom see so many people clearly touched by grief standing together by one grave. These former soldiers, put to the test in many battles and who had stood by many open graves, were not ashamed to show their tears to each other. That event broke all bounds of normal self control because it gripped all of our hearts. Let me reiterate my few short words of farewell at graveside.

Dear Kurt,

Rallied around you, I see a lot of dear, loyal comrades who were at your side in peace, during the war and afterwards. If all who feel a bond with you were to stand here it would constitute a vast crowd. With us belong thousands of comrades who have already been resting in peace for a long time, especially many soldiers of your *12. SS-Panzer-Division*, whom you visited a year ago at the military cemeteries in Normandy. I speak as your former operations officer because we had promised one another that whoever survived would speak the last words for the other one.

I want to tell you what we feel to be your legacy, so that you will continue to live not only in your family, but also in your deeds. You did not build yourself any stone monuments, but you built a shining example in our comrades' hearts. We want to make the effort to be as brave and loyal as you were, to never consider our own profit or advantage.

Your entire being was committed to the fortune and the freedom of our fatherland. Your heart beat for Germany until the very end. We always want to love this fatherland of ours as you did — all the more in time of need and sorrow. As was always your intent, we want to help in creating a great community of free peoples in which the spirit and character of the German people may grow in noble, peaceful competition.

Wherever we are — at home, at work, in pubic or abroad — everyone of us should be able to say with pride: I was a comrade of *Panzermeyer*! In that way, you'll always be with us, your influence will continue on!

We salute you, Kurt!

*

Kurt "Panzermeyer" Meyer

Born: 23 December 1910 in Jerxheim (Province of Braunschweig)

Knight's Cross: 18 May 1941 as an *SS-Sturmbannführer* and the commander of *SS-Aufklärungs-Abteilung 1*

Oakleaves to the Knight's Cross: 23 February 1943 (195th soldier of the *Wehrmacht*) as an *SS-Obersturmbannführer* and the commander of *SS-Aufklärungs-Abteilung 1*

Swords to the Oakleaves of the Knight's Cross: 27 August 1944 (91st soldier of the *Wehrmacht*) as an *SS-Oberführer* and the commander of the *12. SS-Panzer-Division "Hitlerjugend"*

Died: 23 December 1961 in Hagen (heart attack)

Kurt Meyer, 1944, in *Wehrmacht* uniform after his capture. (Jost Schneider)

Imprisonment in Werl.

Below: The reading of the indictment.

Aurich, 1945. Major Arthur Russell of the 4th Battalion of Royal Winnipeg Rifles escorting Kurt Meyer. (Jost Schneider)

Free at last!

Upper right: Field cemetery for *Waffen-SS* grenadiers.

Bottom right: Kurt Meyer's grave in Hagen, Westphalia.

The welcome home.

Kurt Meyer
1910-1961

411

Rank Comparisons

Enlisted

US Army	German Army	Waffen-SS
Private	*Schütze*	*SS-Schütze*
Private First Class	*Oberschütze*	*SS-Oberschütze*
Corporal	*Gefreiter*	*SS-Sturmmann*
(Senior Corporal)	*Obergefreiter*	*SS-Rottenführer*
(Staff Corporal)	*Stabsgefreiter*	(None)

Noncommissioned Officers

US Army	German Army	Waffen-SS
Sergeant	*Unteroffizier*	*SS-Unterscharführer*
Staff Sergeant	*Feldwebel*	*SS-Oberscharführer*
Sergeant First Class	*Oberfeldwebel*	*SS-Hauptscharführer*
Master Sergeant	*Hauptfeldwebel*	*SS-Sturmscharführer*
Sergeant Major	*Stabsfeldwebel*	*(None)*

Officers

US Army	German Army	Waffen-SS
Lieutenant	*Leutnant*	*SS-Untersturmführer*
First Lieutenant	*Oberleutnant*	*SS-Obersturmführer*
Captain	*Hauptmann*	*SS-Hauptsturmführer*
Major	*Major*	*SS-Sturmbannführer*
Lieutenant Colonel	*Oberstleutnant*	*SS-Obersturmbannführer*
Colonel	*Oberst*	*SS-Oberführer* or *SS-Standartenführer*
Brigadier General	*Generalmajor*	*SS-Brigadeführer*
Major General	*Generalleutnant*	*SS-Gruppenführer*
Lieutenant General	*General der Panzertruppen* etc.	*SS-Obergruppenführer*
General	*Generaloberst*	*SS-Oberstgruppenführer*
General of the Army	*Feldmarschall*	*Reichsführer-SS*